# Advances in Spatial Science

## Titles in the Series

*L. Lundqvist* and *L.-G. Mattsson* (Eds.)
National Transport Models
VIII, 202 pages. 2002. ISBN 978-3-540-42426-0

*J. R. Cuadrado-Roura* and *M. Parellada* (Eds.)
Regional Convergence in the European Union
VIII, 368 pages. 2002. ISBN 978-3-540-43242-5

*G. J. D. Hewings, M. Sonis* and *D. Boyce* (Eds.)
Trade, Networks and Hierarchies
XI, 467 pages. 2002. ISBN 978-3-540-43087-2

*G. Atalik* and *M. M. Fischer* (Eds.)
Regional Development Reconsidered
X, 220 pages. 2002. ISBN 978-3-540-43610-2

*Z. J. Acs, H. L. F. de Groot* and *P. Nijkamp* (Eds.)
The Emergence of the Knowledge Economy
VII, 388 pages. 2002. ISBN 978-3-540-43722-2

*R. J. Stimson, R. R. Stough* and *B. H. Roberts*
Regional Economic Development
X, 397 pages. 2002. ISBN 978-3-540-43731-4

*S. Geertman* and *J. Stillwell* (Eds.)
Planning Support Systems in Practice
XII, 578 pages. 2003. ISBN 978-3-540-43719-2

*B. Fingleton* (Ed.)
European Regional Growth
VIII, 435 pages. 2003. ISBN 978-3-540-00366-3

*T. Puu*
Mathematical Location and Land Use Theory, 2nd Edition
X, 362 pages. 2003. ISBN 978-3-540-00931-3

*J. Bröcker, D. Dohse* and R. Soltwedel (Eds.)
Innovation Clusters and Interregional Competition
VIII, 409 pages. 2003. ISBN 978-3-540-00999-3

*D. A. Griffith*
Spatial Autocorrelation and Spatial Filtering
XIV, 247 pages. 2003. ISBN 978-3-540-00932-0

*J. R. Roy*
Spatial Interaction Modelling
X, 239 pages. 2004. ISBN 978-3-540-20528-9

*M. Beuthe, V. Himanen, A. Reggiani* and *L. Zamparini* (Eds.)
Transport Developments and Innovations in an EvolvingWorld
XIV, 346 pages. 2004. ISBN 978-3-540-00961-0

*Y. Okuyama* and *S. E. Chang* (Eds.)
Modeling Spatial and Economic Impacts of Disasters
X, 323 pages. 2004. ISBN 978-3-540-21449-6

Further titles in this series:
http://www.springer.com/series/3302

*L. Anselin, R.J.G.M. Florax* and *S. J. Rey*
Advances in Spatial Econometrics
XXII, 513 pages. 2004. ISBN 978-3-540-43729-1

*R.J.G.M. Florax* and *D. A. Plane* (Eds.)
Fifty Years of Regional Science
VIII, 400 pages. 2004. ISBN 978-3-540-22361-0

*D. Felsenstein* and *B.A. Portnov* (Eds.)
Regional Disparities in Small Countries
VI, 333 pages. 2005. ISBN 978-3-540-24303-8

*A. Reggiani* and *L.A. Schintler* (Eds.)
Methods and Models in Transport and Telecommunications
XIII, 364 pages. 2005. ISBN 978-3-540-25859-9

*H.W. Richardson* and *C.-H.C. Bae* (Eds.)
Globalization and Urban Development
X, 321 pages. 2005. ISBN 978-3-540-22362-7

*G. Arbia*
Spatial Econometrics
XVII, 207 pages. 2006. ISBN 978-3-540-32304-4

*B. Johansson, C. Karlsson, R. Stough* (Eds.)
The Emerging Digital Economy
X, 352 pages. 2006. ISBN 978-3-540-34487-2

*H. Westlund*
Social Capital in the Knowledge Economy
X, 212 pages. 2006. ISBN 978-3-540-35364-5

*A.E. Anderssson, L. Pettersson, U. Strömquist* (Eds.)
European Metropolitan Housing Markets
VI, 363 pages. 2007. ISBN 978-3-540-69891-3

*A.T. Murray, T.H. Grubesic* (Eds.)
Critical Infrastructure
VIII, 311 pages. 2007. ISBN 978-3-540-68055-0

*R. Cooper, K. Donaghy, G. Hewings* (Eds.)
Globalization and Regional Economic Modeling
XIII, 475 pages, 2007. ISBN 978-3-540-72443-8

*R. Capello, R. Camagni, B. Chizzolini, U. Fratesi*
Modelling Regional Scenarios for the Enlarged Europe: European Competiveness and Global Strategies
XX, 337 pages, 2008. ISBN 978-3-540-74736-9

*C. Jensen-Butler, B. Sloth, M.M. Larsen, B. Madsen* and *O.A. Nielsen*
Road Pricing, the Economy and the Environment
XVI, 425 pages, 2008. ISBN 978-3-540-77149-4

*U. Fratesi* and *L. Senn*
Growth and Innovation of Competitive Regions
X, 368 pages, 2009. ISBN 978-3-540-70923-7

Aura Reggiani · Peter Nijkamp
Editors

# Complexity and Spatial Networks

## In Search of Simplicity

Springer

*Editors*

Professor Aura Reggiani
Department of Economics
Faculty of Statistics
University of Bologna
Piazza Scaravilli, 2
40126 Bologna
Italy
aura.reggiani@unibo.it

Professor Peter Nijkamp
Department of Spatial Economics
VU University Amsterdam
De Boelelaan 1105
1081 HV Amsterdam
The Netherlands
pnijkamp@feweb.vu.nl

ISBN 978-3-642-26914-1 e-ISBN 978-3-642-01554-0
DOI: 10.1007/978-3-642-01554-0
Springer Dordrecht Heidelberg London New York



*Cover design:* SPi Publishing Services

Printed on acid-free paper

Springer is part of Springer Science+Business Media (www.springer.com)

# Preface

Complex systems analysis has become a fascinating topic in modern research on non-linear dynamics, not only in the physical sciences but also in the life sciences and the social sciences. After the era of bifurcation theory, chaos theory, synergetics, resilience analysis, network dynamics and evolutionary thinking, currently we observe an increasing interest in critical transitions of dynamic real-world systems in many disciplines, such as demography, biology, psychology, economics, earth sciences, geology, seismology, medical sciences, and so on. The relevance of this approach is clearly reflected in such phenomena as traffic congestion, financial crisis, ethnic conflicts, eco-system breakdown, health failures, etc. This has prompted a world-wide interest in complex systems.

Geographical space is one of the playgrounds for complex dynamics, as is witnessed by population movements, transport flows, retail developments, urban expansion, lowland flooding and so forth. All such dynamic phenomena have one feature in common: the low predictability of uncertain interrelated events occurring at different interconnected spatio-temporal scale levels and often originating from different disciplinary backgrounds. The study of the associated non-linear (fast and slow) dynamic transition paths calls for a joint research effort of scientists from different disciplines in order to understand the nature, the roots and the consequences of unexpected or unpredictable changes in complex spatial systems. Complex dynamics also challenges the findings from conventional equilibrium theory, in particular concerning multi-agent systems. Consequently, the prediction, analysis, and management of non-linear dynamic phenomena in the context of complexity analysis is of great importance for decision making in both the private and the public sector.

In this context, from a methodological viewpoint, in complex systems there is the need for a unifying framework of analysis that embraces the meaning and use of interdisciplinary concepts (such as self-organization, criticality, redundancy, resilience and sustainability). At the same time, the universality and 'simplicity' of network centrality and connectivity laws (such as the entropy and power laws) should be better explored.

The present volume brings together a series of original and innovative contributions in the area of complex spatial dynamics and networks. A wealth of authors – from different disciplines – were invited to write an original piece of work centring around the non-linear dynamic nature of spatial and network systems. This book is the outgrowth of a workshop organized by IPL (Institute Para Limes), the new Institute for frontier research on complex phenomena of a trans-disciplinary nature, based in the Netherlands (for details, see www.paralimes.org). The participants came from all over the world and provided refreshing ideas on the analysis of complexity and non-linear dynamic evolution in space and in spatial networks. Their contributions and various enthusiastic ideas laid the foundation for this publication that aims to be a systematic compilation of carefully selected and refereed papers on interdisciplinary perspectives on spatial complexity and non-linear dynamic network development. The editors wish to thank Jan Wouter Vasbinder, Managing Director of IPL, for his great support in the preparation and organization of our work.

Bologna/Amsterdam,
March 2009

Aura Reggiani
Peter Nijkamp

# Contents

# Contributors

Franz-Josef Bade
Faculty of Spatial Planning, University of Dortmund, Dortmund D 44221, Germany

Michael Batty
Centre for Advanced Spatial Analysis, University College London, 1-19 Torrington Place, London WC1E 7HB, UK

Lucien Benguigui
Solid State Institute and Department of Physics, Technion-Israel Institute Technology, Haifa 32000, Israel

David Bernstein
Department of Computer Science, James Madison University, 800 South Main St. MSC 4103, Harrisonburg, VA 22807, USA

Efrat Blumenfeld-Lieberthal
Environmental Simulation Laboratory, The Porter School of Environmental Studies and the Faculty of Arts, Tel-Aviv University, Tel-Aviv 69978, Israel

Michele Campagna
Dipartimento di Ingegneria del Territorio, Sezione Costruzioni e Infrastrutture, Università degli Studi di Sassari, via De Nicola, Sassari 07100, Italy

Simone Caschili
Dipartimento di Ingegneria del Territorio, Università degli Studi di Cagliari, Piazza d'Armi 16, Cagliari 09123, Italy

Yu Chen
Transport Systems Planning and Transport Telematics, TU Berlin, Salzufer 17-19, Sekr. SG 12, Berlin 10587, Germany

Alessandro Chessa
Dipartimento di Fisica, INFM, Università degli Studi di Cagliari, Complesso Universitario di Monserrato, Monserrato 09042, Italy

Helen Couclelis
Department of Geography, University of California, Santa Barbara, CA 93106-4060, USA

Andrea De Montis
Dipartimento di Ingegneria del Territorio, Sezione Costruzioni e Infrastrutture, Università degli Studi di Sassari, via De Nicola, Sassari 07100, Italy

Giancarlo Deplano
Dipartimento di Ingegneria del Territorio, Università degli Studi di Cagliari, Piazza d'Armi 16, Cagliari 09123, Italy

Kieran P. Donaghy
Department of City and Regional Planning, Cornell University, 106 West Sibley Hall, Ithaca, NY 14853, USA

George Ehrhardt
The Abdus Salam International Center for Theoretical Physics, Strada Costiera 11, Trieste 34014, Italy

Koen Frenken
Urban and Regional Research Centre Utrecht (URU), Faculty of Geosciences, Utrecht University, P.O. Box 80115, Utrecht 3508 TC, The Netherlands

Terry L. Friesz
Department of Industrial and Manufacturing Engineering, The Pennsylvania State University, University Park, PA 16802, USA

Giacomo Galiazzo
School of Public Policy, George Mason University, 4400 University Dr., Fairfax, VA 22030, USA

Dominik Grether
Transport Systems Planning and Transport Telematics, TU Berlin, Salzufer 17-19, Sekr. SG 12, Berlin 10587, Germany

Daniel A. Griffith
School of Economic, Political and Policy Sciences, University of Texas at Dallas, 800 W. Campbell Rd, Richardson, TX 75080-3021, USA

Kingsley E. Haynes
School of Public Policy, George Mason University, 4400 University Dr., Fairfax, VA 22030, USA

Cars Hommes
CeNDEF, Department of Quantitative Economics, University of Amsterdam, Roetersstraat 11, Amsterdam 1018 WB, The Netherlands

Rajendra G. Kulkarni
School of Public Policy, George Mason University, 4400 University Dr., Fairfax, VA 22030, USA

Changyun Kwon
Service Enterprise Engineering Center, The Pennsylvania State University, University Park, PA 16802, USA

Matteo Marsili
The Abdus Salam International Center for Theoretical Physics, Strada Costiera 11, Trieste 34014, Italy

Francesca Medda
Centre for Transport Studies, University College London, Gower Street, London WC1E 6BT, UK

Kai Nagel
Transport Systems Planning and Transport Telematics, TU Berlin, Salzufer 17-19, Berlin 10587, Germany

Peter Nijkamp
Department of Spatial Economics, Faculty of Economics, VU University Amsterdam, De Boelelaan 1105, Amsterdam 1081 HV, The Netherlands

Roberto Patuelli
Institute for Economic Research (IRE), University of Lugano, Via Maderno 24, CP 4361, Lugano CH-6904, Switzerland

Aura Reggiani
Department of Economics, Faculty of Statistics, University of Bologna, Piazza Scaravilli 2, Bologna 40126, Italy

Massimo Ricottilli
Department of Economics, University of Bologna, Piazza Scaravilli 2, Bologna 40126, Italy

Marcel Rieser
Transport Systems Planning and Transport Telematics, TU Berlin, Salzufer 17-19, Sekr. SG 12, Berlin 10587, Germany

Piet Rietveld
Department of Spatial Economics, Faculty of Economics, VU University Amsterdam, De Boelelaan 1105, Amsterdam 1081 HV, The Netherlands

Laurie A. Schintler
School of Public Policy, George Mason University, 4400 University Dr., Fairfax, VA 22030, USA

Kevin Seel
Department of Geography, George Mason University, 4400 University Dr MS1E2, Fairfax, VA 22030, USA

Roger R. Stough
Office of the Provost, George Mason University, 4400 University Dr., Fairfax, VA 22030, USA

Fernando Vega-Redondo
Department of Economics, European University Institute, Villa San Paolo, Via della Piazzuola 43, Florence 50133, Italy

Nigel Waters
Department of Geography, George Mason University, 4400 University Dr MS1E2, Fairfax, VA 22030, USA

Alan Wilson
Centre for Advanced Spatial Analysis, University College London, 1-19 Torrington Place, London WC1E 7HB, UK

# Chapter 1
# Simplicity in Complex Spatial Systems

## Introduction

**Aura Reggiani and Peter Nijkamp**

## 1.1 Moving Worlds

In the past decade, complexity has become an important and fascinating domain for advanced research on nonlinear dynamics, in which a multiplicity of scientific fields are involved (physics, life sciences, social sciences, economics, geography, and so forth). Complex systems analysis refers to research at the dynamic interface of – or the interaction between – small or micro-elements of a system that are interconnected and determine a macro-level of operation of the system that is not just the sum of the micro-elements. As a result of self-organizing forces among interacting micro-units, a dynamic network configuration may emerge that displays its own dynamics, ranging from "butterfly" effects to scale-free evolution, or from bifurcations with unexpected phase transitions to preferential attachment in small-world networks (see Barabási and Albert 1999; Nicolis and Nicolis 2007). The complexity movement has also had far-reaching impacts on dynamics research in the spatial sciences.

The space-economy is often interpreted as a standard well-functioning economic system enriched with the element of space. But space is not just an additional dimension of the economy: it forms an intrinsic feature of any geographic–economic system and may lead to the emergence of complex nonlinear and interactive behaviours and processes in a geographic setting. The foundation for an interpretation of the space economy as an interdependent complex set of economic relationships – at different geographic scales and with a variety of time dimensions involved – can be found in the "first law of geography" formulated by Tobler (1970), who stipulates that everything in space is related to everything else, but near things are more related than distant things. The solidity of this law needs to be reconsidered in the light of recent advances in complexity and network theory. In particular, the latest findings in

A. Reggiani (✉)
Department of Economics, Faculty of Statistics, University of Bologna, Bologna, Italy

A. Reggiani and P. Nijkamp (eds.), *Complexity and Spatial Networks*,
Advances in Spatial Science,
DOI: 10.1007/978-3-642-01554-0_1, © Springer-Verlag Berlin Heidlberg 2009

network theory show how – for certain network typologies – distant things can be related by means of "hubs" or "egos" (preferential nodes/attractors). Spatial networks appear to exert a dynamic impact on an organized space.

One of the striking features in the modern space-economy has been the simultaneous occurrence of spatial dynamics (both fast and slow dynamics) and spatial inertia (for example persistent welfare disparities between regions). Regions and cities are apparently operating in a complex force field, with asynchronously emerging key factors that impact on regional or urban development in different ways and with different growth paces. Examples of such factors are: changing lifestyles, trends towards a highly mobile society, emerging innovations and creativity strategies, shifting views on the competences of policy making, the search for new forms of industrial organization and leadership, the adoption of new technologies, the design of new forms of urban architecture, and so forth.

This rapidly changing scene of regional and urban development has called for new research departures, such as: a reliance on experimental psychology/sociology, design of learning principles for decision makers (based on evolutionary biology), integration of ethical and sociological notions in policies for a multi-cultural society, etc. Consequently, regional and urban research has become richer in scope, with more emphasis on interdisciplinarity, complexity, synergy among research methodologies, conflict management principles, adaptive and evolutionary (notably, learning) behaviour, and increasing interest in the great potential offered by the cognitive sciences. This has also had profound impacts on empirical research, in which computational neural networks, data mining techniques, adaptive and micro-simulation modelling techniques, or analysis of social and ecological externalities are playing a more prominent role. This also calls for a more integrated and interdisciplinary research perspective. One of the fields which has demonstrated a marked shift in focus has been the area of regional growth analysis and management, in which organizational sociology, industrial/urban organization and multi-actor decision making are playing a central role. This is only one example of recent trends that demonstrate a reorientation towards complex dynamics in space, in which a multiplicity of driving forces exert differential dynamic impacts on human and spatial systems.

Textbook economics has paid extensive attention to the conditions under which critical drivers might lead to accelerated growth, with sometimes significant variation among regions or cities (for example increasing returns to scale, product heterogeneity and specialization, etc.). All these elements impact on the welfare and productivity pattern of spatial-economic systems and may be a source of divergent economic achievements by various regions or cities in the space-economy. Nevertheless, the analysis of spatial-economic disparities does often not provide us with a complete picture of all relevant background factors. In other words, many models trying to explain regional growth and spatial differences therein are semantically insufficiently specified. In a number of cases, therefore, economists have resorted to the introduction of complementary explanatory factors, such as X-efficiency factors which refer to often intangible factors (for example personal devotion, altruistic behaviour, concern about the future, social engagement, etc.) and which may offer additional explanations for the performance of various agents (for example regions, administrations, entrepreneurs, employees, etc.).

In addition, the notion of interactive behaviour and processes as evident from modern network theory in open systems challenges the use of conventional ceteris paribus conditions (especially in an interdisciplinary research context). The ceteris paribus condition has become a central tool in economic research so as to draw attention on commonalities in behaviour of economic subjects or agents by assuming that certain contextual (or environmental) factors may be seen as constant across relevant objects of research or over a relevant time horizon. Such factors ensure a certain order or structure which allows for transferability (or even generalization) in an otherwise chaotic world. This methodological approach means essentially an abstraction from a highly varied complex real-world economic system and allows for a focussed investigation of a certain relevant economic phenomenon (see Andersson et al. 2002; Gough et al. 2008).

Recent advances in complexity theory have shown that complex systems evolve in different ways, depending on the type of interdependencies among the components. Thus the identification of system typologies leading to different equilibrium solutions and different evolutionary paths might be useful for a critical review of standard ceteris paribus conditions. In this context, well-known concepts – emerging from different disciplines – such as learning, emergence, scaling, robustness, etc. may also be revisited and unified in the framework of spatial-economic science and network theory.

The trend towards a more thorough analysis of nonlinear dynamic spatial systems and networks finds its parallel in evolutionary thinking concerning spatial complexity, in which in particular interactions among micro-constituents or actors are drivers for the complex dynamics of geographical space (see also van den Bergh 2007; Boschma and Frenken 2006). This book aims to offer a panoramic view of recent advances in spatial complexity, in order to enhance our understanding of complex spatial networks by simplicity in terms of the basic driving forces of systemic impacts, as well as in terms of modelling such systems. Simple models mapping out the evolution of complex networks are undoubtedly a key issue in spatial economic research.

## 1.2 Outline of the Book

Starting from the above considerations, this book on "*Complexity and spatial networks: In search of simplicity*" aims to highlight the "network embedding" implications for modern complexity theory – with reference to spatial-economic analysis – by providing exploratory pathways for novel research lines.

In particular – after this introductory chapter – the volume aims to address three main issues:

- *Part A: Complexity, evolution, and simplicity in space.* This first part investigates evolutionary aspects in spatial economics, by showing how "old" concepts and methods, borrowed from physics, demography and social sciences – such as thermodynamics, rank-size and Kolmogorov complexity – can be interpreted in a "novel" interdisciplinary framework and applied to spatial and urban modelling. In addition, new concepts such as "polyplexity" are proposed to embed social space and time, and may thus be suitable for simplifying the representation of complex social phenomena.
- *Part B: Evolutionary networks in a socio-economic context.* The second part examines the concept of "network embedding" in economic and social science, in the light of planning and policy issues. In particular, learning and noise are explored in complex social networks; the role of constraints and proximity is analysed in the firm's networks, and preferential attachment and space are highlighted in the dynamic network of transport/commodity flows.
- *Part C: Empirical aspects of network complexity in the space-economy.* The third part is devoted to the empirical analysis of network complexity in spatial economics, by considering simulations/applications in transport and urban/regional economics. In this context, simulations and applications are presented with reference to: modal choice in urban transport networks, power generation networks, urban dynamics and its morphology, regional labour markets, and regional transport networks.

A final section, *Part D: Epilogue*, concludes the volume, by offering a synthesis framework and general suggestions for the future research agenda.

As previously mentioned, Part A focuses mainly on spatial evolution and its complexity/simplicity aspects. Part A begins with a contribution by Wilson (Chap. 2), who shows how methods from statistical mechanics can be interpreted and applied to the analysis of urban structure and its evolution, from the perspective of future interdisciplinary research. Benguigui et al. (Chap. 3) address the current debate on the "correct" function that accurately describes the size distribution of entities, by revisiting the rank-size rule and Zipf's law, which – as is well-known – are strictly related to fractal and complexity theory. Based on a multiplicative model of proportionate growth, the authors develop a quantitative comparison to relate the change in the rank-size curves to the change in the real data of Israeli cities during the period 1950–2005. Medda et al. (Chap. 4) deal with spillover and cumulative effects, in the context of urban growth processes. These authors assume a mutual dependence between transportation costs and urban form, and, by applying the Turing morphogenetic algorithm, analyse the dynamic processes induced by this relationship for spatial changes in the city. In particular, they introduce, in their dynamic model, two specific elements: an accumulative trend of the variables, and a diffusion process in their variation. The numerical simulation of an illustrative case study depicts how the entire urban shape can be modified in different ways by an improvement of the transport system. Next, Kulkarni and Stough (Chap. 5) focus on the question of how we formally measure the degree of complexity (simplicity) in spatial systems, by exploring a methodology based on Kolmogorov complexity. The visualization of complex data – by means of comparative complexity/ simplicity measurement is illustrated – and the mapped results are presented. Couclelis (Chap. 6) concludes Part A: this author introduces the notion of "polyplexity" as a new way of approaching the search for the most complex systems. Polyplexity takes into account the possibility that the space and time within which a phenomenon enfolds may themselves be complex.

The issue of evolutionary modelling in complex economic networks is addressed in Part B. Hommes (Chap. 7) discusses complexity models in economics, by considering a model with heterogeneous expectations (in particular, the asset pricing model), in which bounded rationality is disciplined through simple heuristic, adaptive learning and evolutionary selection. The author shows that this model is consistent with learning to forecast laboratory experiments with human subjects, while it also explains observed path-dependent stable and unstable outcomes. Ehrhard et al. (Chap. 8) discuss the issue of homophily and conformity in the evolution of complex social networks. They propose a stylized model in which agents are involved in a local coordination game with their neighbours in a co-evolving network. The social process of network formation exhibits sharp transitions, hysteresis, and equilibrium multiplicity. The robustness of these conclusions is tested by introducing some persistent noise which may disturb both the establishment of links and adjustments of actions. The role of constraints is further explored by Ricottilli (Chap. 9); the author shows that constraints might improve the acquisition of technological capabilities, in a framework of firms interacting in a cognitive network.

In subsequent chapters, the relevance of space in economic networks is highlighted. Frenken (Chap. 10) models social networks emerging from the mobility decisions of entrepreneurs moving between product divisions within or between firms, and within or between cities, by focusing on the concept of proximity as an essential contribution for interpreting these interdependencies. Friesz et al. (Chap. 11) consider a dual time-scale transportation planning model in which demand evolves on a day-to-day time-scale and traffic flows on arcs of the transportation network fluctuate on a within-day time-scale. They show how one may incorporate in such a model a non-traditional sub-model of demand growth based on the paradigm of preferential attachment familiar from network science and the notion of learning dynamics from evolutionary game theory. Finally, Donaghy (Chap. 12) discusses methodological challenges for modelling an economy as a spatial system oriented towards complexity. In this context, the author sketches a prototype model, by specifying a non-cooperative dynamic game between shippers and carriers, with reference to the evolution of commodity flows in the Midwest of the United States.

Part C is devoted to empirical or operational contributions in complex spatial networks. Grether et al. (Chap. 13) present a transport network approach where modal choice is integrated into a multi-agent simulation of travel behaviour and traffic flows. This approach is tested on the basis of a real-world base case for the city of Zurich (Switzerland). A sensitivity analysis concerning the "disutility" of travelling by a non-car mode is also carried out, by highlighting the advantage of the microscopic nature of the model. Seel and Waters (Chap. 14) discuss the complexity issue, by critically reviewing past studies of complexity in the social sciences and in geography. In particular, the authors pay attention to the problem of spatializing system dynamics, by developing a system dynamics model of the Forrester type, in which spatial variations are incorporated. They adopt this model in order to offer insight into how the deregulated (complex) market for electrical power in Alberta (Canada) will evolve. The authors emphasize how a spatially explicit, system-dynamics approach allows an understanding and remediation of undesirable aspects of the price responses of this market system. Several forms of investor behaviour are modelled. An unanticipated result of the modelling exercise shows that boom-and-bust oscillations may be avoided through the introduction of capacity payments that are transparent to electricity users. Urban pattern formation in network dynamics is examined by Schintler and Galiazzo (Chap. 15). Here the authors assume that urban areas and their development patterns are inherently spatial networks of people, firms and institutions, which interact and adjust vis-à-vis one another over time. To explore this urban network dynamics, they introduce an approach based on a combination of GIS and graph-theoretic techniques. Specifically, a Kriging method is used to create a continuous surface of the phenomenon of interest, for example population density, and raster analysis is adopted to characterize the underlying network topology. Next, graph-theoretic techniques (such as betweenness) measure the structural importance of each part of the network: for instance, population density versus employment density. The authors apply this approach to three metropolitan areas in the USA to explore how patterns of population and employment in these cities have changed in relation to one another over the last couple of decades.

The next three chapters refer to empirical analyses in spatial commuting networks. Griffith (Chap. 16) raises the issue of spatial autocorrelation effects within different spatial interaction model specifications (specifically, unconstrained, singly-constrained and doubly-constrained spatial interaction models). Spatial filtering methodology is used in this analysis. The empirical experiments carried out for the 2000 Texas journey-to-work data corroborate the notion that positive spatial autocorrelation biases the estimation of distance decay effects uncovered with geographic flows models that fail to account for it. The author concludes his chapter by suggesting that the theme of negative spatial autocorrelation should be explored in order to gain a better understanding of the complexity of spatial interaction data. Next, De Montis et al. (Chap. 17) carry out a comparative analysis of the commuting networks of the two main Italian islands (Sardinia and Sicily), by exploring the interplay between topological, traffic and demographic characteristics of the two networks by means of network analysis tools. The authors also highlight the necessity to investigate four specific research areas in the framework of a possible research agenda: (a) integration of GIS and complex network modelling; (b) study of the evolution of networks over time; (c) analysis of comparable and non-comparable networks, and (d) detection of communities on networks. Patuelli et al. (Chap. 18) then analyse the evolution of commuting networks in Germany from two perspectives: space and connectivity. The results of their empirical experiments – concerning the home-to-work commuting flows among 439 German districts for the years 1995 and 2005 – aim to identify, among the main German attraction districts, the most "open" and connected ones. These emerging districts can be considered as future spatial-economic attractors, as well as network interconnectors, that is "hubs".

Finally, Reggiani (Chap. 19) concludes the book, with an "interdisciplinary" synthesis of the methodological relationships between spatial economics and network science. The focus here is on similarities and connections concerning the theoretical foundations, approaches, and functional forms between these two frameworks, in the light of a unifying conceptualization. Reflections on a new research agenda for both theoretical and empirical research – with the aim of jointly exploring the two fields of analysis (spatial economics and networks) – are also offered. All these complexity studies indicate the need to deepen – in a cross-disciplinary approach – theoretical, methodological, and empirical investigation into the field of complexity, evolution and networks in space, an observation also made by Baofu (2007).

## References

Andersson C, Rasmussen S, White R (2002) Urban settlement transitions. Environ Plann B 29: 841–865

Baofu P (2007) The future of complexity. Imperial College Press, London

Barabási A, Albert R (1999) Emergence of scaling in random networks. Science 286:509–512

Boschma R, Frenken K (2006) Why is economic geography not an evolutionary science? J Econ Geogr 6(3):273–302

Gough I, Runciman G, Mace R et al. (2008) Darwinian evolutionary theory and the social sciences. 21st Century Soc 3(1):65–86

Nicolis G, Nicolis C (2007) Foundations of complex systems. Imperial College Press, London

Tobler W (1970) A computer movie simulating urban growth in the Detroit region. Econ Geogr 46(2):234–240

van den Bergh JCJM (2007) Evolutionary thinking in environmental economics. J Evol Econ 17:521–549

# Part A
# Complexity, Evolution, and Simplicity in Space

# Chapter 2
# The "Thermodynamics" of the City

## Evolution and Complexity Science in Urban Modelling

**Alan Wilson**

## 2.1 Introduction

The primary objectives of this chapter are twofold: first, to offer a review of progress in urban modelling using the methods of statistical mechanics; and second, to explore the possibility of using the *thermodynamic* analogy in addition to statistical mechanics. We can take stock of the "thermodynamics of the city" not in the sense of its physical states – interesting though that would be – but in terms of its daily functioning and its evolution over time. *We will show that these methods of statistical mechanics and thermodynamics illustrate the contribution of urban modelling to complexity science and form the basis for understanding the evolution of urban structure.*

It is becoming increasingly recognised that the mathematics underpinning thermodynamics and statistical mechanics have wide applicability. This is manifesting itself in two ways: broadening the range of systems for which these tools are relevant; and seeing that there are new mathematical insights that derive from this branch of Physics. Examples of these broader approaches are provided by Beck and Schlagel (1993) and Ruelle (1978, 2004). The recognition of the power of the method and its wider application goes back at least to the 1950s (Jaynes, 1957, for example) but understanding its role in complexity science is much more recent. However, these methods are now being seen as offering a major contribution. In general, the applications have mainly been in fields closely related to the physical sciences. The purpose of this chapter is to demonstrate the relevance of the methods in a field that has had less publicity but which is obviously important: the development of mathematical models of cities. The urban modelling field can be seen, in its early manifestation, as a precursor of complexity science; and, increasingly, as an important application within it (Wilson 2000).

A. Wilson
Centre for Advanced Spatial Analysis, University College London, London, UK

A. Reggiani and P. Nijkamp (eds.), *Complexity and Spatial Networks*, Advances in Spatial Science,
DOI: 10.1007/978-3-642-01554-0_2, 

There have been two main phases of development in this branch of urban science and a third now beckons. The first was in the direct application of the methods of statistical mechanics in urban analysis in the modelling of transport flows in cities (Wilson 1967). These models were developed by analogy though it was soon recognised that what was being used was a powerful general method. A family of spatial interaction models was derived and one of these was important as the beginnings of locational analysis as well as the representation of flows in transport models (Wilson 1970).

The second phase extended the locational analysis to the modelling of the evolution of structures, with retail outlets providing an archetypal model (Harris and Wilson 1978). This was rooted in the developments in applied nonlinear dynamics in the 1970s and not directly connected to statistical mechanics. The equations were largely solved by computer simulation, though some analytical insights were achieved. This provides the beginnings of a method for modelling the evolution of cities – the urban analogue of the equivalent issue in fields such as developmental biology. It is a powerful example of the possibility of modelling evolution within complexity science.

The emerging third phase reconnects with statistical mechanics. It was shown in the evolution modelling that there could be sudden changes in structure at critical values of parameters. Are these analogues of phase transitions in statistical mechanics? There was always the possibility that analogies with Ising models in Physics and their progeny – concerned with the properties of molecules on a lattice – would offer further insights since these represented a kind of locational structure problem and some interesting mathematics were associated with these models. Statistical mechanics is now handling much more complex structural models and there is a much fuller understanding of phase transitions. This makes it worthwhile to pursue the analogy again.

The chapter is structured as follows. In Sect. 2.2 we present two archetypal models – first the transport model and second the retail model – to represent urban systems of interest. In each case, we combine the description of the models with a thermodynamic interpretation. In Sect. 2.3, we show how the retail model can be extended to be an archetypal model of the evolution of urban structure and, again, the associated thermodynamics. In Sect. 2.4, we explore future challenges.

## 2.2 The Thermodynamics of Spatial Interaction

### *2.2.1 Introduction*

In this section, we combine presentations of some archetypal models of cities which have been, or can be, rooted in concepts that are in common with those of statistical mechanics – representing transport flows and flows to retail centres. We intersperse these presentations with explorations of thermodynamic and statistical mechanical analogies. In Sect. 2.2.2, we introduce the systems of interest and define the key variables. In Sect. 2.2.3, we present the transport model and in Sect. 2.2.4, the model of flows to retail centres.

### 2.2.2 *The Archetypal Submodels*

Transport planners have long needed to understand the pattern of flows in cities and a core scientific task is to model these flows both to account for an existing situation and to be able to predict the consequences of change in the future – whether through, for example, population change or through planned transport investment and network development. The models in principle provide the analytical base for optimising transport policy and investment.

Assume that the city can be divided into a set of discrete zones, labelled 1, 2, 3, . . . , $N$. Then the core of the modelling task is to estimate the array $\{T_{ij}\}$, where $T_{ij}$ is the number of trips from zone $i$ to zone $j$. This pattern obviously depends on a whole host of variables: trip demand at $i$ (origins, $O_i$), trip attractions at $j$ (destinations, $D_j$), the underlying transport network and associated congestion effects, and so on. The network is handled through a matrix of generalised travel costs, $\{c_{ij}\}$. We describe the core model in Sect. 2.2.3.

Suppose we now focus on retail trips alone, represented by a matrix $\{S_{ij}\}$. These might be proportional to the spending power at $i$ ($e_iP_i$, with $e_i$ as per capita expenditure, $P_i$, the population) and the attractiveness of retail facilities in $j$ (which we designate as $W_j$). The model can then predict a locational vector $\{D_j\}$ which is $\Sigma_i S_{ij}$, the sum of the flows into a retail centre attracted by $W_j$. An ability to predict $\{D_j\}$ is valuable for planning purposes, whether in the private (retail) sector or for public facilities such as hospitals and schools. This model is elaborated in Sect. 2.2.4. We can use what might be called phase 1 methods to estimate $\{S_{ij}\}$, but this shows the phase 2 task to be the modelling of the dynamics of the structural vector $\{W_j\}$. We indicate an approach to this in Sect. 2.3.

### 2.2.3 *The Transport Model*

Transport flows were initially modelled on the basis of an analogy with Newtonian physics – the so-called gravity model. We use the notation introduced in Sect. 2.2.2.

$$T_{ij} = KO_iD_jc_{ij}^{-\beta}, \tag{2.1}$$

where $k$ and $\beta$ are constants. This proved unsatisfactory and various factors were added to improve the fit to reality. The breakthrough (Wilson 1967) was to recognise that these had a resemblance to statistical mechanics' partition functions.

To show the connection and to facilitate later analysis, we introduce some of the core concepts of statistical mechanics here. The simplest Boltzmann model is represented by the microcanonical ensemble. This is a set of copies of the system each of which satisfies some constraint equations which describe our knowledge of the macro system. It is assumed that each copy can occur with equal probability but Boltzmann's great discovery was to show that one distribution occurs with overwhelming probability. This distribution can be found by maximising an appropriate probability function which then turns out to be, essentially, the entropy function.[1] For a perfect gas with a fixed number of articles, $N$ and fixed energy, $E$, if $n_i$ is the number of particles with energy $\varepsilon_i$, then the most probable number of particles in each energy level – the most probable distribution – is obtained by maximising the entropy:

$$S = -\Sigma_i n_i \log n_i, \tag{2.2}$$

subject to

$$\Sigma_i n_i = N, \tag{2.3}$$

$$\Sigma_i n_i \varepsilon_i = E, \tag{2.4}$$

to give

$$n_i = N \exp(-\beta\varepsilon_i)/\Sigma_i \exp(-\beta\varepsilon_i), \tag{2.5}$$

where

$$\beta = 1/kT. \tag{2.6}$$

$T$ is the temperature and $k$ is Boltzmann's constant.

It is convenient to define the partition function as:

$$Z = \Sigma_i \exp(-\beta\varepsilon_i). \tag{2.7}$$

It is useful for a future point in the argument to note here that we can link thermodynamics and statistical mechanics through the free energy, $F$ (and here we follow Finn 1993) defined in terms of the partition function as:

$$F = -NkT \log Z, \tag{2.8}$$

and all thermodynamic properties can be calculated from this.

[1]The detailed justification for this is well known and not presented here.

The post-Newton, Boltzmann-like, transport model can then be developed on the basis of such a microcanonical ensemble. Now, instead of a single state label, $i$, representing energy levels, there is a double index, $(i,j)$, labelling origin–destination pairs. The constraint equations then become:

$$\Sigma_j T_{ij} = O_i, \tag{2.9}$$

$$\Sigma_i T_{ij} = D_j, \tag{2.10}$$

$$\Sigma_i T_{ij} c_{ij} = C. \tag{2.11}$$

The “number of particles” constraint – a single equation in physics – is replaced by the sets of constraints (2.9) and (2.10). $C$ is clearly the urban equivalent of “energy” for this system and the $c_{ij}$, measures of the cost of travel from $i$ to $j$, are the equivalent of energy levels. If $c_{ij}$ is measured in money units, then $C$ is measured in money units also. Then, maximising a suitable “entropy”[2]

$$S = -\Sigma_i T_{ij} \log T_{ij} \tag{2.12}$$

gives, subject to (2.9)–(2.11), the so-called doubly-constrained model:

$$T_{ij} = A_i B_j O_i D_j \exp(-\beta c_{ij}). \tag{2.13}$$

The parameter $\beta$ measures the “strength” of the impedance: if $\beta$ is large, trips are relatively short, and vice versa. It can be determined from (2.11) if $C$ is known, but in practice it is likely to be treated as a parameter of a statistical model and estimated from data. $A_i$ and $B_j$ are balancing factors to ensure that (2.9) and (2.10) are satisfied. Hence:

$$A_i = 1/\Sigma_j B_j D_j \exp(-\beta c_{ij}), \tag{2.14}$$

and

$$B_j = 1/\Sigma_i A_i O_i \exp(-\beta c_{ij}). \tag{2.15}$$

The inverses of $A_i$ and $B_j$ are the analogues of the partition functions. However, they do not translate easily (or at all) into thermodynamic form.

It is generally recognised that to make the models work, $c_{ij}$ should be taken as a generalised cost, a weighted sum of elements like travel time and money cost. To

[2]There are many possible definitions of entropy that can be used here, but for present purposes, they can all be considered to be equivalent.

take the thermodynamic analogy further, we do need a common unit and, as noted earlier and to fix ideas, we take "money" as that unit. These will then be the units of "energy" in the system.[3] Given that the units are defined, then the $\beta$ parameter, together with the definition of a suitable Boltzmann constant, $k$, will enable us to define temperature through:

$$\beta = 1/kT. \tag{2.16}$$

We are accustomed to estimating $\beta$ through model calibration. An interesting question is how we define $k$ as a "universal urban constant" which would then enable us to estimate the "transport temperature" of a city. Note that with $c_{ij}$ having the dimensions of money, then $\beta$ has the dimensions of (money)$^{-1}$ and so from (2.16), $kT$ would have the dimensions of money. If $k$ is to be a universal constant, then $T$ would have the dimensions of money. It is also interesting to note that it has been proved that (Evans 1973), in the transport model, as $\beta \rightarrow \infty$, the array $\{T_{ij}\}$ tends to the solution of the transportation problem of linear programming in which case $C$, in (2.11), tends to a minimum. This is the thermodynamic equivalent of the temperature tending to absolute zero and the energy tending to a minimum.

### 2.2.4 *Retail Systems: Interaction Models as Location Models*

The next step is to introduce a spatial interaction model that also functions as a location model. We do this through the singly-constrained "retail" model that is, retaining a constraint analogous to (2.9), but dropping (2.10). We begin with the conventional model and introduce a new notation to distinguish it from the transport model. We use the notation introduced in Sect. 2.2.2

The vector $\{W_j\}$ can be taken as a representation of urban structure – the configuration of $W_j$s. If many $W_j$s are non-zero, then this represents a dispersed system. At the other extreme, if only one is non-zero, then that is a very centralised system. A spatial interaction model can be built for the flows on the same basis as the transport model. Maximizing an entropy function:

$$-\Sigma_{ij} S_{ij} \log S_{ij}, \tag{2.17}$$

we find

$$S_{ij} = A_i e_i P_i W_j^{\alpha} \exp(-\beta c_{ij}), \tag{2.18}$$

[3] For simplicity, we will henceforth drop the quotation marks and let them be understood when concepts are being used through analogies.

where

$$A_i = 1/\Sigma_k W_k^\alpha \exp(-\beta c_{ik}), \tag{2.19}$$

to ensure that

$$\Sigma_j S_{ij} = e_i P_i, \tag{2.20}$$

$$\Sigma_{ij} S_{ij} \log W_j = X, \tag{2.21}$$

and

$$\Sigma_{ij} S_{ij} c_{ij} = C. \tag{2.22}$$

Equation (2.21) represents a new kind of constraint. It is inserted to generate the $W_j^\alpha$ term in (2.18), but the form of this equation shows that log $W_j$ can be taken as a measure of size benefits to consumers using $j$ and $X$ an estimate of the total. $\alpha$ is a parameter associated with how consumers value "size" of retail centres – and is actually the Lagrangian multiplier that goes with the constraint (2.21). In thermodynamic terms, as we will see shortly, $X$ can be taken as another kind of energy. As in the transport model, $C$ is the total expenditure on travel. $\beta$ measures travel impedance as in the transport model and is the Lagrangian multiplier that associated with (2.22).

Because the matrix is only constrained the origin end, we can calculate the total flows into destinations as:

$$D_j = \Sigma_i S_{ij} = \Sigma_i e_i P_i W_j^\alpha \exp(-\beta c_{ij})/\Sigma_k W_k^\alpha \exp(-\beta c_{ik}), \tag{2.23}$$

and this is how the model also functions as a location model.[4]

$W_j^\alpha$ can be written:

$$W_j^\alpha = \exp(\alpha \log W_j). \tag{2.24}$$

If we then assume, for simplicity and for illustration, $W_j$ can be taken as "size" and that benefits are proportional to size, then this shows explicitly that log $W_j$ can be taken as a measure of the utility of an individual going to a shopping centre of size $W_j$ but at a transport cost, or disutility, represented by $c_{ij}$. The significance of this in

[4]This model, in more detailed form, has been widely and successfully applied.

the thermodynamic context is that $\alpha$ can be seen (via another Boltzmann constant, $k'$) as related to a different kind of temperature, $T'$:

$$\alpha = 1/k'T'. \tag{2.25}$$

It was originally shown in Wilson (1970), following Jaynes (1957), that this argument can be generalised to any number of constraints and hence any number of temperatures. It can easily be shown, as in Physics, that if two systems are brought together with different temperatures, then they will move to an equilibrium position at an intermediate temperature through flows of heat from the hotter to the colder body. This also means, therefore, that in this case, there can be flows of different kinds of heat. In this case, the flow of heat means that more people "choose" destinations in the "cooler" region.

### 2.2.5 *Deepening the Thermodynamic Analogy*

In order to learn more from the thermodynamic analogy, we need to remind ourselves of some of the core concepts. The two key laws of thermodynamics, the first and second, are concerned with (a) the conservation of energy and (b) the fact that a system's energy cannot be increased without an amount of work being done on the system which is greater than or equal to the energy gain.

There are a number of so-called thermodynamic functions of state and we briefly note those needed for our ongoing argument. The internal energy (which we will equate with our "$C$") is particularly important. It normally appears in differential form, for example as:

$$dU = dQ + \Sigma_i X_i dx_i, \tag{2.26}$$

where $dQ$ is the flow of heat and $\Sigma_i X_i dx_i$ represents the work done on the system by various external forces, $\{X_i\}$. The $\{x_i\}$ are system descriptors – variables – so that $dx_i$ measures the change in the variable from the application of the force. Essentially, the increase in the internal energy is the sum of the heat flow in and the work done. For example, there may be a change in volume, $V$, an $x$-variable, from the application of pressure, $P$, an $X$-force.

We can introduce entropy, $S$, for the first time (in its thermodynamic form) by defining it through:

$$dQ = TdS, \tag{2.27}$$

so that

$$dU = TdS + \Sigma_i X_i dx_i. \tag{2.28}$$

The second law can then be formulated as:

$$TdS \geq 0 \tag{2.29}$$

(or, "entropy always increases"). For a fluid of volume, $V$, and pressure, $P$, the work done can be represented by $PdV$, and

$$dU = TdS - PdV \tag{2.30}$$

(there is a negative sign because the work done on the system produces a reduction in volume and so the minus sign turns this into a positive contribution to work). In other cases, the $X$ and the $x$ might be the degree of magnetisation brought about by a magnetic field, for example. The general formulation in (2.26) and (2.28) is particularly important for our discussion of cities below: the challenge then is to identify the $\{X_i\}$ and the $\{x_i\}$ in that case.

We can introduce the free energy, $F$, as:

$$F = U - TS, \tag{2.31}$$

and in differential form as:

$$dF = -SdT - PdV, \tag{2.32}$$

or in a more general form, from (2.28) and (2.31), as:

$$dF = -SdT + \Sigma_i X_i dx_i. \tag{2.33}$$

F can be specified as a function of $T$ and $V$ and then all other properties can be deduced.

The free energy (Pippard 1957, p. 56) is a measure of the work that can be done – a decrease in $F$ – by a system in an isothermal reversible change. Given the second law, it is the *maximum* amount of work that can be done by a system. We can also note that by inspection of (2.31), the principle of maximizing entropy, which we will invoke below, is equivalent – other terms being kept constant – to minimizing free energy. This notion has been very interestingly exploited by Friston [for example – see Friston et al. (2006) and Friston and Stephan (2007)] in a way that we will examine briefly later in Sect. 2.4.

In the case of spatial flow models, we need to recognise two kinds of change through work being done on the system (or heat flowing). In terms of the transport elements of either of our archetypal models, this can be a $\delta C$ change or a $\delta c_{ij}$ change. The former is a whole system change that means, for example, there is a greater resource available for individuals to spend on transport – and this will decrease $\beta$ and hence increase the temperature; the latter would probably be produced by a network change – say the investment in a new link. Even with fixed $C$, if this leads to a reduction in cost, we would expect it to generate an increase in temperature. In terms of the Physics analogy, a positive $\delta C$ change is equivalent to an increase in energy. It would be possible in principle to define an external coordinate, $x_i$, and a generalised force, $X_i$, so that $X_i \delta x_i$ generated $\delta C$. It is less easy to find a Physics analogue for $\delta c_{ij}$ changes – because that would involve changing energy levels.

This analysis enables us to interpret the principal laws of thermodynamics in this context. "Work done" on the system will be manifested through either $\delta C$ or $\delta c_{ij}$ changes. Essentially, what the laws tell us is that there will be some "waste" through the equivalent of heat loss. Note that an equivalent analysis could be offered for the retail model for $\delta W_j$ or $\delta X$ changes.

We should now return to the basics of the thermodynamic analogy and see if there are further gains to be achieved – particularly by returning to the $\Sigma_i X_i dx_i$ terms [from (2.2)]. It is worth noting that a system of interest is described by variables that divide into two sets: the *extensive* variables, that are dependent on size, and the *intensive* variables that are system properties that are not size dependent. The volume of a gas, $V$, is an example of the first; its temperature, $T$, and pressure, $P$, are examples of the second. It is a task of thermodynamics analysis to seek state equations that relate the key variables. For an ideal gas, there is Boyle's Law:

$$PV = nRT, \tag{2.34}$$

where $n$ is a measure of the number of particles and $R$ is a universal constant.

In the urban case, we have available to us a temperature through the parameter $\beta$ (actually $1/kT$, an inverse temperature). The next step is to explore whether there is an $x_i$ which is the equivalent of a volume, $V$. The volume of a gas is the size of the container. In this case, for simplicity for this initial exploration, we can take the area, $A$, of the city as a measure of size.[5] This would then allow us to work with the free energy as a function of $T$ and $V$ – or in the urban case, $\beta$ and $A$: $F(\beta, A)$, say. We can then explore the idea of a state equation and it seems reasonable to start with Boyle's law since people in cities are being modelled on the same basis as an ideal classical gas. This suggests, by analogy with (2.34) that:

$$PA = NRT, \tag{2.35}$$

where $N$ is the total population and $R$ is a constant. In terms of $\beta$, this becomes:

$$P = NR/\beta kA, \tag{2.36}$$

where we have taken $A$ to the other side of the equation.[6] There are, of course, two constants, $R$ and $k$, in this equation which cannot be obtained in the same way as in Physics, but let us assume for the moment that they can be estimated. Then, (2.36)

---

[5] We should explore whether we can determine a measure of $A$ from the topology of the $\{c_{ij}\}$.

[6] Note that $P$ appears to have the dimensions of "density" x'money'.

gives us a definition of an urban "pressure". It has the right properties intuitively: it increases if $A$ or $\beta$ decreases or $N$ increases (in each case, other variables held constant).

The final step in this exploration of a deepening analogy is to link the thermodynamics with the statistical mechanics that generated the flow models. In physics, this is achieved by connecting the free energy to the partition function of the system of interest. We saw in the transport case that while we could find analogues of partition functions, the analogy was not exact.[7]

In the retail case we have dropped one set of "number" constraints and this suggests that the inverse of the $A_i$ term will function as a partition function. Consider

$$Z_i = \Sigma_k \exp(\alpha \log W_k - \beta c_{ik}). \tag{2.37}$$

This looks like a partition function, but as a function for each zone $i$ rather than for the system as a whole. This is because the consumers leaving a zone can be treated as an independent system.[8] It is perhaps then not too great a leap to make the heroic assumption that an appropriate partition function for the system is:

$$Z = \Sigma_i Z_i = \Sigma_{ik} \exp(\alpha \log W_k - \beta c_{ik}). \tag{2.38}$$

We can then seek to work with the free energy and the model at (2.16) and (2.26). Then, using (2.8):

$$F = -[N/\beta] \log Z. \tag{2.39}$$

We can also explore the standard method of calculating state functions from the free energy:[9]

$$P = -(\partial F/\partial A)_T, \tag{2.40}$$

$$S = -(\partial F/\partial T)_A, \tag{2.41}$$

or, using (2.16):

$$S = -k\beta^2(\partial F/\partial \beta)_A. \tag{2.42}$$

[7] Can we take $A_i B_j$ as an $i$–$j$ partition function? Can we work backwards and ask what we would like the free energy be for this system? If (2.11) specifies the energy and $\beta$ (=$1/kT$) the temperature, then $F=U-TS$ becomes $F=C-S/k\beta$? Then if $F=NkT \log Z$, what is $Z$?

[8] ter Haar (1995, p. 202) does show that each subsystem within an ensemble can itself be treated as an ensemble provided there is a common β value.

[9] The following equations can be derived from (2.31) with A substituted for $V$ and $T=1/k\beta$.

And, with $U=C$, using (2.23),

$$U = F - (\partial F/\partial T)_A = -T^2(\partial/\partial T \cdot F/T)_A = (\partial/\partial\beta \cdot k\beta F)_A. \qquad (2.43)^{10}$$

In this formulation, $A$ does not appear in the partition function. We might consider $A$ to be defined by the topology of the $\{c_{ij}\}$ and possibly the spatial distribution of the $W_j$ and this should be explored further. Indeed, more generally we might write (2.40) as:

$$X_i = -(\partial F/\partial x_i)_T. \qquad (2.44)$$

It might be particularly interesting to look at the concepts of specific heat. "Heat" flowing into a city will be in the form of something like investment in the transport system and this will increase $T$ and hence decrease $\beta$ but each city will have a specific heat and it will be interesting to look at how different cities can effectively absorb investment. This should connect to cost–benefit analysis, possibly through NPVs. The standard formulae for specific heats can be transformed into the urban formalism as follows:

$$C_V = (\partial U/\partial T)_V \to -1/k\beta^2(\partial U/\partial\beta)_A, \qquad (2.45)$$

and

$$C_P = (\partial U/\partial T)_P + P(\partial V/\partial T)_P = [-1/k\beta^2(\partial U/\partial\beta)_P + P(\partial V/\partial\beta)_P]. \qquad (2.46)$$

It remains a challenge to calculate these in the urban case.

We should also examine the possibility, noted earlier, of examining some of these concepts at the level of a zone within city – building on ter Haar's concept of subsystems.[11]

It remains to ask the question of whether there could be phase changes in spatial interaction systems.[12] This seems intuitively unlikely for the spatial interaction models: smooth and fast shifts to a new equilibrium following any change is the likely outcome. If the model is made more realistic – and more complicated – by adding different transport modes, then the position could be different. There could then be phase changes that result in a major switch between modes at some critical parameter values (see, for example, Wilson, 1976). However, there is the possibility of significant phase changes in the structural model and it is to this that we now turn.

---

[10] What does this produce for $U$? And is it possible to do all the calculations implied by (2.40)–(2.46)?

[11] It is possible to introduce a $\beta_i$ rather than a $\beta$ which reinforces this idea.

[12] We elaborate the notion of phase changes in the next section. Essentially, in this case, they would be discrete "jumps" in the $\{T_{ij}\}$ or $\{S_{ij}\}$ arrays at critical values of parameters such as $\beta$.

## 2.3 Urban Structure and its Evolution

### *2.3.1 The Model*

We have presented an archetypal singly-constrained spatial interaction model, representing (among other things) flows to the retail sector. We can now add a suitable hypothesis for representing the dynamics (following Harris and Wilson 1978):

$$dW_j/dt = \varepsilon(D_j - KW_j)W_j, \tag{2.47}$$

where $K$ is a constant such that $KW_j$ can be taken as the (notional) cost of running the shopping centre in $j$.[13] This equation then says that if the centre at $j$ is profitable, it grows; if not, it declines. The parameter $\varepsilon$ determines the speed of response to these signals.

The equilibrium position is given by:

$$D_j = KW_j, \tag{2.48}$$

which can be written out in full, using (2.23), as:

$$\Sigma i\{e_i P_i W_j \exp(-\beta c_{ij})/\Sigma_k W_k \exp(-\beta c_{ik})\} = KW_j. \tag{2.49}$$

The (2.47) are analogous to Lotka–Volterra equations – in the form of species competing for resources. In this case, we have retail developers competing for consumers. Because this model combines Boltzmann's statistical mechanics (B) and Lotka's and Volterra's dynamics (LV), these have been characterised as BLV models and it has been shown that they have a wide range of application (Wilson 2008).

What is clear to the present time is that it is possible to characterise the kinds of configurations that can arise for different regions of $\alpha$ and $\beta$ space: for larger $\alpha$ and lower $\beta$, there are a smaller number of larger centres; and vice versa.[14] This can be interpreted to an extent for a particular zone, say $j$, by fixing all the $W_k$, for $k \neq j$. A key challenge is to solve this problem with all the $W_j$s varying simultaneously. There are many procedures for solving (2.49) iteratively but we constantly need to bear in mind the sensitivity to the initial conditions.

The zonal interpretation is shown in Fig. 2.1. The left and right hand sides of (2.49) are plotted separately and of course, the intersections are the possible equilibrium points. If $\alpha \leq 1$, there is always a possible equilibrium point, but if - $\alpha > 1$, there are three possible cases: only zero as an equilibrium; one additional non-zero stable state; and the limiting ($\alpha = 1$) case that joins the two. The $\beta$ value also

---

[13] $K$ could be $j$-dependent as $K_j$ (and indeed, usually would be) but we retain $K$ for simplicity of illustration.

[14] Clarke and Wilson (1985).

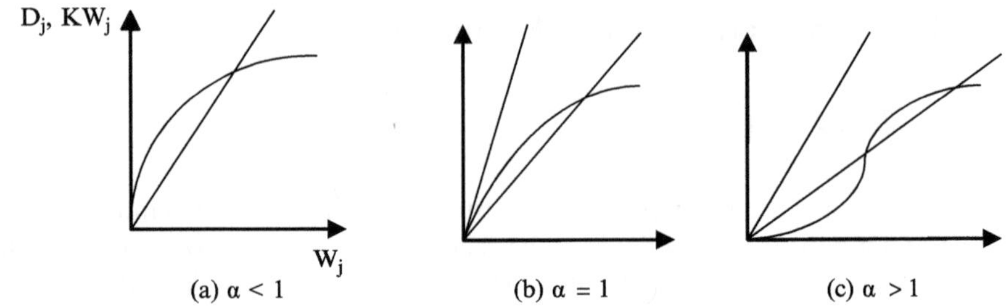


**Fig. 2.1** Zonal analysis of phase transitions

determines the position of the equilibria. This analysis shows a number of properties that are typical of nonlinear dynamical systems: multiple (system) equilibria and strong path dependence – that is, sensitive dependence on initial conditions. It also shows that as the parameters $\alpha$ and $\beta$ (and indeed any other exogenous variables) change slowly, there is the possibility of a sudden change in a zone's state – from development being possible to development not being possible, or vice versa [as depicted by the two $KW_j$ lines in Fig. 2.1b, c]. These kinds of change can be characterised as phase transitions – in this case at a zonal level, but clearly there will be system wide changes of this kind as well. This analysis is the basis of a very powerful tool for identifying complex phase transitions. We return to this in the Sect. 2.4.6.

Recall that this analysis is dependent, for a particular $W_j$, on the set $\{W_k\}$, $k \neq j$, being constant. It is almost certainly a good enough approximation to offer insight, but the challenge is to address the problem of simultaneous variation. The system problem is to predict equilibrium values for the whole set $\{W_j\}$ and the trajectories through time, recognizing the points at which phase changes take place. This is where newer statistical mechanics models potentially can help.

This analysis exemplifies characteristics of models of nonlinear complex systems: multiple equilibrium solutions, path dependence and phase transitions and so demonstrate the contribution of urban modelling to complexity science.

### *2.3.2 The Thermodynamics of Structural Change*

We have seen that the spatial interaction model, whether in its doubly-constrained (transport) form, $\{T_{ij}\}$, or singly-constrained (retail) form, $\{S_{ij}\}$, is best represented by a microcanonical ensemble and we can reasonably assume a rapid return to equilibrium following any change. We have offered an equation representing the dynamics of $\{W_j\}$ evolution but we can now work towards an interpretation of this model in a statistical mechanics format. It will be represented by a canonical ensemble. This differs from a microcanonical ensemble in that the energy is allowed to vary. The return to equilibrium after a disturbance is likely to be much slower: it takes developers much longer to build a new centre than for individuals to adjust their transport routes for example. What is more, the two systems are linked because the structural variables $\{W_j\}$ are exogenous variables in the retail model (and there is an equivalent vector in the transport model). In the case of the structural model, we will have to assume some kind of steady state independent rapid-return-to-equilibrium for the interaction arrays. We have indicated in the previous section that there will be discrete changes. We now explore the possible statistical mechanics' bases to see if these are in fact phase changes.

In Physics, the energy is most generally represented in a Hamiltonian formulation and so we denote it by $H$. We can now construct a canonical ensemble in which each element is a state of the system with potentially varying energy. For each "system" energy, that part of the ensemble will be a copy of the corresponding microcanonical ensemble. That is, it can be shown (Wilson 1970, Appendix 2) that the microcanonical distribution is nested in the canonical distribution for each energy value in the latter.

We can again work with probabilities, but we denote them by $P_r$ since they relate to the probability of the *system* state occurring – and we label a particular state r. We can then maximise a system entropy to get the result that:

$$P_r = \exp(-\beta H_r)/\Sigma_r \exp(-\beta H_r). \tag{2.50}$$

Physicists have modelled the distribution of states of particles on a lattice. An early example was the Ising model which is concerned with spin systems and the alignment of spins at certain temperatures that produce magnetic fields. The interactions in the Ising model are only with nearest neighbours and there are no phase transitions. However, when it is extended to two and three dimensions, there are phase transitions, but it is very much more difficult to solve. In our case, of course, we are interested in interactions that extend, in principle, between all pairs. Such models have been explored in statistical mechanics and, below, we explore them and seek to learn from them – see Martin (1991).

Locations in urban systems can be characterised by grids and urban structure can then be thought of as structure at points on a lattice. We can consider zone labels $i$ and $j$ to be represented by their centroids which can then be considered as the nodes of a lattice. The task, then, is to find a Hamiltonian, $H_r$, as a function of the structural vector $\{W_j\}$. We can then write (2.50) as:

$$P_r = \exp(-\beta H_r(\{W_j\}))/\Sigma_r \exp(-\beta H_r(\{W_j\})), \tag{2.51}$$

and we have to find the $\{W_j\}$ that maximises $P_r$. Since the denominator is the same for each $r$, this problem becomes:

$$L\left[\{W_j^{\text{opt}}\}\right] = \text{Max}_r \exp(-\beta H_r(\{W_j\})). \tag{2.52}$$

So the immediate issue is to decide on the Hamiltonian. Suppose we take the measure of profit used in (2.47). Then:

$$H = \Sigma_j (D_j - KW_j), \tag{2.53}$$

and the problem becomes:

$$L\left[\{W_j^{opt}\}\right] = \text{Max}_r \exp[-\beta\Sigma_j(D_j - KW_j)], \tag{2.54}$$

where $K$ is a unit cost for retailers and $D_j$ can be obtained in the usual way. Substitution then gives:

$$L\left[\{W_j^{opt}\}\right] = \text{Max}_r \exp[-\beta\Sigma_j(\Sigma_i\{e_iP_iW_j \exp(-\beta c_{ij})/\Sigma_k W_k \exp(-\beta c_{ik})\}KW_j)], \tag{2.55}$$

which shows what a formidable problem this appears to be. However, scrutiny of the right hand side shows that we maximise $L$ by maximizing the exponent and because of the first negative sign, this is achieved by minimising

$$\Sigma_j(\Sigma_i\{e_iP_iW_j \exp(-\beta c_{ij})/\Sigma_k W_k \exp(-\beta c_{ik})\} - KW_j), \tag{2.56}$$

which then suggests that the equilibrium value for $\{W_j\}$ occurs when this expression is a minimum. However, by inspection, we can see that this happens when each term within $\Sigma_j$ is zero:

$$\Sigma_i\{e_iP_iW_j \exp(-\beta c_{ij})/\Sigma_k W_k \exp(-\beta c_{ik})\} = KW_j, \tag{2.57}$$

which is, of course, simply the equilibrium condition (2.48) [or (2.49)]. This then seems to indicate that a statistical mechanics exposition produces an equivalent equilibrium condition for the $\{W_j\}$.

What we know from the analysis of Fig. 2.1 is that at a zonal level, there are critical values of $\alpha$ and $\beta$, for example, beyond which only $W_j = 0$ is a stable solution for that zone – that is, the expression inside $\Sigma_j$. So we know that there are critical points at a zonal level at which, for example, there can be a jump from a finite $W_j$ to a zero $W_j$ (see Dearden and Wilson 2008, for a simulation of this). This implies there is a set of $\alpha$ and $\beta$ at which there will be critical changes somewhere in the system. This is particularly interesting when we compare this situation to that in statistical mechanics. There, we are usually looking for critical temperatures for the whole system at which there is a phase transition. Here, there will be many more system phase transitions, but in each case consisting of a zonal transition (which then affects the system as a whole – since if a $W_j$ jumps to zero, then other $W_k$s will jump upwards – or vice versa). It would be interesting to see whether the set of critical $\alpha$s and $\beta$s form a continuous curve. If we further add, say, $K$ and the $\{e_iP_i\}$, then we are looking for a many-dimensional surface. It will also be interesting to see whether there are other systems – ecosystems? – that exhibit this kind of phase change.

To take the argument further at the system level, we need to construct an order parameter. In Physics, at a phase change, there is a discontinuity in the order parameter and hence indeterminacy in some derivatives of the free energy. An obvious example in Physics is in magnetism: an ordered system has particles with spins aligned – ordered – and there can be phase transitions to and from disordered states. In these cases, the order parameters are straightforward to define. In the urban case, intuition suggests that it is the nature of the configurations of $\{W_j\}$ that we are concerned with. A dispersed system with many small centres can be considered less ordered than one with a small number of large centres. This suggests that we should examine $N[W_j > x]$ – the number of $W_j$ greater than some parameter $x$. If $x$ is set to zero, this will be a measure of ubiquity of centres and we know that there will be transitions at $\alpha = 1$. Or we could set $x$ to be large and seek to identify configurations with a small number of large centres to see whether they are achieved through phase transitions as parameters vary.[15] There is also the interesting possibility that entropy is used as a measure of dispersion and so $-\Sigma_j W_j \log W_j$ could be used as an order parameter.[16]

### *2.3.3 An Alternative Thermodynamic Formulation for the $\{W_j\}$*

In this analysis so far, we have assumed that $\{W_j\}$ can be obtained by solving the equilibrium equations. It is interesting to explore the possibility of a suboptimal $\{W_j\}$ via entropy maximizing – something more like a lattice model with each $W_j$ as an occupation number. We can use the same argument that generates conventional spatial interaction models and differentiates them from the transportation problem of linear programming (and, of course, as we noted earlier, it has been shown that as $\beta \to \infty$, the spatial interaction model solution tends to the linear programming limit). We can proceed as follows. Assume $\{S_{ij}\}$ is given.[17] Then maximise an entropy function in $\{W_j\}$ subject to appropriate constraints.

$$\text{Max } S = -\Sigma_j W_j \log W_j, \tag{2.58}$$

such that

$$\Sigma_{ij} S_{ij} \log W_j = X, \tag{2.59}$$

[15] It would be interesting to calculate the derivatives of the free energy – the $F$-derivatives – to see whether there is a way of constructing $N[W_j > x]$ out of $F$. Are we looking at first or second order phase transitions?

[16] I am grateful to Aura Reggiani for this suggestion.

[17] It can be shown that we can carry out an entropy maximizing calculation on $\{S_{ij}\}$ simultaneously and that leads to a conventional a spatial interaction model and the same model for $\{W_j\}$. The implication of this argument is that if we obtain a $\{W_j\}$ model with the method given here, we should then recalculate $\{S_{ij}\}$ from an spatial interaction model and then iterate with $\{W_j\}$.

and

$$\Sigma_i S_{ij} = kW_j + Y, \tag{2.60}$$

$X$ and $Y$ are constants – $X$ determining the total amount of benefit that consumers derive from size (or attractiveness) and $Y$ the extent to which the equilibrium condition (2.48) is being treated as suboptimal. The Lagrangian for this problem is:

$$L = -\Sigma_j W_j \log W_j - \mu\Sigma_i[S_{ij} - KW_j - Y] - \alpha(\Sigma_i S_{ij})/W_j, \tag{2.61}$$

and setting

$$\partial L/\partial W_j = 0 \tag{2.62}$$

gives, with some re-arrangement,

$$\log W_j + \alpha D_j/W_j = \mu K_j \tag{2.63}$$

(where we have substituted $D_j$ for $\Sigma_i S_{ij}$ without loss of generality since we are taking the $\{S_{ij}\}$ as fixed). These equations could be solved numerically for $\{W_j\}$ – and indeed graphically.

It is then interesting to interpret (2.63) and then to look at the $\alpha\rightarrow\infty$ limit. Write (2.63), by dividing by $\mu$, as follows:

$$(1/\mu)\log W_j + \alpha D_j/\mu W_j = K_j. \tag{2.64}$$

The left hand side is clearly cost per square foot. The first term on the right hand side is a measure of scale benefits; the second term is revenue per square foot modified by the factor $\alpha/\mu$.

By analogy with the linear programming version of the transport model, as $\alpha\rightarrow\infty$, we would expect the normal equilibrium condition to be satisfied and hence $Y\rightarrow 0$. Equation (2.64) then suggests that as $\alpha\rightarrow\infty$, we must have $\mu\rightarrow\infty$ in such a way that $\alpha/\mu\rightarrow 1$. The first term in (2.64) then clearly tends to 0 and the equation then becomes equivalent to (2.48).

## 2.4 Ongoing Challenges

### *2.4.1 Introduction*

We noted at the outset that there have been three phases of relevant work in urban science:

- spatial interaction models – and associated location models – rooted in statistical mechanics, which work very effectively;
- models of developing and evolving structures, including the recognition of urban phase transitions;
- the use of newer methods in statistical mechanics to accelerate our understanding of development and evolution.

Within this framework, we have aimed to add thermodynamic interpretations to the findings of each phase – an area that has been raised in the past but far from fully developed. How do we now move forward?

### 2.4.2 Spatial Interaction

These are the models about which we can feel most confident in practice. Only archetypal models have been presented here, but by now they have been fully disaggregated and tested in a wide variety of circumstances. However, it is clear from the argument presented here that there remain possibly interesting areas of interpretation which can be developed through the thermodynamic analogy. In particular, it would be valuable to seek an understanding of the urban partition functions that arise from the more complex "particle number" constraints that are introduced. There is also scope for a fuller exposition of the thermodynamics of these models. A start has been made in this chapter but it would be useful, for example, to expound more fully the external variables that underpin changes in these systems.

### 2.4.3 Development and Evolution

The structural evolution issues have only been explored to date with archetypal models. There is a case for exploring, for example, phase transitions with more realistic disaggregated models and also exploring (in the case of the retail model) alternative revenue and production (that is, cost) functions to see whether new kinds of phase transitions would emerge. We should also recognise that there are other submodels that demand different formulations: for example, economic input–output models, flows of goods, even energy – leading to real thermodynamics! Finally, huge progress has been made in the development of comprehensive models and these also should be explored for phase transitions.

### 2.4.4 The "New" Thermodynamics and Statistical Mechanics

We noted at the outset that authors such as Ruelle (2004) and Beck and Schlagel (1993) – and many others – are presenting the mathematics of thermodynamics and statistical mechanics in a more general format and demonstrating in principle the applicability to a wider range of systems. It would be valuable systematically to translate the models presented in this chapter into these formats to explore the extent to which further advances are possible. We have only scratched the surface of possibilities in this chapter and further research is to be encouraged.

### *2.4.5 Models in Planning*

Urban models have long had uses in various forms of planning – public and commercial – through their forecasting capabilities and, to a lesser extent, through being embedded in optimisation frameworks. Our understanding of nonlinearities now puts a bound on forecasting capabilities but in an interesting way. While forecasting may be impossible in terms of structural variables over a long time scale – because of path dependence and phase transitions – what becomes possible is the identification of phase transitions that may be desirable or undesirable and then one aspect of planning is to take actions to encourage or avoid these as appropriate.

There is a potential new interest, that we have alluded to briefly earlier, which brings statistical mechanics to bear on urban planning, and that is Friston's (et al.'s) work on free energy and the brain. It is interesting to place this model into a planning framework. The essence of Friston's argument – for the purposes of building the analogy – is to model the brain and its environment as interacting systems, and that the brain, through its sense mechanisms, builds a model of the environment and that it handles environmental uncertainties through free energy minimisation. If the brain is replaced by "urban planning system" and its environment by "the city", then Friston's argument resonates with Ashby's (1956) law of requisite variety – essentially in this case that the planning system has to model the city in order to have a chance of success.

### *2.4.6 Concluding Comments*

There have been some spectacular successes in the application of statistical mechanics to urban modelling and some insights have been achieved in adding thermodynamic interpretations. However, it is also clear that the potential benefits of combining the tools from different disciplines in the urban modelling context have not been fully worked out. What is needed is a coming together of skills, the building of new interdisciplinary teams: urban modellers, statistical physicists and the mathematicians who have been generalizing the thermodynamic and statistical mechanics formalisms. It might also be valuable to add the skills of those who have been using these tools for modelling in other fields, such as neuroscience. In some cases, neuroscience being an example, the emphasis has been on taking a statistical (Bayesian) view and this complements the mathematical one in an interesting way. It may be helpful to conclude, therefore, with some indications of what remains to be achieved – but which intuition suggests is achievable!

The urban "big picture" needs to be completed – for example through the specification of suitable "external variables" and generalised forces – the $X_i\delta x_i$ terms – for cities. We perhaps made the beginnings of progress by introducing "area" as a quasi "volume" measure but leaving open the question of whether a better measure could be found from $\{c_{ij}\}$ topology. We also had some difficulty in specifying "urban" partition functions because of the usual nature of the "number of particles"/origin-destination constraints. This is a research question to be resolved.

Potentially the biggest advances to come lie in the modelling of the evolution and emergence of structures – the $\{W_j\}$ in the archetypal model. Again, intuition suggests that the methods down being applied to solids in statistical physics – and their mathematical generalisations, should enable us to make more progress than we have achieved so far. However, that progress is not inconsiderable: it has allowed us to identify phase transitions in urban evolution and to show that they are of a different character to the most obvious ones in Physics.

# References


Ashby WR (1956) Design for a brain. Chapman and Hall, London

Beck C, Schlagal F (1993) Thermodynamics of chaotic systems. Cambridge University Press, Cambridge

Clarke M, Wilson AG (1985) The dynamics of urban spatial structure: the progress of a research programme. Trans Inst Br Geog 10: 427–451

Dearden J, Wilson AG (2008) An analysis system for exploring urban retail system transitions, Working Paper, 141 Centre for Advanced Spatial Analysis, University College London

Evans SP (1973) A relationship between the gravity model for trip distribution and the transportation model of linear programming, Transp Res 7:39–61

Finn CBP (1993) Thermal Physics. Nelson Thornes, Cheltenham

Friston K, Mattout J, Trujillo-Barreto N et al. (2006) Variational free energy and the Laplace approximation. Neuroimage 34:220–234

Friston K, Stephan KE (2007) Free energy and the brain. Synthèse 159:417–458

ter Haar D (1995) Elements of statistical mechanics (3rd ed.). Butterworth-Heinemann, London

Harris B, Wilson AG (1978) Equilibrium values and dynamics of attractiveness terms in production-constrained spatial-interaction models. Environ Plan A 10:371–88

Jaynes ET (1957) Information theory and statistical mechanics. Phys Rev 106:620–630

Martin P (1991) Potts models and related problems in statistical mechanics. World Scientific, Singapore

Pippard AB (1957) The elements of classical thermodynamics. Cambridge University press, Cambridge

Ruelle D (1978/2004) Thermodynamic formalism, 2nd edn. Cambridge University Press, Cambridge

Wilson AG (1967) A statistical theory of spatial distribution models. Transp Res 1:253–69

Wilson AG (1970) Entropy in urban and regional modelling. Pion, London

Wilson AG (1976) Catastrophe theory and urban modelling: an application to modal choice, Environ Plan A 8:351–356

Wilson AG (2000) Complex spatial systems: the modelling foundations of urban and regional analysis. Prentice Hall, Harlow

Wilson AG (2008) Boltzmann, lotka and volterra and spatial structural evolution: an integrated methodology for some dynamical systems. J R Soc Interface 5:865–871

# Chapter 3
# Macro and Micro Dynamics of the City Size Distribution

## The Case of Israel

**Lucien Benguigui, Efrat Blumenfeld-Lieberthal, and Michael Batty**

## 3.1 Introduction

Complex systems evolve and grow from the bottom up. Their key characteristic is emergence in that the actions of the system's basic elements are uncoordinated yet their effects at greater scales appear organized. Hence we say that a complex system exhibits order at higher scales which is usually measurable using some scale-free characteristics. In city systems for example, it is clear that there is a hierarchy of sizes and that these sizes follow a scaling law which can be approximated by a power law. Within cities, different types of centre also follow such scaling not only in terms of their sizes but also in terms of their frequency and spacing. Such systems are sometimes said to exhibit self-similarity which means that if the system is examined at different scales, it appears the same; that is if a system has a certain pattern at one scale, this pattern can be transformed to another scale by enlargement or contraction so that it is impossible to see the difference between the two scales. Self-similarity is a key feature of geometries that are said to be fractal and in terms of cities, such fractal patterns have been widely observed (Axtell 2001). In this chapter we will exploit this fact by examining the pattern of city sizes which have a characteristic signature which is a power law. This signature which is sometimes referred to as the rank size rule is one of the most fundamental features of complexity in that many systems in the physical, natural and social world exhibit such scaling.

The statistics of sizes is a topic studied in many disciplines across the natural and the social sciences (Batty 2008; Buldyrev et al. 2003; Carvalho and Penn 2004; Duranton 2006; Chattopadhyay and Mallick 2007). One of the key problems is how to describe mathematically the function which describes the sizes of the objects or entities that compose such distributions. The two most common distributions

M. Batty (✉)
Centre for Advanced Spatial Analysis, University College London, London, UK
e-mail: m.batty@ucl.ac.uk

A. Reggiani and P. Nijkamp (eds.), *Complexity and Spatial Networks*, Advances in Spatial Science,
DOI: 10.1007/978-3-642-01554-0_3, 

in the literature are the power law and the lognormal distribution (Laherrere and Sornette 1998; Blank and Solomon 2000; Limpert et al. 2001) but it is clear that in many cases, neither of these distributions replicate the shape or form of functions in a satisfactory way. In fact, more accurate distributions lie somewhere between these two options (Limpert et al. 2001; Benguigui and Blumenfeld-Lieberthal 2006).

Among the difficulties concerning the choice of distribution is the problem of the lower tail. In the past, many applications have simply dealt with the largest entities usually because data has been available only for the largest, or sometimes because it is assumed that the most important entities are those that are the largest. The lower tail, or long tail as it is sometimes called, of the frequency distribution of sizes is often disregarded – cut off, and it is clear that by changing the size of the lower tail of the entities' size distribution, the function which fits the distribution will also change. It is not easy to choose a criterion to define the cut off for the lower tail and very often it is chosen arbitrarily.

It is thus a major problem in exploring size statistics to relate the properties of a particular set of entities (for example, incomes, stock market values, populations of cities, frequencies of words and letters that comprise languages, etc.) to a function that describes accurately their size distribution. Several models attempt to solve this problem by relating the entities' size distributions to some hypotheses concerning their behavior (Benguigui and Blumenfeld-Lieberthal 2006). These models, however, usually examine the size distribution at a particular time, thus grounding the analysis in comparative statics, often beginning with some arbitrary initial distribution at the starting point for simulation and iterating the model until some equilibrium state is reached.

The purpose of this chapter is twofold; in the first part, we investigate the *dynamics* of the size distribution (without searching for an equilibrium) and show how it is possible to relate the change in the distribution to the properties of the entities. For that, we use an approach which defines a distribution using a new exponent $\alpha$ (Blank and Solomon 2000) presenting a simulation based on a model we have recently developed (Benguigui and Blumenfeld-Lieberthal 2006). In the second part, we study the dynamics of the Israeli system of cities, first at the macro level and then at the micro, comparing the micro-dynamics in the real data and in the simulation.

Most of the existing work in the field looks for an agreement between the empirical distributions and that given by the models through measures of the accuracy of the model. This, in fact, seems sufficient in the case of a static model. When adding dynamics to the model, however, one has to consider not only the change in the distribution but the temporal variation of the entities as well. Recently, Batty (Sornette 2000; Batty 2006) proposed the rank-clock representation to study the micro dynamics of systems with entities that appear to generate homogeneous rank size distributions at the macro scale. This representation follows the dynamics of individual entities within the system and here we use it to check the validity of our work in terms of micro dynamics.

In this chapter, we use the Israeli system of cities as our case study in examining the validity of our simulation. In the first part of the chapter, we compare the macro dynamics of the Israeli system of cities from 1950 to 2005 at seven snapshots of time showing that the temporal change in the distribution is related to the variation of the number of cities in Israel through time. In the second part of the chapter, we compare the micro dynamics of the simulation with the Israeli system of cities from 1950 to 2005. Sections 3.2–3.6 concern the macro dynamics of the simulation and the real system of cities while Sect. 3.7 presents their micro dynamics.

In the next section, we present the new exponent $\alpha$ and its correlation with the distributions. In Sect. 3.3, we then cover the data concerning the cities of Israel followed by an outline of the model (in Sect. 3.4) and its application to Israeli cities (in Sect. 3.5). In Sect. 3.6, we present the results of the model at the macro level while Sect. 3.7 concerns the micro dynamics of the simulation in comparison to the real system of Israeli cities. Finally we discuss the results of the model, introducing the rank clock analysis which makes comparisons between the model and the real data. Surprisingly, the model provides a rather good description of the distributions at the macro level, but fails to give a sufficiently accurate analysis of the individual changes in the entities at the micro level.

## 3.2 A New Exponent

Recently, we proposed a phenomenological approach (Limpert et al. 2001) when analyzing the size distribution of entities. We based our approach on the three equivalent representations of the size distribution (Benguigui and Blumenfeld-Lieberthal 2007):

- The density function $D(S)$ which gives the number of entities with size between $S$ and $S + \mathrm{d}S$.
- The cumulative function $P(S)$ which gives the number of entities with a size larger or equal (or smaller) than a given $S$. These two functions $D(S)$ and $P(S)$ can also be expressed in relative terms. The two relative functions do not give the number of entities but rather the percentage (or probability) of the total number of entities.
- The rank size representation which is transformed into the logarithm of the size ($y = \ln \mathrm{Size}$) and plotted as a function of the logarithm of the rank ($x = \ln \mathrm{Rank}$). The function $y(x)$ is called the rank-size curve. When the relationship between the size and rank of the entities can be expressed by the function $y \sim x^n$ where $n$ is a power of the function, the distribution is referred to popularly as Zipf's law after Zipf (Batty 2007) who presented a graphical and rather trenchant summary of such relationships This relationship is represented pictorially by a linear equation plotted as a double logarithmic graph.

Our proposed equation concerns the function $y(x)$. We propose to analyze the rank-size curve for a system of entities using the following expression:

$$y = y_0 - H(1 - \alpha)\mu[b + H(1 - \alpha)x]^{\alpha}, \tag{3.1}$$

where $y = y(x)$. $H(1 - \alpha)$ is the Heaviside function,[1] equal to $-1$ if $\alpha < 1$ and to 1 if $\alpha \geq 1$. $y_0$, $b$, and $\mu$ are parameters and the exponent $\alpha$ can be smaller, equal or larger than 1. In the case $\alpha = 1$, Zipf's law is recovered, that is, there is a linear relation between $x$ and $y$. When $\alpha \neq 1$, the curve $y(x)$ has different shapes following the value of $\alpha$ which we call the "shape exponent." This means that each value of $\alpha$ defines a particular distribution.

## 3.3 The Cities of Israel

We have analyzed the rank size distribution of all the settlements in Israel from 1950 to 2005 (Zipf 1949) and found that qualitatively, they all have the same shape. In Fig. 3.1 we present the rank-size curves of all the settlements in three selected years 1961, 1983, and 2005. These three curves have the same shape and the only observed differences are the shifts of the curves upwards with time. The distribution demonstrates discontinuity around the value of population equal

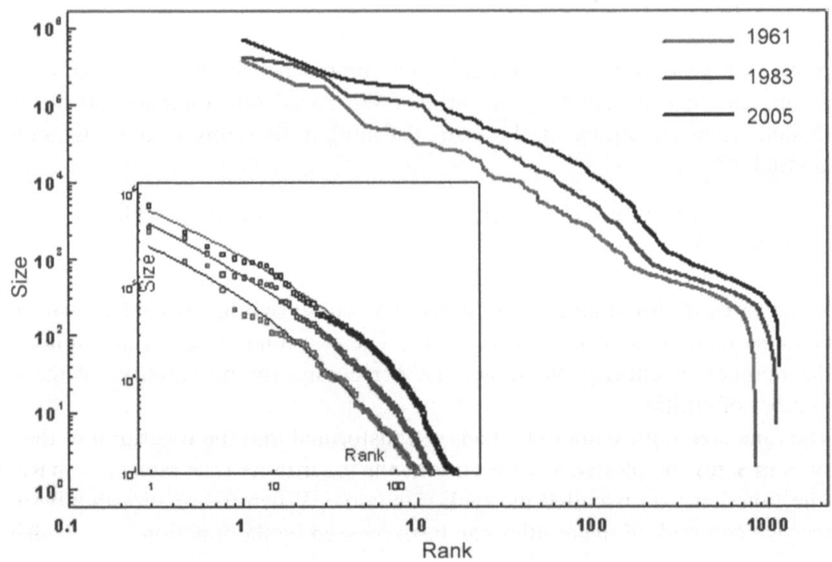


**Fig. 3.1** The rank-size distribution of all the settlements in Israel in the years: 1961, 1983, and 2005. In the inset: the rank size distribution of all the settlements in Israel with population larger than 1,000

[1] A Heaviside function is a discontinuous step function which is equal to zero for a negative variable and one for a positive variable.

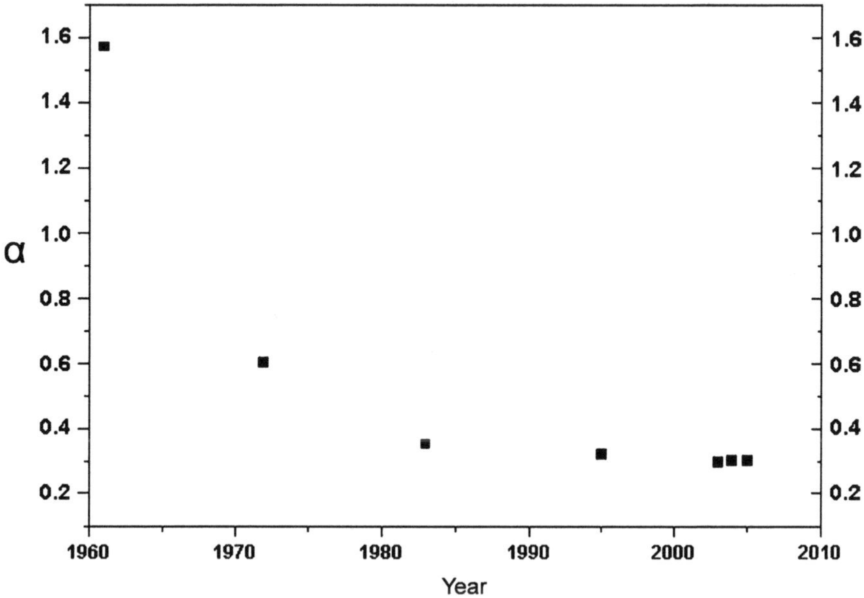


**Fig. 3.2** The variation of the "shape exponent" α between the years 1961 and 2005 for the Israeli system of cities

to 1,000. In other words, the distributions can be divided into two parts around this value. Based on the above, we defined the lower tail of the distribution for all settlements with population smaller than 1,000 for the entire period. The inset in Fig. 3.1 presents the rank size curves for the years 1961, 1983, and 2005, after the exclusion of their lower tails (based on settlements with populations smaller than 1,000).

Similar to several other cases (Limpert et al. 2001), the fit of the distributions to (3.1) is very good ($R^2 > 0.97$) as demonstrated in Fig. 3.1. Figure 3.2 presents the variations of the exponent α with time. The exponent α is larger than 1 in 1961 and changes to values lower than 1 in the following years. In Fig. 3.3 we show the change in the number of cities (settlements with population larger than 1,000) as a function of time. The data fits a quadratic equation with good precision.

## 3.4 The Model

The model we have used is based on a computer simulation that we have recently presented (Limpert et al. 2001; Benguigui and Blumenfeld-Lieberthal 2006). We begin with $N_0$ cities with population equal to 1. At the first stage, each city grows by

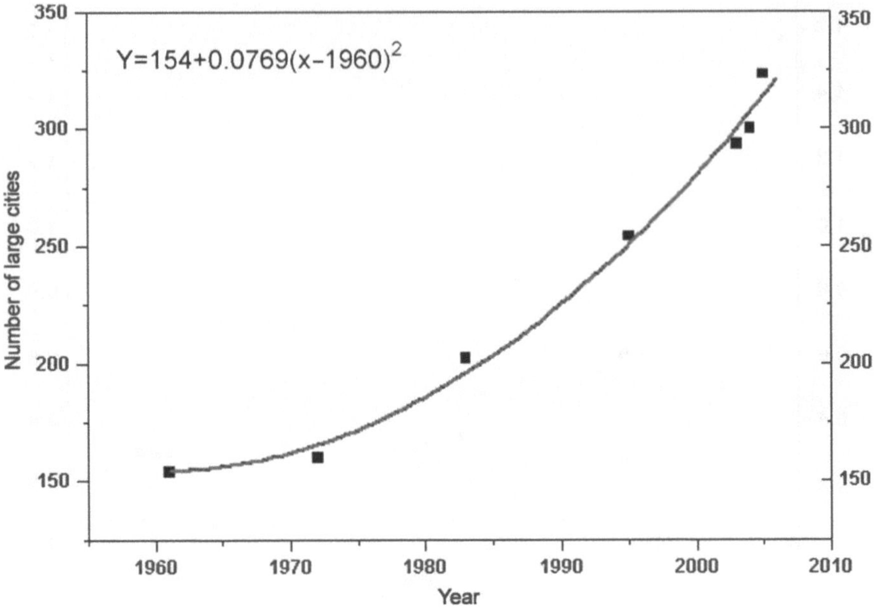


**Fig. 3.3** The change in the number of cities (with population larger than 1,000) between the years 1961 and 2005 for the Israeli system of cities

random multiplicative growth: one city is chosen at random and its population is changed from step $T$ to step $T+1$ as:

$$S(T+1) = \gamma\, S(T), \tag{3.2}$$

where $\gamma$ is a random variable uniformly distributed between $\gamma_m < 1$ and $\gamma_M > 1$ such that the mean

$$\frac{\gamma_m + \gamma_M}{2}$$

is fractionally greater than 1. At the second stage, if the population of a city decreases below 1, the city disappears from the system. At the third stage, a new city is added to the system with the population equal to 1 after $K$ steps. As for the initial cities, if the size of the new city decreases below 1, it disappears from the system.

The distributions that this model yields are dependent on the number of steps $T$ (or in other words, the total growth time) and on the rate of introducing new cities $K$ (for a given $\gamma$ distribution). If $K > 300$ and is constant, the exponent $\alpha$ changes as a step function when, for small $T$, it is smaller than 1 and for large T it changes to values larger than 1. For smaller values of $K$, the exponent $\alpha$ is larger than 1 for small $T$ and becomes equal to 1 for larger $T$. A significant property of the model involves its statistical basis. For given values of $T$ and $K$, the results of the model do not always yield the same parameters. When fitting the resulting rank size curves of the model to (3.1), a distribution of such parameters emerges.

An important issue in the model is the definition of the time $t$. It is not equal to the number of steps ($T$) since the number of steps (on average) separating two consecutive choices of the same city in the growth process is dependent on the number of cities. The unit of time is chosen as the mean number of steps separating two consecutive choices of the same city. For a number of steps $\Delta T$, the interval of time $\Delta t$ is equal to $\frac{\Delta T}{N}$ or if we consider the continuum limit:

$$\mathrm{d}t = \frac{\mathrm{d}T}{N}. \tag{3.3}$$

If we add to this the rate of creating new cities, we get:

$$\frac{\mathrm{d}N}{\mathrm{d}T} = \frac{1}{K}. \tag{3.4}$$

If $K$ is constant, it is not difficult to show from (3.3) and (3.4) that the variation of the number of cities with time is exponential, and is given by

$$N = p_s N_0 \exp\left(\frac{t}{K}\right)$$

where $p_s$ is the probability of a new city surviving, with $p_s$ approximately equal to 0.27.

## 3.5 Application of the Model to Israeli Cities

In this section, we show how we use the model to interpret the system of cities in Israel. Our goal is to use the model in order to determine the city size distribution or more precisely the rank-size function for Israel's real system of cities. Since we know the function $N(t)$ is quadratic, we had to find the function $K(T)$ which fits this result. For that, the following system of equations needs to be solved:

$$\frac{\mathrm{d}t}{\mathrm{d}T} = \frac{1}{N}, \tag{3.5a}$$

$$\frac{\mathrm{d}N}{\mathrm{d}T} = \frac{1}{K}, \text{ and} \tag{3.5b}$$

$$N = N_0 + N_1(t - t_1)^2. \tag{3.5c}$$

Analytic solutions to this system of equations are complicated, and thus we propose to find an approximate solution using a heuristic approach. After some systematic trial and error iteration, we generated the following expression:

$$K(T) = K_0 + K_1 T^{1/3},$$

which yields good results, as indicated below. Based on this expression, it is possible to show that $N(t)$ is given by:

$$N = \left(N' - \frac{6K_0^2}{K_1^3}\right) + \left(\frac{3}{2K_1}\right)\left(T^{1/3} - \frac{K_0}{K_1}\right)^2 + \left(\frac{3K_0^2}{K_1^3}\right)\ln\left(K_0 + K_1 T^{1/3}\right), \quad (3.6)$$

where $N'$ is an integration constant. Considering that the log term in (3.6) changes very slowly in its dependence on $T$, it is possible to add it to the constant term and get a good approximation using the following simple expression for $N$:

$$N = N_0 + N_1\left[T^{1/3} - (T_1)^{1/3}\right]^2, \quad (3.7)$$

where

$$T_1^{1/3} = K_0/K_1.$$

The constants $N_0$ and $N_1$ in (3.7) are dependent on the coefficients $K_0$ and $K_1$ and also on the integration constant.

To find the relation between the time $t$ and the number of steps $T$, we integrate (3.5a) and (3.7) to get:

$$t = \frac{u}{N_1} + B\theta^{-1}\left[u\sqrt{\frac{N_1}{N_0}}\right] + C\ln\left(N_0 + N_1 u^2\right) + t', \quad (3.8)$$

where

$$u = \left(T^{1/3} - T_1^{1/3}\right).$$

In (3.8), $t'$ is an integration constant, and the coefficients $B$ and $C$ are dependent on $K_0$, $K_1$, and $N_0$, $N_1$.

It is possible to consider the second and third terms on the right hand side of (3.8) as constants: the log term because it changes very slowly in its dependence on $N$ and hence on $T$; and the $\theta^{-1}$ term because its argument is larger than 1 (considering the real values of the parameters). The final result suggests that $t$ is linearly dependent on $T^{1/3}$. One can choose the integration constant such the model can be written as follows:

$$t = D\left[T^{1/3} - (T_1)^{1/3}\right]. \quad (3.9)$$

From (3.7) and (3.9), it can be deduced that the dependence of $N$ on $t$ is indeed the quadratic equation we expected.

In the following steps, we consider $K$ as constant, choose an initial state for which the number of cities is roughly 150 (see Fig. 3.3) and select the exponent $\alpha$ as approximately 1.6 (corresponding to the values of $\alpha$ in 1961 as seen in Fig. 3.2). We found that this corresponds to a state with 50 initial cities, $T = 45{,}000$ steps and $K = 80$. For larger values of $T$, we used the following function to describe $K$ as a function of $T$:

$$K = K_0 + K_1 T^{1/3}$$

such that the value of

$$T_1 = (K_0/K_1)^3$$

is near the value of $T$ in the initial state. Then, we ran the simulation for several values of $T$ where for each $T$ we found the value of $N$. We also determined the time by graphical integration of the function $N^{-1}$ vs. $T$, and we plotted the rank-size curve determining the shape exponent $\alpha$ by fitting the curve using (3.1).

## 3.6 Macro Analysis of the Model: The Rank Size Curve and the Number of Cities

The outputs of the model are presented in Figs. 3.4–3.8; Fig. 3.4 presents the relation $N$ vs. $T^{1/3}$ where it is clear that the relation in (3.7) is indeed verified. The relation $t$ vs. $T^{1/3}$ is presented in Fig. 3.5 where a linear relationship between

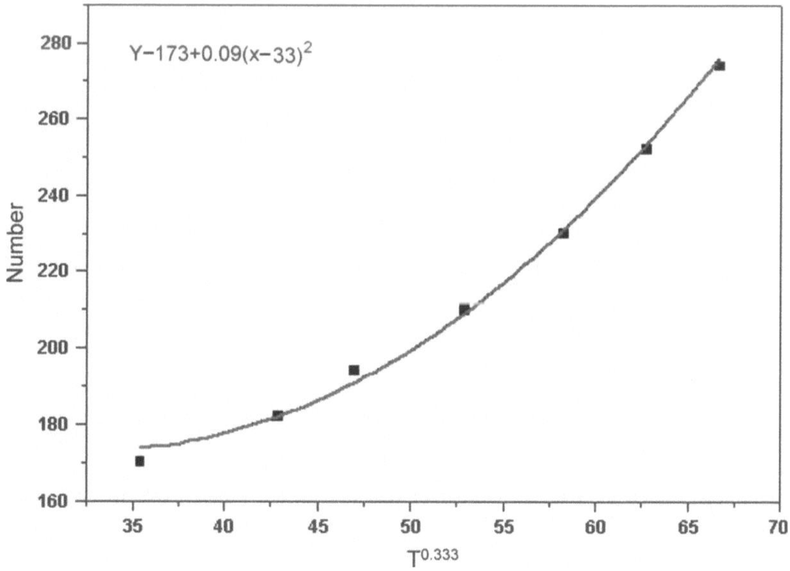


**Fig. 3.4** Results of the model: the number of cities $N$ vs. $T^{1/3}$, where $T$ represents the number of steps

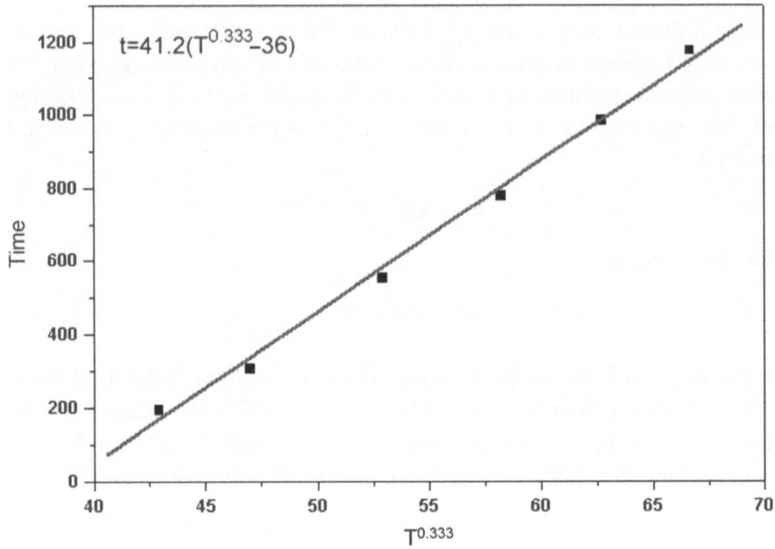


**Fig. 3.5** Results of the model: time $t$ vs. $T^{1/3}$, where $T$ represents the number of steps

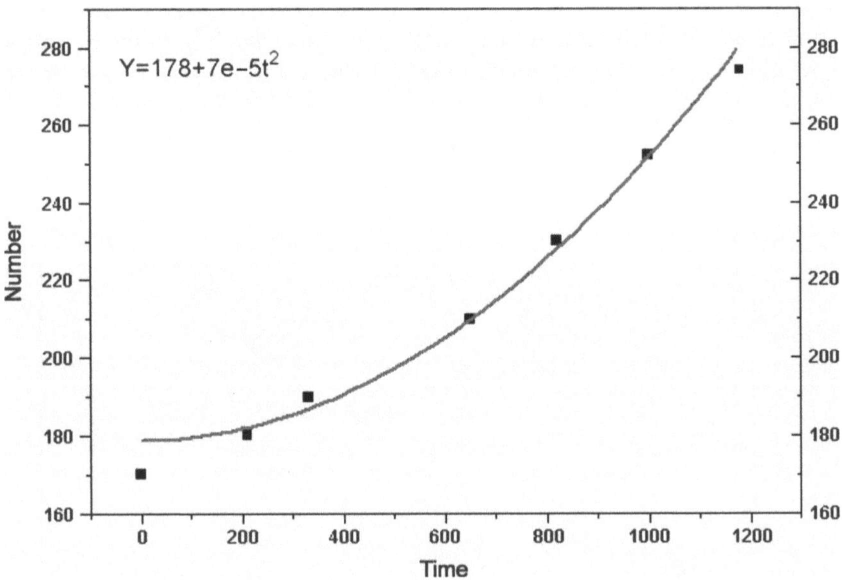


**Fig. 3.6** Results of the model: the number of cities $N$ vs. the time $t$. Note that the data fits a quadratic equation

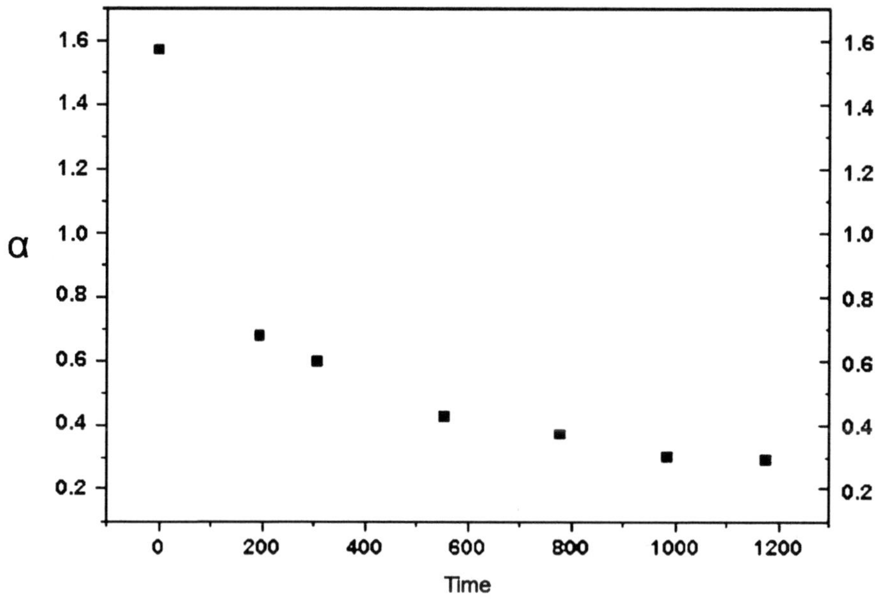


**Fig. 3.7** Results of the model: the "shape exponent" α vs. the time $t$

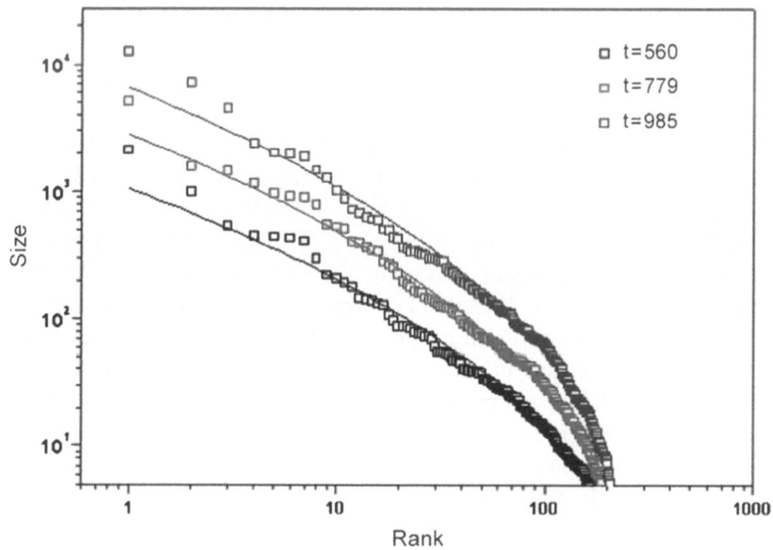


**Fig. 3.8** Results of the model: the rank size distribution of the cities in the model on a log–log graph

time and $T^{1/3}$ is also verified with the same value of $T_1^{1/3}$ found in Fig. 3.4. Figure 3.6 shows $N$ vs. $t$ where the quadratic relation is also found. Note that the values of $N$ and of the coefficients $N_0$ and $N_1$ found in the model, are very close to the ones in the real data. In Fig. 3.7 we present the exponent $\alpha$ vs. $t$. Here too, there is a very good quantitative agreement between the model and the real data. Finally, we show the rank-size curves of the model in Fig. 3.8. These curves are qualitatively very similar to the ones that resulted from the real data. Based on the above, we think the model provides a good description of the evolution of the Israeli system of cities (but clearly only when considering cities with populations larger than 1,000). More particularly, the qualitative change in the distribution (that is in the exponent $\alpha$) is a direct consequence of the variation of the number of cities with time.

## 3.7 Microdynamics: The Rank-Clock Analysis

So far, we have focused only on the macro dynamics of the Israeli system of cities. This means that we ignored the changes that appear within the positions (or ranks) of individual cities with time, and looked only at the rank size distribution of the entire system. In this section, we follow Batty (2006) and present an initial analysis of the micro dynamics of both the Israeli system of cities and the above simulation model.

We start with the real data of Israeli cities (and settlements) with populations larger than 5,000 from 1950 to 2005 (Batty 2007). The number of these settlements increased from 34 in 1950 to 172 in 2005. Figure 3.9 presents the rank clock of Israel for these years. The cities are colored according to their rank and the time they first entered the system with red representing cities which enter first through a spectrum of color – red to yellow to green to blue – for the cities that enter last. The micro dynamics of this system of cities presents little irregularity, is mostly stable in structure and shows a system that is rapidly growing with cities rising rapidly up the ranks but few cities falling out. We can conclude all this in rather impressionistic terms by simply viewing the color balance of the clock and comparing this to a system like the US where there is much greater volatility into and out-of the top ranked cities. There are only a few cases where cities move toward the center of the rank clock over this period while the cities that existed in the early stages of the development from 1950 remain the largest cities in the system (with cities entering earlier nearer the center of the rank clock). Few cities that were introduced in later years manage to increase significantly and move to the center of the clock. We can define a number of measures or parameters that characterize the clock, hence the system of cities. First, the rank shift is a parameter which indicates the stability of the dynamics of individual entities in the system. It is defined as:

$$d_i(t) = |r_i(t) - r_i(t-1)|, \tag{3.10}$$

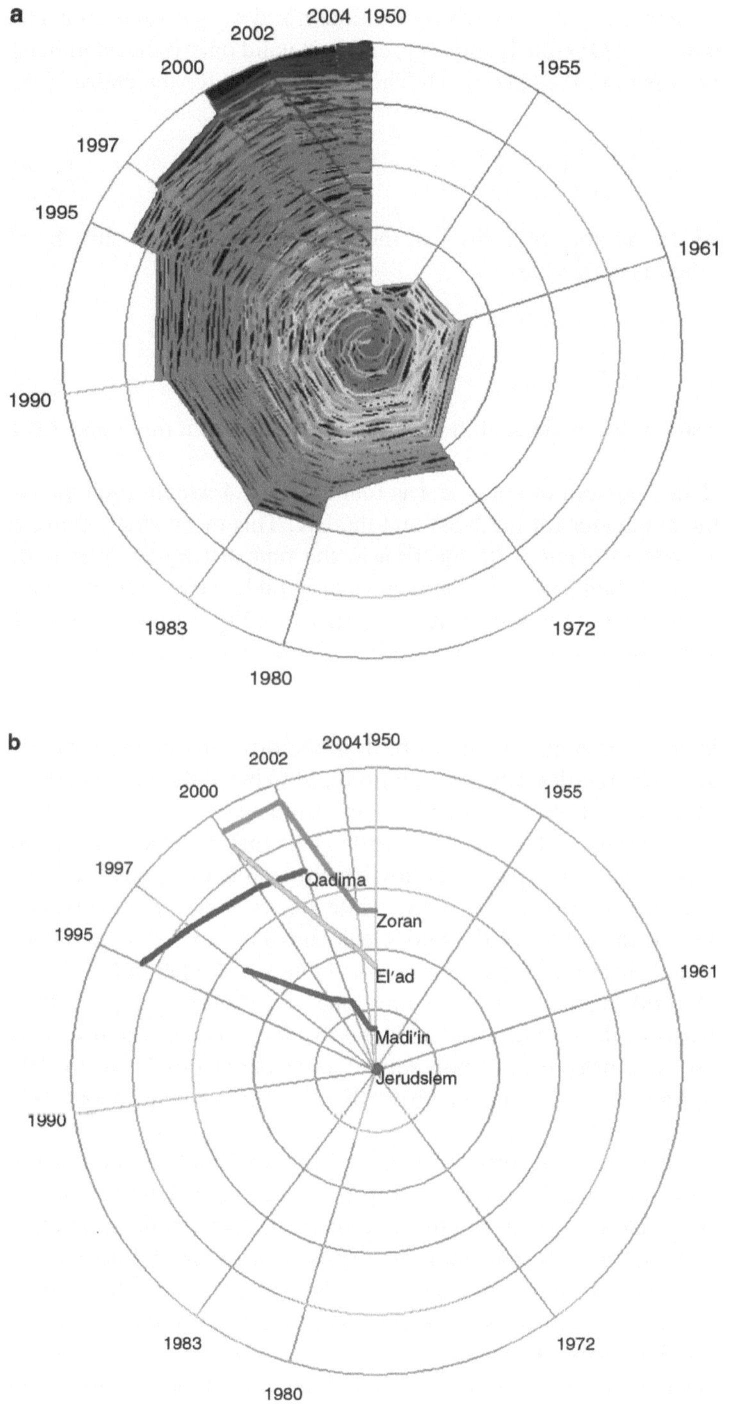


**Fig. 3.9** The rank clock representation of the Israeli system of cities (a) between the years 1950 and 2005 and (b) with some specific cities also plotted

where $r_i(t)$ represents the rank of city $i$ at time $t$, and $r_i(t-1)$ represents the rank of city $i$ at time $t-1$. Obviously, this expression is valid only if the examined city is in the system at both times $t$ and $t-1$. The average shift for the entire system is:

$$d(t) = \frac{\sum_i d_i(t)}{N}, \tag{3.11}$$

where $N$ is the number of entities in the system. The average shift for the entire period studied is defined as:

$$d = \frac{\sum_t d(t)}{T}, \tag{3.12}$$

where the sum is the average shift of the system at different times and $T$ is the entire period.

In the Israeli system of cities, $d$ was found to be 5.4 which resembles the values of the same parameter for the USA and the UK. This means that on average, each city in the system changes its location in the rank list by 5.4 places during the studied period. When analyzing these changes in rank, we can see that most cities in Israel presented very few changes in their ranks (see Fig. 3.9), while a small number of cities changed their ranks considerably. These cities can be divided into three groups, based on the reasons for their growth; the first group consists of orthodox-religious settlements, characterized by high annual growth rates (~10%), which can be explained by the high volume of birth in the orthodox community. El'ad is an example for such a settlement. It was introduced to the system only in the year 2000 and within the next 4 years changed its rank from 144 to 64 ($d_i(t) = 20$).

The second group includes several settlements that were united (by government decision) into one municipality. Following this unification, some of these settlements disappeared from the system, while others increased systematically and moved towards the center of the rank clock showing rapid change reflected in the steepness of their slope. Zoran which was united with Qadima in the year 2003, changed its rank from 152 to 87 within one year ($d_i(t) = 65$!). Lastly, the third group consists of the city of Modi'in, the only city in Israel which was completely planned (in terms of its located population) to be a large city between Tel Aviv and Jerusalem. Modi'in's rank changed from 88 to 23 between the years 1997 and 2005 ($d_i(t) = 9.3$).

When analyzing the results of the model for the Israeli system (see Sect. 3.5), one finds $d = 14.8$ which is considerably larger that the value of $d$ for the real data describing the Israeli system of cities. Figure 3.10 presents the rank clock for this model and based on this, the findings suggest that the model does not provide a good description for the micro dynamics of the system. Even before calculating the value of $d$, we can see that the rank clock is different from the one of Israel's real data as the colors in the clock are mixed and present no organized pattern. However, the calculated value of $d = 14.8$ is similar to the case of the top city populations in the world data set from classical times to the modern day. This value is relatively

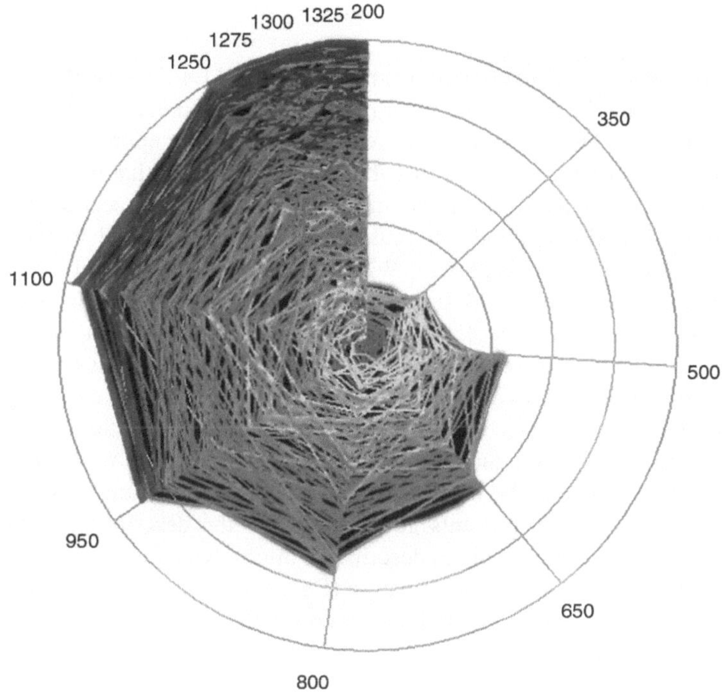


**Fig. 3.10** The rank clock representation of the model results at equivalent time points to the real evolution in Fig. 3.9

high and indicates that the individual entities of the system presented great changes in their sizes and ranks over time.

The importance of this comparison is that it shows that even though the model provides a very good description for the macro dynamics of the Israeli system of cities, it does not explain their micro dynamics. It appears that further work needs to be done in order to develop a model that will provide a good description for both the macro and micro dynamics of a system and this would probably have to include many more specific spatial factors which characterize the urban development of Israel during these years. The reason for that lies in the fact that the micro dynamics of the Israeli system of cities were affected by various reasons (such as government regularities, as described earlier) that cannot be imitated by the current model. The model, which is a complex system, is based on many random variables that are all dependent on one another. In its current version, it is very difficult to depict the exact variable that controls each aspect of the changes in the micro level of the system. Thus further work is needed in order to calibrate the model to fit the micro dynamics of the Israeli system of cities.

## 3.8 Conclusions

We have presented an adaptation of a simulation model for the growth of entities in the Israeli system of cities so that we might examine the dynamics of the distribution through time. Our approach is different from most other approaches to city size distributions in particular and size distributions in general in that our focus is upon the micro dynamics as well as the macro statics of cross sectional city size distributions. We also applied our multiplicative growth model to the cities of Israel and were partially successful. When considering the ensemble of cities at the macro scale, that is their rank size distributions, we get a convincing explanation of the time variation of these distributions which is dependent upon the rate of creating new cities. However, when studying the micro dynamics of the system, that is the evolution of individual cities over time, using the rank-clock representation, it is clear that the relative variation in size and rank of the cities (average shift) is considerably larger in the model than in the real data. Hence, we believe there is an inconsistency between the macro and micro aspects of the analysis.

We believe the model presents good results for the description of the macro dynamics of the system but fails to describe its micro dynamics, thus it needs to be extended. One option, suggested by Havlin (Private communication), is to consider the growth of cities with interactions or correlations among themselves. Such extensions would take the model to one dealing with systems of cities and their interactions which have meaning in terms of trade and other transportation flows. In the current model the growth of each city follows (3.2) alone, that is each city grows independently of every other. The proposal which we will follow in future research is to introduce enough but not too rich a set of interactions between cities such that the growth of any one city will be dependent on the growth of others.

Finally, we wish to emphasize that even the known "static models" present some evolution until they obtain the desired distribution. Until recently (Batty 2006), the evolution of the individual entities was hardly investigated but it seems necessary to do so in order to understand the evolution of systems of entities as a whole. In other words, we believe that dynamics has to be introduced in all models that study the distribution of sizes. In this chapter, we have cast doubt on the standard scaling that represents the key feature of complex systems, elaborating this to produce a more general model of size distributions. There are still many puzzles involved in the search for key signatures of city systems but we have illustrated that the basic principles of scaling still hold notwithstanding that the models we have generated go beyond the simplest form of power law scaling.

**Acknowledgments** The authors thank George Kun from the Israel Central Bureau of Statistics for providing the data concerning the cities of Israel and for useful discussions.

## References

Axtell RL (2001) Zipf distribution of U.S. firm sizes, Science 293:1818–1820
Batty M (2007) Visualizing creative destruction, CASA working papers 112 http://www.casa.ucl.ac.uk/publications/workingPaperDetail.asp?ID=112
Batty M (2006) Rank clocks, Nature 444:592–596
Batty (2008) The size, scale, and shape of cities, Science 319:769–771 352
Benguigui L, Blumenfeld-Lieberthal E (2006) From lognormal distribution to power law: A new classification of the size distributions, Int J Mod Phys C 17(10):1429–1436
Benguigui L, Blumenfeld-Lieberthal E (2007) A dynamic model for city size distribution beyond Zipf 's law, Physica A 384(2):613–627
Buldyreva SV, Dokholyana NV, Erramilli S, Hong M, Kim YJ, Malescio G, Stanley HE (2003) Hierarchy in social organization, Physica A 330:653–659
Carvalho R, Penn A (2004) Scaling and universality in the micro-structure of urban space, Physica A 332:539–547
Chattopadhyay KA, Mallick SK (2007) Income distribution dependence of poverty measure: A theoretical analysis, Physica A 377(1):241–252
Duranton G (2006) Some foundations for Zipf's law: Product proliferation and local spillovers, Regional Science and Urban Economics, 36(4), 542-563
Havlin S (2008) Private communication, Department of Physics, Bar Ilan University (havlin@ophir.ph.biu.ac.il)
Israel Bureau of Statistics (2008) www.cbs.gov.il
Laherrere J, Sornette D (1998) Stretched exponential distributions in nature and economy: "fat tails" with characteristic scales, Euro Phys J B 2:525–539
Limpert E, Stahel WA, Abbt M (2001) Log-normal distributions across the sciences: Keys and clues. Bioscience 51: 341-352
Sornette D (2000) Critical phenomena in natural sciences. Springer, Berlin
Zipf GK (1949) Human behavior and the principle of least effort. Addison-Wesley, Cambridge, MA

## References

# Chapter 4
# A Morphogenetic Perspective on Spatial Complexity

## Transport Costs and Urban Shapes

**Francesca Medda, Peter Nijkamp, and Piet Rietveld**

## 4.1 Introduction

A modern city is a complex entity characterized by a plurality of behaviour, volatility of interactions, and mobility of residents. It is in a permanent state of flux due to dynamic forces that impact on its functional structure and its spatial configuration (Ingram 1998). Urban dynamics often mirrors fundamental changes in a transport system and its spatial spillovers (Crane 2000; Handy 1996). The externality dimensions of urban growth often relate to congestion and detrimental environmental effects due to car usage (air pollution, noise, accidents); for this reason, a proper investigation of evolving urban forms, and their patterns of change, could potentially be a means of understanding and combating urban sprawl, reducing automobile dependence, increasing the use of alternative transport modes, and supporting pedestrian mobility.

In the literature we see that the relationship between transport and urban form has been studied extensively. A number of analyses (Cervero and Gorham 1995; Friedman et al. 1994; Newman and Kenworthy 1989) investigate the interactions between urban form and transport by using aggregate indicators or measures such as urban density or urban land rent in relation to trip frequency or average trip lengths. Their approaches give rise to significant results between an urban transport system and a general characterization of urban form, and may therefore support land use policies which might effectively lead to different overall travel patterns in the city. Nonetheless, they neither convincingly address the problem of how specific characteristics of urban forms correlate with different travel patterns nor do they illustrate how urban form influences individual decisions. For example, multivariate regression in disaggregate models (Boarnett and Crane 2001), which considers socio-economic and travel characteristics of individuals, yields mixed

---

F. Medda (✉)
Centre for Transport Studies, University College London, London, UK

A. Reggiani and P. Nijkamp (eds.), *Complexity and Spatial Networks*, Advances in Spatial Science,
DOI: 10.1007/978-3-642-01554-0_4, 

results on the relationships between urban form and transport, implying that a modification of the urban form (for example from pre-World War II traditional communities to post-World War II dispersed communities) does not always significantly correspond to realized or anticipated changes in travel behaviour. When investigating whether there are possible benefits a city derives from an improvement in urban transport systems in relation to land rents, Mohring (1993) concludes "*regrettably, the answer is very little*".

The relationship between urban form and transport, and in particular travel behaviour, is markedly complex, because it depends on the characteristics of the urban form (functional–geographic structure of the city, activity-based zoning, etc.) and the characteristics and purposes of the travel under scrutiny (working, shopping, by car, by mass transit system, etc.). The objective of this work is to analyse the relationship between urban form and collective transport systems by considering the behaviour of two types of transport costs: the external cost of transport and the private transport cost. Both costs influence individual choices of households in relation to location, and thus will impact on the morphologic structure and dynamics of the city and its shape.

Our proposed methodological–conceptual approach applies the essentials of the morphogenetic algorithm based on Turing (1952), which we deploy to study the effects of transport costs on city shape changes. Alan Turing, who encrypted secret codes during World War II, and who is one of the founders of modern computer science, defined near the end of his career an algorithm that analyses the formation of spatial concentration patterns which occur due to different diffusion rates of considered "substances". What interests us about this formulation is his finding that, contrary to our intuition, diffusion is no longer associated with smooth processes, but is instead related to the creation of peaks of concentrated "substances". The model we develop in our study assumes a linear city where distance as such is not relevant in the household's choice of a residential location. These modifications of the standard urban economic model have led us to the definition of a model in which time – and spillover effects of the variables – take precedence in the urban form process. Although our model as defined here is mathematically more complicated than the standard urban economics approach, it can nevertheless adequately capture economic processes such as transport cost interactions, and provide further insights into the dynamics of this urban phenomenon.

Because we are developing a dynamic model, the variables determining growth must have an accumulative trend and must therefore increase over time. We also assume that the variables which activate growth not only have an accumulative trend, but that they also determine the increase of the variables which *inhibit* growth. Stability is achieved when economic variables, which are increasing and therefore determining growth, are constrained by the inhibitor variables. Importantly, if a variable inducing growth is defined in the morphogenetic algorithm of reaction-diffusion as the *activator* of growth, the definition of the inhibitor may not be so straightforward. The difficulty of the definition arises because the inhibitor does not inherently inhibit growth; it is a variable that may or may not inhibit growth through its interaction with other functions. This complexity aspect is fundamental in our urban economics context, whereby the same variables may both stimulate and halt growth.

In our analysis we first propose the urban dynamic model based on the morphogenetic algorithm and then simulate how the interrelationship between external and private transport costs have direct effects on urban form.

## 4.2 Transport and Urban Morphology

As in the model defined in Medda et al. (2006), we study the formation of the urban shape but we assume a linear city. We divide the city into $i=1, \ldots, p$ districts. Households in the city are identical, each has an income $Y$, and each chooses a quantity of housing space of which the rent, $R$, is an aggregate compound function of the number of people living in the district. All households are assumed to travel along the city from district to district. Their total number is equal to $N$; and the maximum number of people living in each district is equal to $N/p$, the starting point of our simulation. We assume a fixed urban space occupancy per person, and the maximum density in all districts is fixed and equal to $D$. If at some stage a district were to attract and hence have to accommodate a number of people greater than $N/p$, the district would require a larger area for its residents in order to maintain the same density level $D$, and thus it would need to expand. We assume that the growth of our linear city is outward and in one direction.

We assume a mutual dependence between transport cost and population distribution and will now discuss the composition of these transportation costs. We assume that the External Transport Cost (congestion, environmental costs, safety costs) (ETC) and the Private Transport Cost (time and money) (PTC) together form the Total Transport Cost (TTC).

The External Transport Cost function is:

$$\mathrm{ETC}_i = K_i + f(n_{i,t}) \tag{4.1}$$

where $K_i =$ a fixed external sunk cost related to air and noise pollution and other intrusion and disturbances caused by the collective use of the transport system; $f(n_{i,t}) =$ the congestion cost; this cost comprises the variable travel cost related to using the transport system when the number of people living in a particular district $i$ is $n_i$ (at time $t$). Travel cost increases when the number of people in the district increases; it therefore includes a congestion cost component.

The Private Transport Cost function is:

$$\mathrm{PTC}_i = H_i + c(F(n_{i,t})) \tag{4.2}$$

where $H_i =$ the fixed costs (for example fare or tax) related to use of the collective transport system; $c(F(n_{i,t})) =$ the cost, without congestion, (the congestion cost is incorporated in (4.1)), of the total travel time for the number of people living in a given district $i$, including waiting time. This cost is an indirect function of the standard travel time or frequency $F$ of the transport service, offered by the city. We assume a supply response system, which means that the higher the number of people living in a particular district $i$ at time $t$, the higher will be the supply of infrastructure or the frequency of transport services. This implies that, as the transport system supply or the frequency of transport increases, total travel time will be lower, and thus the total cost related to travel time will decrease.

The household in a specific district $i$ will minimize its total transport cost under a given budget constraint as follows:

$$\text{Min TTC}_i = \text{ETC}_i + \text{PTC}_i \tag{4.3}$$

subject to: $Y_i > \text{TTC}_i + R(n_{i,t})$, where $R(n_{i,t})$ is the rent value, a direct function of the households living in the given district $i$ at time $t$. The higher the number of people living in the district, the higher will be the rent cost for the household.

The relationship between the two transport costs will now be examined through the analytical form embodied in the Turing algorithm, which was developed to study the spontaneous generation of spatial patterns in morphogenetics. The algorithm, more commonly known as the reaction–diffusion model, describes how two variables operating in an antagonistic way produce a spatial pattern formation. We may interpret our model from this morphogenetic point of view and observe that the two types of transport costs act as "activator" and "inhibitor" of urban growth.

$$\frac{\partial T_1}{\partial t} = R_1(T_1, T_2) + D_1\nabla^2 T_1 \tag{4.4}$$

and

$$\frac{\partial T_2}{\partial t} = R_2(T_1, T_2) + D_2\nabla^2 T_2 \tag{4.5}$$

where $T_1 =$ External Transport Cost (ETC) (inhibitor); $T_2 =$ Private Transport Cost (PTC) (activator); $R_1(T_1,T_2)$ and $R_2(T_1,T_2)$ identify the mutual dependencies between the external transport cost (inhibitor) and the private transport cost (activator). Both costs are functions of the number of people living in the district, $n_{i,t}$, thus $R_1$ and $R_2$ are functions of $n_{i,t}$. Additionally, $R$ indicates how, by changing the activator (PTC), the inhibitor will react; and because the effect will be a movement of people along the city, the antagonistic behaviour of these two costs represents the origin of urban form change. $D_1$ and $D_2$ are the diffusion coefficients which account for the effects of the transport costs in the city, $D_2 \gg D_1$; the Laplacian operator, $\nabla^2 \equiv \partial^2/\partial x^2$, describes the processes of diffusion in space.

After having defined the model in the form of an inhibitor–activator system, we must ensure its stability. The initial condition is a homogeneous steady-state pattern. If this status is perturbed by a change in transport costs, which is illustrated more precisely by sinusoidal perturbations, we need to know whether the "sinusoidal perturbations will *die away* and dampen back to the flat, spatially homogeneous steady-state, or if they will amplify and create a high-amplitude pattern" (Kauffman 1993). By imposing the conditions of Lyapunov's stability theory, we can verify that the eigenvalues of the system are positive, and are therefore associated with a sinusoidal perturbation which grows until it manifests a spatial pattern formation.[1] In particular, an increase in private transport cost in a specific district of the city will have a direct effect on the number of people leaving that district. The movement of people will trigger the external transport cost change, which will have a corresponding impact on the people deciding whether to stay or to leave this specific district. Since we have assumed a constant urban density, these two effects will modify the shape of the district. Moreover, the two costs behave as two positive "waves" that diffuse in the city; their effect will have impacts on all the urban districts, thus modifying the whole urban shape. These two diffusion waves will decay over time, reaching a new urban form.

## 4.3 Numerical Solutions

The interactions between the variables and, in particular the two transport costs, can be described through the sand dune paradox. "Naively, one would expect that the wind in the desert causes a structure-less distribution of the sand. However, wind, sand and surface structure together represent an unstable system. Sand deposits more rapidly behind a wind shelter. This increases the wind shelter which, in turn, accelerates the deposition of more sand – a self-enhancing process' (Meinhardt 1998). The simulations of our urban complexity model based on transportation and land use are conducted through the use of the software program SP (Meinhardt 1998).

The equilibrium conditions are at $t=0$, the Total Transport Cost (TTC) is given and constant for all districts and the rent value is equal across all districts. At $t=1$, the fixed Private Transport Cost $H_i$ in district $i$ (at the extreme left corner of the linear city) is assumed to increase (Fig. 4.1).

Therefore, our urban system tends to move away from the original equilibrium states, while the two transport costs will respond in mutually opposing ways. In Fig. 4.1 we have assumed that $H_i^0 = 0.06$ increases to $H_i^1 = 0.08$[2] in the specific district $i$ which is at the far left side of the linear city (we will assume in all three simulations that the shock resulting from changes in the transport cost always

[1] See for a complete proof, Babloyantz and Hiernaux (1975), Nicolis and Prigogine (1977).

[2] We start by considering a perturbation which reaches after a period of time a stable state of an homogenous periodic pattern in the linear city. In this case the following variables have values: $H^0 = 0.06$; $K^0 = 0.00$; $D_1{}^0 = 0.01$; $D_2{}^0 = 0.35$.

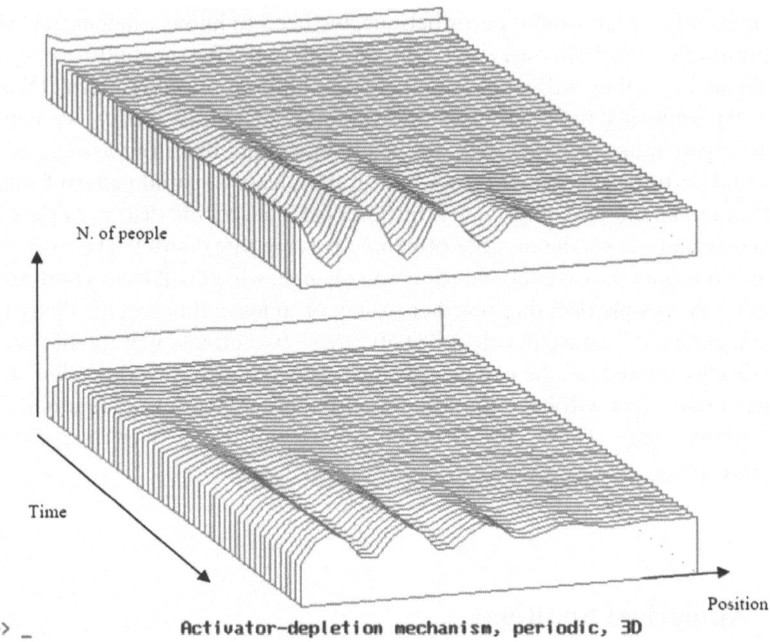


**Fig. 4.1** Dynamic pattern due to the increase of the fixed Private Transport Cost (PTC). The horizontal axis in both graphs represents the location of each district in the linear city. The vertical axis on the top graph represents the number of people living in the district in relation to the activator, that is PTC. The vertical axis of the bottom graph represents the people living in the district in relation to the inhibitor, that is External Transport Cost (ETC)

happens in this specific district $i$). The top graph in Fig. 4.1 identifies how, with an increase in the fixed Private Transport Cost, people will relocate over a period of time in new districts, and in particular create four major expansions in the city. The bottom graph in Fig. 4.1 is the reaction mechanism of the change in PTC. The number of people in each district is a function of the External Transport Cost (ETC) in this case. The urban formation related to ETC is antagonistic to the urban formation determined by PTC. As we can observe, the pattern does not diffuse in the entire city, but the impact of the increase of the fixed Private Transport Cost affects only one part of the city.

Let us now assume an increase of the fixed PTC, $H_i$, and also an increase in the fixed ETC, $K_i$, that is, $H_i^0 = 0.06$ increases to $H_i^1 = 0.08$ and $K_i^0 = 0.00$ increases to $K_i^1 = 0.2$. In this case we have multiple interactions, the households move along the linear city, and the effect, rather than to impact only a few districts, is in this simulation diffused throughout the city (Fig. 4.2).

We have assumed that the rate of diffusion of the inhibitor, that is the External Transport Cost, is higher than the rate of diffusion of the activator, the Private Transport Cost. We now assume a decrease in the value of the rate of diffusion of the activator from $D_1^0 = 0.01$ to $D_1^1 = 0.003$, while maintaining the value of the rate

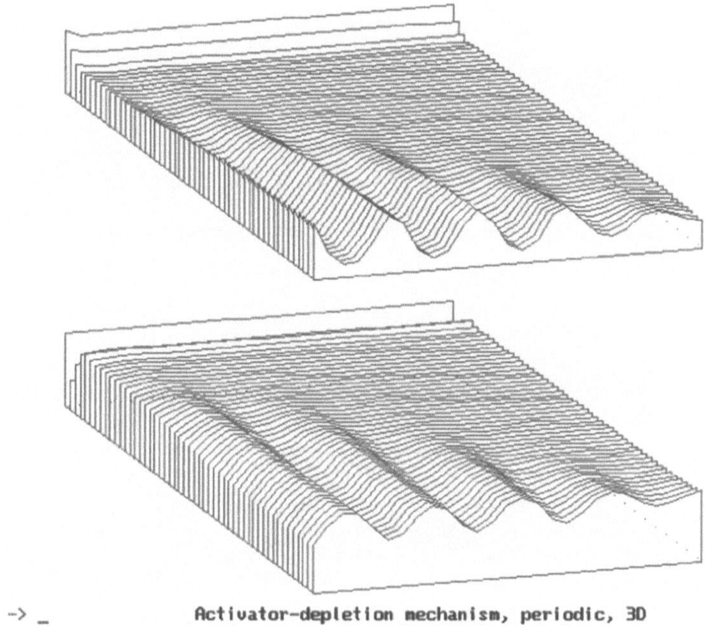


**Fig. 4.2** Dynamic pattern due to the increase of the fixed Private Transport Cost (PTC) and of the fixed External Transport Cost (ETC). The horizontal axis represents the location of all the districts in the linear city. The vertical axis in the top graph represents the number of people living in the district in relation to the activator, that is PTC. The vertical axis of the bottom graph represents the people living in the district in relation to the inhibitor, that is External Transport Cost (ETC)

of diffusion of ETC, that is, $D_2{}^0 = 0.35 = D_2{}^1$. In this case we observe in Fig. 4.3 that the number of people living in certain districts increases, reaching maximum at the expense of the districts nearby. The sharpness of these urban growth peaks resulting from PTC is offset by urban growth in the "troughs" of the peaks due to the ETC.

These illustrative examples show how a transport improvement can determine a direct impact upon an entire urban shape. Since we analyse a variation in the number of people living in the district, we are assuming a consequent change in urban land use. Our modelling experiment highlights three consequences, the first being that a transport improvement can, according to the hypotheses of our model, determine effects not only in the area where the improvement is located, but also through spill over effects in distant areas. The second consequence we can derive from the model is that transport improvements in different locations in the city can determine variations in the initial urban shape and thus the formation of a new urban shape. The third consequence is the connection between the two transport

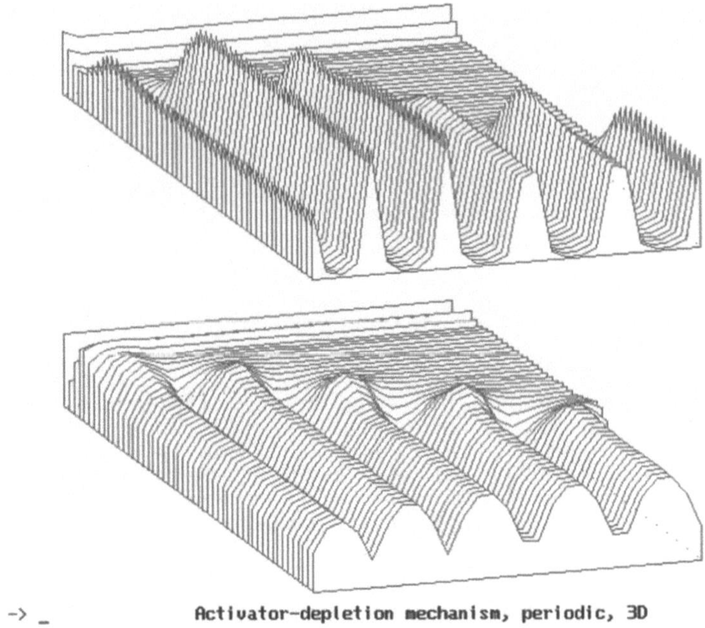


**Fig. 4.3** Dynamic pattern due to the increase of the fixed Private Transport Cost (PTC) and of the fixed External Transport Cost (ETC), and a decrease of the rate of diffusion of the activator. The horizontal axis represents the location of all the districts in the linear city. The vertical axis in the top graph represents the number of people living in the district in relation to the activator, that is PTC. The vertical axis of the bottom graph represents the people living in the district in relation to the inhibitor, that is the External Transport Cost (ETC)

costs, in particular the reactive mechanisms of the ETC in relation to changes in PTC. An increase in PTC determines an increase in the external cost related to travel.

## 4.4 Conclusion

We have argued that the city is analogous to other systems in biology, mathematics and chemistry, in that it is a complex self-organized system which evolves towards order through the modification of its components. By order we mean a status achieved through *balance* and *collaboration* of the various components.

If we consider the urban shape as a result of a *selective* process, we can study this process by means of a dynamic urban growth model that examines the spontaneous generation of urban spatial patterns. The simple model we have developed depicts urban shape changes under the impact of transport costs. We have analysed a linear city, subdivided into various districts, and one (collective) transport system.

The innovative aspect of our urban model in relation to the more standard urban economics approach is the dynamic framework through which we examine the problem of spatial urban growth and the two specific mechanisms that relate the pertinent variables: an accumulative trend of the variables and a diffusion process in their variation. These two elements, which were lacking in previous urban economics approaches, assume a fundamental role when we consider the type of impact that a collective transport system improvement can generate. We have demonstrated that such an impact not only occurs in the surrounding area where we have the improvement, but also in areas distant from the improvement point. The impact area of a transport improvement is therefore not limited to a pre-defined area calculated by iso-transport cost curves, but actually includes the entire city. The final stable status is the result of a dynamic process where antagonistic effects, that is the activator and the inhibitor, operate in all points throughout the city. The numerical simulation experiments of three illustrative simple case studies have highlighted this selective process.

To achieve these results we have imposed very restrictive assumptions, but despite these restrictions, we have nevertheless been able to formulate a model that reflects interesting and relevant urban economic mechanisms and offers us a fuller picture of the process of urban pattern formation. We are confident that such an approach can be the basis for further methodological and empirical inquiry into the relationship between economic processes and urban spatial structure.

**Acknowledgments** We are grateful to Professor Aura Reggiani and Sir Alan Wilson for useful discussions and comments. Any mistake or misinterpretation is, however, ours and should not be attributed to any of the institutions with which we are affiliated.

## References

Babloyantz A, Hiernaux J (1975) Models for cell differentiation and generation of polarity in diffusion-governed morphogenetic fields. Bull Math Biol 37:637–645.

Boarnett M, Crane R (2001) Travel by design: the influence of urban form on travel. Oxford University Press, Oxford.

Cervero R, Gorham R (1995) Commuting in transit versus automobile neighbourhoods. J Am Plann Assoc 61(2):210–225.

Crane R (2000) The influence of urban form on travel: an interpretive review. J Plann Lit 15(1):3–23.

Friedman B, Gordon SP, Peers JB (1994) Effect of neo-traditional neighborhood design on travel characteristics. Transp Res Rec 1466:63–70.

Handy S (1996) Methodologies for exploring the link between urban form and travel behaviour. Transp Res D 1(2):151–165.

Ingram G (1998) Patterns of metropolitan development: what have we learned? Urban Stud 35(7):1019.

Kauffman SA (1993) The origins of order: self-organization and selection in evolution. Oxford University Press, New York.

Medda F, Nijkamp P, Rietveld P (2006) Dynamic effects of transport costs on urban shape. In: Reggiani A, Nijkamp P (eds) Spatial dynamics, networks and modelling. Edward Elgar, Cheltenham.

Meinhardt H (1998) The algorithmic beauty of sea shells. Springer.
Mohring H (1993) Land rents and transport improvements: some urban parables. Transportation 20:267–283.
Newman P, Kenworthy J (1989) Cities and automobile dependence: an International sourcebook. Gower, London.
Nicolis G, Prigogine I (1977) Self-organization in nonequilibrium systems. Wiley-Interscience New York.
Turing AM 1952 The chemical basis of morphogenesis. In: Saunders PT (ed) 1992. Morphogenesis: the collected works of A.M. Turin. North-Holland, Amsterdam.

# Chapter 5
# Algorithmic Complexity and Spatial Simplicity

**Rajendra G. Kulkarni, Roger R. Stough, and Kingsley E. Haynes**

## 5.1 Introduction

Recently the field of complexity science has emerged as an amalgamation of many different areas borrowing ideas and attracting researchers from the physical, biological and social sciences (Holland 1992; Bak 1996; Kohonen 1997; Fabian 1998; Wolfram 1994; Kauffman 2000). Of late, much of this interdisciplinary research has been facilitated by ideas and tools borrowed from another field, namely, computer science.

Merging of complexity with computer science has provided researchers with a variety of tools to test new ideas and theories and carry out simulations that have offered greater insight into a variety of properties of how complex adaptive behavior evolves and how simple rules guiding interactions at the micro level give rise to complex macro behavior as well as to identifying the properties of self-organization and emergent behaviors.

In this chapter we discuss another aspect of computer science that is slowly making inroads into complexity research, namely the field of Algorithmic complexity also known as Kolmogorov complexity (Chaitin 1966). Applications of Algorithmic complexity may offer new insights into problems such as:

1. How to measure complexity?
2. Is such a metric fixed or does it change? And in that case
3. Is there a continuum between complexity and simplicity?

Of the three questions posed in the Introduction, two are addressed in this chapter. The third question of determining if there is a continuum between complexity and simplicity is much more difficult and beyond the scope of the current chapter. However, we may be able to point toward a path that addresses this issue and we hope to respond to that question in the future.

R.G. Kulkarni (✉)
School of Public Policy, George Mason University, Fairfax, VA, USA

A. Reggiani and P. Nijkamp (eds.), *Complexity and Spatial Networks*, Advances in Spatial Science,
DOI: 10.1007/978-3-642-01554-0_5, © Springer-Verlag Berlin Heidlberg 2009

## 5.2 Background

Let us begin with an example from the field of astronomy that illustrates in an *exaggerated* sense how a seemingly complex problem becomes much more accessible and hence simpler. Much before the age of the renaissance, for more than a millennium and half, the field of astronomy was dominated by a complex dichotomous belief system of sublime heavens and material (corrupt) earth. The mechanics of planetary motions was described by a very convoluted system of concentric shells with a stationary earth at the center (http://abyss.uoregon.edu/~js/glossary/kepler.html).

However that complex and inscrutable geocentric view changed into a much more accessible and hence simpler view with a series of discoveries by Copernicus, Kepler, Galileo and Newton. The field of astronomy was transformed from a mysterious, very complex one into a field that gave rise to unified theories of planetary motions, a field that was more accessible and hence simpler. Newton's laws of motion based on a few principles (postulates) have universal applicability although under a given set of constraints. When the relative speed of objects approaches the speed of light, a different set of laws and constraints formulated by Einstein become applicable. And in the realm of the subatomic universe quantum laws come into play.

The main point of the above illustration is to stress how a complex system can change into a less complex one with the introduction of empirically tested theories that are based on a set of axioms; and laws derived from these axioms. A brief discussion on this point follows.

## 5.3 Complexity to Simplicity

Suppose one discovers (invents) a solution 's' for a complex problem 'p' based on a theory 'T'. The theory 'T' has been specified by a set 'A' of axioms, a set of rules 'R' derived from these axioms. Thus with the help of theory 'T' a complex problem may become less so. The process is similar to discovering or inventing algorithms to solve a specific problem.

However, it may happen that one discovers new aspects of problem 'p' that cannot be solely explained with theory 'T,' that is, 'T' appears to be incomplete and suddenly the less complex problem 'p' appears to be more complex. To address the new issues, theory 'T' needs to change. This could be accomplished by adding new axioms and rules or by completely abandoning this theory. In the later case, one can devise a new theory 'T1' with a new set 'A1' of axioms and rules 'R1.' Based on the new theory 'T1' the complexity of the problem may be reduced but subject to adding new axioms. The modified 'T1' is similar to modification or discovery of inventing a better algorithm to solve a set of problems. Essentially there is no one theory (or algorithm) that can explain all possible outcomes of a problem.

In fact there is long history of research going back to Russell's *paradoxes* based on Cantor sets (Stanford Encyclopedia of Philosophy; http://plato.stanford.

edu/entities/rusell-paradox), Godel's discovery of *incompleteness* (Chaitin 1966) and Turing's *halting* problem (Chaitin 1966), that in effect show that any formal system is incomplete as well as inconsistent to prove or disprove truth of some phenomenon whose truthfulness is intuitively obvious. Further, the only way to prove/disprove such phenomena is to modify the formal system to include new axioms.

From the above discussion it can be argued that complexity is not static but dynamic in nature. And if it is dynamic, it raises further questions such as "is there a continuum between complexity and simplicity?" and "is such a continuum linear or non-linear?" Further, how does one measure complexity and changes in the levels of complexity? Kolmogorov (Algorithmic) complexity offers some tantalizing answers to these questions (Haynes et al. 2007).

## 5.4 Kolmogorov Complexity

The basic idea behind Kolmogorov complexity is simple, if a physical system can be described in a concise manner then such a system is less complex compared to one that is verbose.

For example, consider the problem of describing an object or a phenomenon 'M'. Note that in this chapter any *object* or *phenomenon* are to be considered to be semantically equivalent and thus all further description applies equally to phenomena.

There may be many different ways to describe an object and yet there is one description that offers the most concise and complete description of that object. The compressed description has made that object less complex. On the other hand, it may be that there is no concise and complete description and the only way to describe the object is to give the full description. The description in the later case cannot be compressed. What has been discussed here is the complexity of object description, thus a compressed description is equivalent to a less complex object and vice versa. Obviously any compression prioritizes some dimensions of an object or concept over others. Hence there are always alternative compression possibilities, it is usually the more general ones that are of greater interest and more useful but clearly any compression and increase in simplicity is done at a cost often in terms of time but it is expected that the increase in simplicity will offset that cost.

However, these descriptions (compressions) are subject to interpretations (Zhao and Stough 2005). In that case is complexity of an object subjective? No, if we could use or do use a *formal descriptive language* based on binary alphabets.

### 5.4.1 *Formal Descriptive Language*

Assume that every letter, word, symbol and expression is described in terms of strings made of binary alphabets of '0' and '1.' With this assumption one can build a lexicographical ordering, a sort of lookup table or an interpreter that generates the strings for each letter, word and symbol. Thus a description of an object now consists of a series or string of ‘0’s and ‘1’s. Here is an example of a lookup table that has lexicographical ordering,

$$\mathrm{U} = [(\varepsilon; 0), (0; 1), (1; 2), (00; 3), (01, 4), (10, 5), \ldots], \tag{5.1}$$

where, the first symbol of the pair of symbols in side the parenthesis represents binary equivalent string of the second symbol.

Coming back to systems with concise descriptions, if such a system description is based on formal descriptive language that is concise then this system is less complex. In fact, a concise description is direct result of order or regularity in a physical system.

What about irregular systems? Descriptions of irregular systems are verbose. Due to lack of order, such descriptions are akin to very large (infinite) random sequences.

In fact Kolmogorov complexity has its roots in the theories of randomness, especially the notion of random infinite sequences. It appears that there are an infinite number of sequences that do not have concise description.

For example, for each $n > 0$, there are $2^n$ binary strings of length $n$ however there are only $2^{(n-1)}$ strings of length less than length $n$ (Ming and Vitanayi 1997). Based on ‘pigeon hole principle’, there is at least one string of length $n$ that cannot be compressed. Because $n$ can have infinite values, there are an infinite number of incompressible strings. As the value of $n$ increases, the difference

$$\mathrm{diff} = (2^{\mathrm{n}} - 2^{(\mathrm{n}-1)}) \tag{5.2}$$

increases as

$$2^{(n-1)} \sim= O(2^n) \tag{5.3}$$

For example, consider a *toy* problem of strings constructed from $n = 6$ binary digits. There are $2^6$ possible strings, however only $2^5$ strings are compressible. That leaves $2^5$ strings that cannot be compressed. As the value of $n$ increases, the number of incompressible stings increases as shown in (5.2).

Of course there are exceptions, for example, objects such as square root of 2 or value of pi. These concisely expressed objects have binary expansions that are seemingly random and infinitely long. In a way, existence of such objects illustrates the concepts behind Godel’s *incompleteness* theorem or Turing’s *halting* problem.

**Definition 1**. The length of a string $s_i$ given by $l_i = |s_i|$, that is, sum of counts of ‘1’s and ‘0s’.

Now if ‘$M$’ has $n$ different descriptions then the description with the smallest length ‘$l_{si}$’ would be the most concise and precise. Or there may be a case where all $n$ descriptions are of same length ‘$l$’ and third case may be that $n = 1$, that is, there is only one possible description with length ‘$L$’. However, this is a crude description since in any and all of these cases there might be patterns inside the string suggesting that one could compress the string and thus reduce the length of the string. If there is no pattern then the description essentially consists of random ‘0s’ and '1s', an example may be (011010010...). On the other hand if there are patterns then it is possible to compute how different the patterns are from being random.

Our discussion so far has led us to describe objects in terms of strings of binary alphabets. Further, such descriptions for patterns or the lack of it allows one to find out whether these strings are random or regular.

**Definition 2**. The complexity of an object can be expressed as the length of the smallest description expressed in binary alphabet.

Based on the above definition we could express the complexity of object 'M' in terms of its description as the length of the smallest binary string.

Below is a toy example on how to compute complexity. The process of describing this example will further formalize the discussion above.

*Example 1*. Consider a coin tossing experiment. For each toss, if the outcome is a head, we write '1', otherwise we write '0'. The following is one of the possible outcomes of a large number of coin tosses.

$$s = 0010100011101000010 \cdots 000110101000. \tag{5.4}$$

Though the frequency of number of tails and heads (zeros and ones) turns out to be nearly 1/2 as the number of coin tosses tends to infinity, the series can never be predicted to reproduce itself exactly in the same way as is shown above. Hence the randomness of this series is measured in terms of the length of the series $\sim |s|$, or the number of bits needed to specify the series '$s$'. On the other hand, if with a biased coin we get only heads (all ones) or only tails (all zeros), obviously a non-random sequence, its randomness is simply the number of bits needed to specify the number '$n$'. For example, a series with random heads and tails of length 99 would need 99 (binary) bits to specify that series, while a series of 99 ones (or zeros) can be specified by a maximum of seven binary bits, that is, 99 can be expressed as 0110011 in binary and the length $l = |0110011|$ is 7.

Zurek (1998) calls it a measure of *algorithmic randomness* and describes it as, '... given by the size, in bits, of the most concise message...' The concise message that Zurek is referring to essentially is a computer algorithm.

**Definition 3**. Kolmogorov complexity also known as Algorithmic complexity measures the complexity of a problem by the size of the algorithm that solves the problem.

Thus Algorithmic complexity $k$ of a sequence $s$ is measured in terms of the number of bits of the smallest program that can regenerate the sequence (Solomonoff 1997).

$$k(s) = |s^*| \tag{5.5}$$

where the right hand side of the above equation measures the length of the program $s^*$. Note that a program $s^*$ includes the binary code and the binary data to produce the sequence $s$.

Of course, if there is no such a program then the length of the entire sequence '*s*' in bits becomes the complexity of the sequence and this is so because the entire sequence is random.

It should be mentioned that there exist programs such as the one that emulates *linear congruential generator algorithm* which produces a seemingly random sequences that cannot be compressed and yet its algorithmic complexity is small. This is similar irrational numbers and 'Pi', discussed in the previous section, whose expansion is essentially incompressible.

*Example 2*. Let us apply the idea behind Kolmogorov complexity to the following traffic situation on a free way (Kulkarni et al. 2000). Imagine a single lane segment of a freeway with a traffic sensor installed somewhere near the middle of this segment. Let the output of the traffic sensor be fed to a processor (computer) for further analysis of traffic patterns, just as the outputs from all other sensors from different sections of a freeway road network are fed to the processor. At any instant the traffic sensor detects the presence or absence of a vehicle. Let the presence of a vehicle be coded as '1' and the absence as '0'. If the lane segment has near free flow traffic, then we may observe a series of '1's and '0's such as the one shown below:

$$0000000000110000100100001 \cdots 00001. \tag{5.6}$$

The series in (5.6) shown above has no pattern and appears as a random sequence of zeros and ones. Next let us imagine extremely congested traffic on the same one lane link. One of the possible series of observations is given by:

$$11111111111111111111 \cdots 11111. \tag{5.7}$$

The series of '1's is clearly not random. To describe the series in (5.7), all one needs to do is to count the number of '1's $= n \times 1 = n$, an integer number. On the other hand there is no way to describe the series in (5.6), but to reproduce the entire sequence as is. To get maximum information from the series in (5.7),

All we need to know is the number 'n', representing the number of ones, while for the series in (5.6), the only way to gain information is to look at the entire sequence.

This is analogous to the description of Kolmogorov randomness. To quote Chaitin, "...A series of numbers is random if the smallest algorithm capable of specifying it to a computer has about the same number of bits of information as the series itself." (Chaitin 1966). Thus the computer would process the series in (5.7) in a single step, while the series shown in (5.6) would need as many steps as there are ones and zeros in the series. In fact, as was shown in Example 1, the series in (5.6) appears similar to the outcomes of a series of un-biased coin tosses, while series in (5.7) is similar to a biased coin toss that invariably produces a series of heads (ones).

## 5.5 Complexity and Spatially Distributed Phenomena

In this part of the chapter we explain how to compute complexity of spatially distributed phenomena that can be visualized by mapping the datasets associated with such phenomena.

Consider 100 same sized square shaped tiles laid in $10 \times 10$ grid fashion. If all these tiles are of same color then the description of the $10 \times 10$ grid is: $10 \times 10$ sized grid of color X. Coding each color tile with symbol '0' and laying each row of 10 such symbols next to each other would produce a series that has one hundred '0' symbols as follows:

$$00000000000 \cdots\cdots 00000000000 \tag{5.8}$$

Suppose each of the odd numbered tiles is replaced with a tile of color Y. Coding each Y colored tile with symbol '1' now would result in alternating '1' and '0's, such as:

$$10101010101 \cdots\cdots 01010101010 \tag{5.9}$$

Next, suppose we replace randomly one fourth of the 100 tiles, with tile of color X or Y. One of the possible outcomes of the resulting pattern would be:

$$1101000101011101 \cdots .00101101010 \tag{5.10}$$

Among the three sequences, sequence (5.8) is less complex than sequence (5.9), which in turn is simpler than sequence (5.10). In terms of a binary formal system, the complexity of (5.8) < complexity of (5.9) < complexity of (5.10).

Now consider a $10 \times 10$ grid where the tiles are of all different color. If we can represent colors as binary numbers then each tile can be assigned a color binary number, for example, just 6 binary digits are sufficient to represent 64 colors, starting with 000000 and ending with 111111. Generating a sequence starting from top-left tile, one could in theory get a 6,000 digit binary number whose complexity will be defined in terms of how random or regular the colors are in the $10 \times 10$ grid. Obviously the level of complexity here is much higher than any of the sequences listed in (5.8), (5.9) and (5.10).

Next consider visualization of complex data that consists of a very large number of spatial locations, each of which has attributes whose values can be mapped. To gain more insight into the spatial patterns one may use attributes associated with these spatial locations such as demographics, income, businesses, poverty, water usage, pollution levels, etc. Using attribute values as weights one can generate density/interpolation (also known as raster) maps to study patterns of spatial distribution. Analysis of raster maps involves digitizing density/interpolation values at each location. The entire process, although somewhat routine creates a complex visual dataset with intricate patterns. How can one determine the complexity

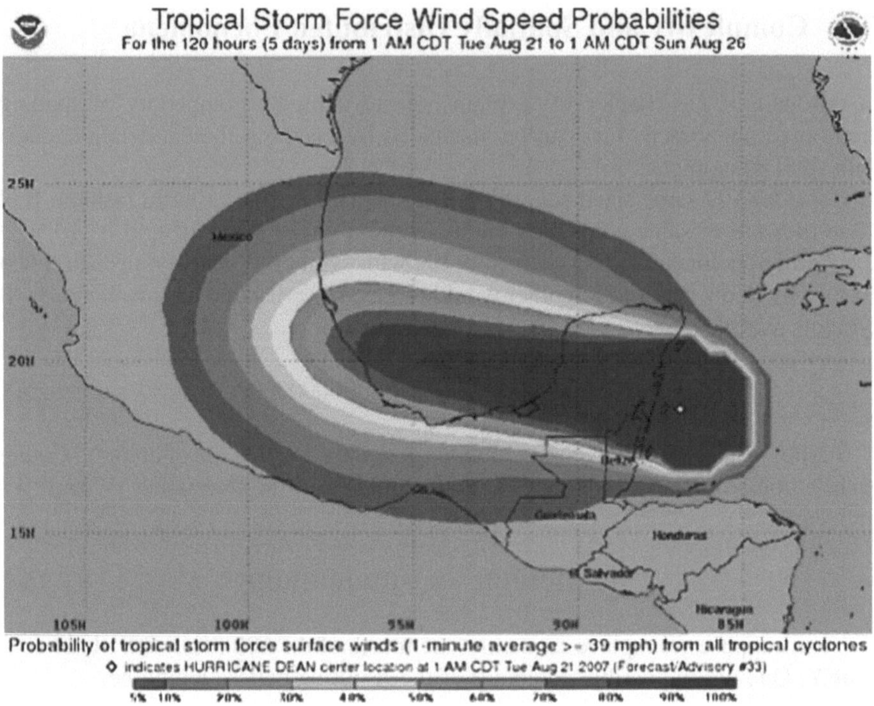


**Fig. 5.1** Atlantic hurricane season lasts from June 15 through November 15 with the peak occurring somewhere between the second week of August and the 2nd week of September

of these patterns? It is possible to do this based on concepts from Kolmogorov theory and using compression algorithms which is described below.

*Example 3*. Consider the problem of *predicting* hurricane paths. Although there are many scientific explanations of how hurricanes form, including how small weather disturbances over Ethiopia become the seeds for future hurricanes (NOAA), there is no single accepted theory that is able to predict the paths hurricanes follow once they form over the waters of the Atlantic and meander in a west-north westerly direction towards North America. Like many weather related phenomena the best one can do is produce a short term forecast of possible paths illustrated as a cone of probabilities (Fig. 5.1).

The poorly understood and possibly non-linear interaction between very a large number of known and unknown dynamic variables related to atmospheric and sea conditions serves as an explanation of the complexity of hurricane path forecasting.

Not unlike the radio astronomers who due to a lack of scientific theory do the next best thing and that is to study patterns of innumerable very high energy X-ray and $\gamma$-ray sources in the sky (http://imagine.gsfc.nasa.gov/docs/introduction/xray_information.html); one could study patterns formed by paths of all of the past hurricanes over the last several decades. These spaghetti like tracks shown in Map 5.1 are of all of the Atlantic hurricanes between 1851 through 2004 (http://

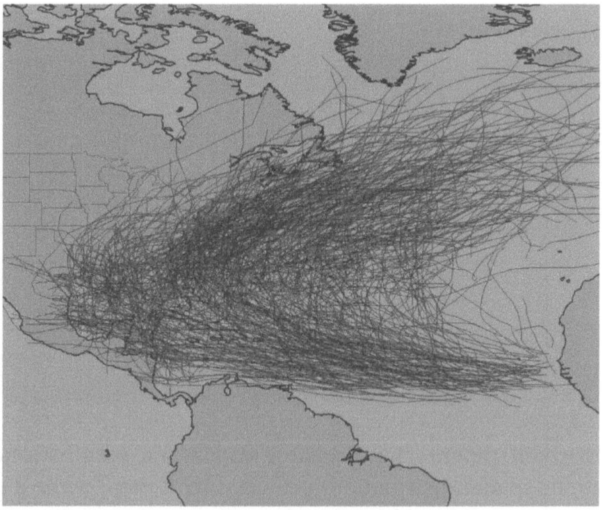

**Map 5.1** Atlantic hurricane tracks between 1851 and 2005

**Map 5.2** Density map of hurricane paths from Map 5.1

www.nhc.noaa.gov). So how does one measure how complex these patterns are? (Maps 5.2 and 5.3).

We could express the digitized map by matrix $M_{(p,q)}$, of $p$ rows and $q$ columns where each pixel value is coded in binary (0s and 1s) based on the color scheme which itself is a binary number whose values represent the hue and intensity. Note that depending on the resolution, one could create a finer grid of digitized map.

Here are more examples of spatially distributed phenomena that can be mapped and used to study the related complexity.

*Example 4*. Consider a map of EPA's (Environmental Protection Agency) toxic substance discharge in to rivers and streams by location (Map 5.4 in Appendix). The map has over 11,000 locations, each of which has a record of tons of toxic discharge

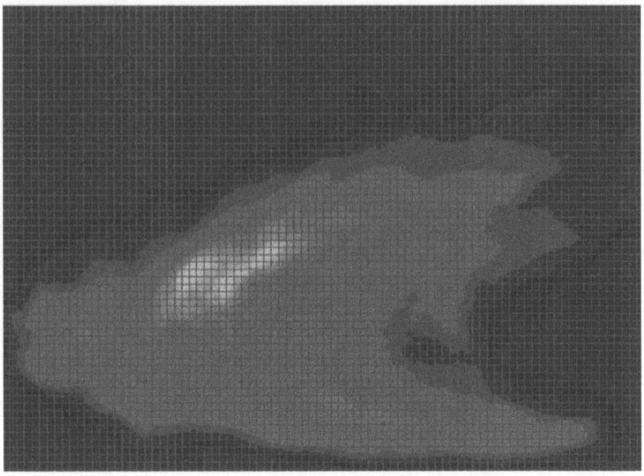

**Map 5.3** Map 5.3 shows a digitized version of the hurricane density Map 5.2

into the nearest river/stream. The discharge data when seen as a table lacks the visual information (spatial distribution) provided by a map, while a map of points that show locations does not add much value to our understanding of the data. On the other hand if we generate a density map weighted by toxic discharge levels, it starts highlighting problem areas (Map 5.5 in Appendix).

*Example 5.* Consider another example of the distribution of all businesses in the core of the Washington, D.C., metro region in 2006. There are over 173K businesses that employ more than 2.5 million people across 13 jurisdictions (Map 5.6 in Appendix). The spatial distribution of these 173 k business locations appears on the surface to have a near uniform distribution. However a density map weighted by number of employees by location offers a different picture (Map 5.7 in Appendix).

## 5.6 Inferences from above Examples

In each of these visualization examples, we have generated a complex digitized data matrix $M_{(p,q)}$, of $p$ rows and $q$ columns where each pixel value is coded in binary (0s and 1s) based on the color scheme which itself is a binary number whose values represent the hue and intensity. Matrix $M_{(p,q)}$ can be easily converted into a sequence $R$ of length $p*q$ by appending all the rows. When $R$ is fed to a compression algorithm, the output $T$ will depend on the compression algorithm (a fixed value $C$) and the sequence $R$. The process of digitizing maps/objects can be considered as a fixed size program (algorithm) just like the compression algorithm and thus does not contribute in determining the degree of complexity.

Using the Kolmogorov entropy definition,

$$\mathrm{K}(\mathrm{C}+R) = |\mathrm{C}+R| = T. \tag{5.11}$$

For a given compression algorithm, $T$ is determined entirely by input $R$.

$$T \propto \mathrm{K}(R). \tag{5.12}$$

**Table 5.1** Comparison of compression ratios based on compression algorithms

| Size in Mbyte | bzip2 $S_2$ | gzip $S_2$ | Commrcl $S_2$ | Uncompressed $S_1$ |
|---|---|---|---|---|
| Hurricane paths | 1.46 | 2.36 | 2.30 | 15.36 |
| WashDC business | 1.52 | 2.38 | 2.34 | 15.38 |
| EPA discharge | 1.72 | 2.65 | 2.62 | 15.84 |
| | bzip2 $C_R = S_2/S_1$ | gzip $C_R = S_2/S_1$ | Commrcl $C_R = S_2/S_1$ | $S_2 = S_1 \rightarrow C_R = S_2/S_1$ |
| Hurricane paths | 0.096 | 0.154 | 0.149 | 1 |
| WashDC business | 0.099 | 0.155 | 0.152 | 1 |
| EPA discharge | 0.112 | 0.172 | 0.170 | 1 |

Another way to express the degree of complexity is as follows: Let $S_1$ represent the length in bits of sequence $R$. One could test for degree of regularity/order or the lack of it by compressing this sequence using a standard compression algorithms. Let $S_2$ represent the length of the compressed sequence $R' = T$. Let the ratio between two sequences be expressed as

$$C_R = S_2/S_1. \tag{5.13}$$

A value closer to 1 indicates that the sequence $S_1$ is irregular or random and hence the object represented by that sequence is irregular/random or complex. On the other hand if the value is closer to 0 then the sequence is highly regular and hence it represents an object that is less complex (more simple).

Table 5.1 shows results of applying three different compression programs, namely gzip, bzip2 and a common commercially available version of the compression program and the ratio $C_R$ between compressed and uncompressed file sizes.

## 5.7 Conclusions and Future Research

Based on the current results, different compression algorithms give different output determined by the distribution patterns of the phenomena. For the current set of examples, it appears bzip2 offers better compression than either gzip or commercially available compression program and in general hurricane paths appear to be less complex than the distribution of Washington, D.C., business locations which in turn appears to be less complex than the EPA discharge locations. However, it is conceivable that given examples from different domains, the compressed results and the $C_R$ ratio may vary from one program to the next. It is also important to note that the data preparation phase which is a proxy for the discovery phase may affect the results obtained. It is worth mentioning that there are certain number sequences that cannot be compressed using the standard compressions algorithms. For exmple, the seemingly random number sequence generated by the *linear congruential generator algorithm* is *incompressible* using standard compression algorithms. And yet, as was noted earlier, the program to generate such a sequence is very small and with small complexity which is not detected using the process outlined by (5.11) through (5.13).

In this chapter we have presented a methodology to determine the complexity of objects. Given a set of similar objects (such as spatial distributions across a region) one may use the proposed complexity measure to determine levels of complexities of these objects, which is equivalent to determining the degree of their simplicity.

Future research will involve the issue of determining for a set of objects, if a continuum exists between simplicity and complexity and also the degree to which data preparation phase influences the end results.

## 5.8 Appendix

See Maps 5.4–5.7

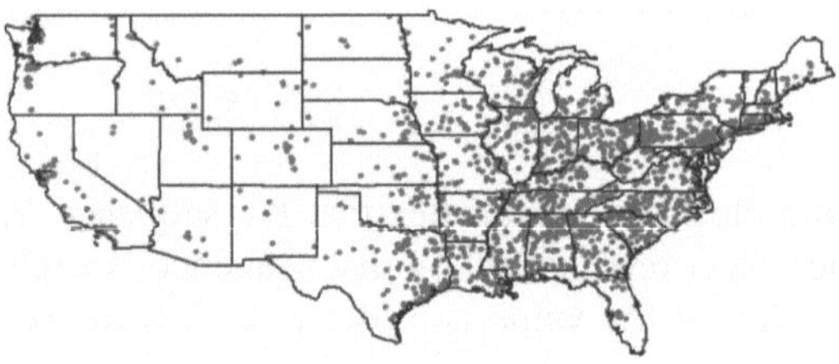

**Map 5.4** Spatial distribution of EPA's toxic release inventory sites across the U.S.

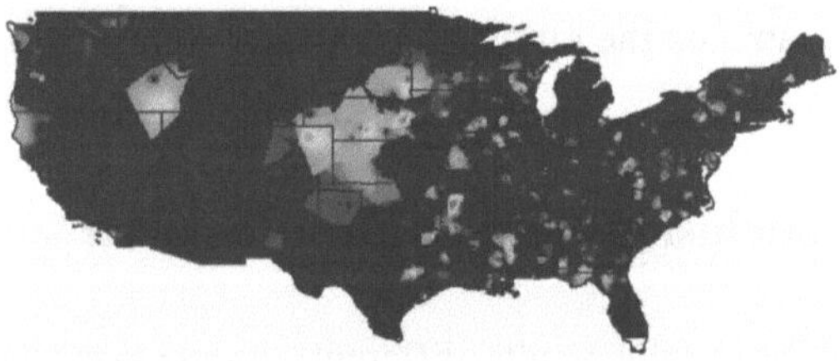

**Map 5.5** Digitized map of EPA's toxic release inventory sites across the U.S.

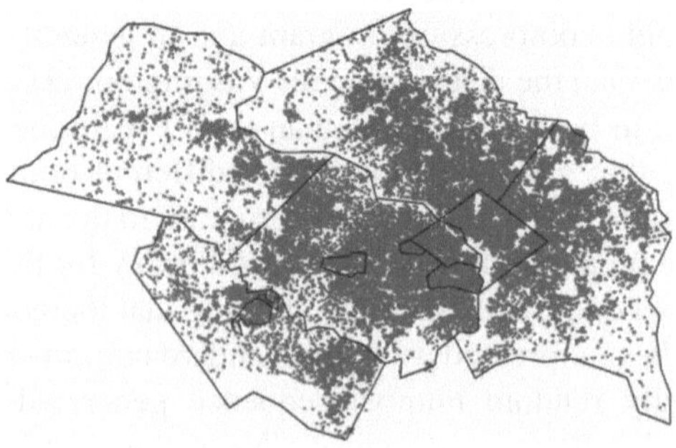

**Map 5.6** Spatial distribution of business locations in the Washington, DC metro region

**Map 5.7** Digitized map of distribution of business locations in the Washington, DC metro region

## References


Bak P (1996) How nature works. Springer, New York

Chaitin GJ (1966) ACM 13:547–569

Chaitin GJ (1987a) Algorithmic information theory. Cambridge University Press, Cambridge

Chaitin GJ (1987b) Information, randomness and incompleteness. World Scientific, Singapore

Chaitin GJ (1988) Randomness in arithmetic. Sci Am 259:1

Environmental Protection Agency (EPA). http://www.epa.gov (Date visited Oct 25, 2008)

Fabian AC (ed) (1998) Evolution. Cambridge University Press, Cambridge, MA

Haynes KE, Kulkarni RG, Stough RR (2007) Traffic grammar and algorithmic complexity. Network Spatial Econ 7:333–351

Holland JH (1992) Adaptation in natural and artificial systems. MIT, Cambridge, MA

Kauffman S (2000) Investigations. Oxford University Press, Cambridge, MA

Kelpler's Universe. http://abyss.uoregon.edu/~js/glossary/kepler.html (Date visited Oct 25, 2008).

Kohonen T (1997) Self-organizing maps. Springer, Berlin

Kulkarni RG, Stough RR, Haynes KE (2000) A formal language of Urban Freeway Traffic Patterns. Proceedings of JCIS (Joint Conference on Information Science), vol. 1, pp. 771–717

Ming L, Vitanyi P (1997) An introduction to Kolmogorov complexity and its applications, Springer, Berlin

NASA, Greenbelt, MD USA, http://imagine.gsfc.nasa.gov/docs/introduction/xray_information.html (Date visited Oct 25, 2008)

National Hurricane Center, USA http://www.nhc.noaa.gov (Date visited Oct 25, 2008)

Solomonoff R (1997) The discovery of algorithmic probability. J Comput Syst Sci 55(1):73–88

Stanford Encyclopedia of Philosophy http://plato.stanford.edu/entities/rusell-paradox (Date visited Oct 25, 2008)

Wolfram S (1994) Cellular automats and complexity (Collected Papers). Addison-Wesley, Reading, MA

Wolfram S (2002) The new science. Addison-Wesley, Reading, MA

Zurek H (ed) (1998) Complexity entropy and the physics of information, vol. III. SFI studies in the science of complexity. Addison-Welsley, Reading, MA

Zhao Z, Stough RR (2005) Measuring similarity among various shapes based on geometric matching. Geogr Anal 37:371–383

# Chapter 6
# Polyplexity

## A Complexity Science for the Social and Policy Sciences

**Helen Couclelis**

*An ant, viewed as a behaving system, is quite simple. The apparent complexity of its behavior over time is largely a reflection of the complexity of the environment in which it finds itself.* (Simon 1969)

## 6.1 Introduction

Simon's famous ant metaphor points to the possibility of two alternative representations for the same complex phenomenon: the ant's convoluted path on the beach may be described as complex behaviour against a simple background, or as simple behaviour against a complex background (or as a little of both, of course). The metaphor also supports the intuition that complexity is largely in the eye of the beholder – a fruitful philosophical position to take, as it encourages the observer to seek the representation that is the most useful for the purpose at hand rather than engage in a wild goose chase for "the" correct kind of representation. However, the ant-on-the-beach scenario falls short in one important respect: it views phenomena as consisting of a system of interest and an environment, whereas in fact every system description also involves a (usually tacit) underlying spatio-temporal framework.

I propose the notion of *polyplexity* as a new way of approaching the study of the most complex of systems, that is the systems studied in the social and policy sciences. Polyplexity goes one step further than most conventional approaches to complex systems by taking into account the possibility that the space and time within which a phenomenon enfolds may themselves be complex. It proposes a "divide and conquer" modelling strategy based on apportioning the apparent complexity of a phenomenon among the three major constituent parts of any system

H. Couclelis
Department of Geography, University of California, Santa Barbara, CA, USA

A. Reggiani and P. Nijkamp (eds.), *Complexity and Spatial Networks*, Advances in Spatial Science,
DOI: 10.1007/978-3-642-01554-0_6, © Springer-Verlag Berlin Heidlberg 2009

representation: the system of interest itself, its environment, and its spatio-temporal context. Polyplexity suggests that the widely acknowledged greater complexity of social relative to natural science phenomena may be seen to be due in part to more complex underlying space–time frameworks. Should this be the case, accounting for spatio-temporal complexity in addition to system environment complexity in social science modelling may help simplify the representation of certain systems of interest.

Social scientists embraced the complexity paradigm fairly early on, making major contributions of their own along the way. However, despite increasingly sophisticated models of complex socio-spatial dynamics and agent-based systems, social science has adopted more or less unquestionably the Cartesian framework of the natural sciences. The result is in many ways a more elaborate form of "social physics", with models such as those simulating the emergent behaviour of growing sand piles replacing the planetary "gravity model" metaphors of the 1950s and 1960s. On the whole, the space and time of social science remain monotonously flat. The shortcomings of the current homogeneous, isotropic space–time assumptions may be especially evident in the attempts of geographers and others to model information-age phenomena such as the "death of distance" or the "extensible individual". It is conceivable that these taken-for-granted Cartesian assumptions are hampering progress in a much broader spectrum of social science and policy research. After several decades of achievements in complex system modelling, I believe that the field is mature enough to consider exploring approaches more specifically tailored to the challenges of the "difficult" (as opposed to "hard") sciences. One possible direction would be to focus on notions of social space and time and their potential role in simplifying the representation of complex social phenomena. This emphasis seems to makes sense because, as Nigel Thrift notes, "complexity theory is preternaturally spatial" (cited in O'Sullivan 2004, p. 284). Polyplexity is meant to be an early wobbly step in that direction.

Not surprisingly, complexity is itself a complex notion. There are several different complexity paradigms highlighting its different aspects: discontinuous change under smooth parameter variation, self-organization, emergence, path dependence, feedback, deterministic unpredictability, and so on. These include Thom's (1975) catastrophe theory, Prigogine's (1980) bifurcation theory, Haken's (1983) synergetics, chaos theory, and a host of related computational approaches among which agent-based simulation and cellular automata modelling are especially popular in the Anglo-American world. Less well explored outside its field of origin is one of the oldest complexity paradigms, that deriving from Turing's work on the mathematical theory of computation (see Copeland 2004). Through its two major branches of automata theory and formal language theory, the theory of computation contributes the notion that complex representations can be built gradually from simpler ones through the systematic expansion of the domains of the operands and operators considered. Polyplexity hopes to capitalize on this principle though the details are still nebulous.

Going back to the issue of a complexity science for the social and policy sciences, there are a number of desiderata, most of which are not very well served by more traditional approaches to complexity. For example, it would be really nice if we were able to handle the following kinds of problems with something like the power and elegance possible for the description of complex physical processes:

- The description of social processes and events, which involve reasons (telic considerations) as well as causes
- The representation within the same general framework of multiple perspectives on – and interpretations of – the same social process or event
- The modelling of emerging institutional structures that are not simply the result of bottom-up interactions
- The representation of individual decision and choice in highly complex environments
- The support of decision making in planning and policy under deep uncertainty and conflict
- Etc. (add your own wish list here)

An overarching desideratum would be the development of a unified perspective on complex system modelling in the social and policy sciences for handling and integrating the above kinds of issues.

As an agenda for polyplexity, this sounds extravagant to the point of foolishness – but who knows? The time may be right for confronting tentative, high-risk ideas of this kind, such as the notion that polyplexity could perhaps simplify the representations of social phenomena and policy problems of interest by relegating some of their apparent complexity to suitably complex but still manageable spatio-temporal structures. To be useful though these structures should first be integrated within some more general and systematic framework. For example, Simon's ant-on-the-beach metaphor could be generalized to the "principle of consistently optimizing behaviour", stating that "every choice is an optimal choice when examined against the appropriate background of empirical, logical and spatio-temporal assumptions".

In the following pages I discuss the three main components of the notion of polyplexity. Complex time and complex space are examined in the next section, and then the notion of "prior structure" is presented as a perspective on modelling that might conceivably support the philosophical ambitions of polyplexity. The conclusion, which is by necessity sober and brief, mentions some of the challenges of pursuing such a program, and summarizes numerous open questions that this chapter leaves in its wake.

## 6.2 Complex Time, Complex Space

This is not the right place to review the achievements of complex systems research in social science. Several of the field's protagonists are represented in this volume and can speak for themselves. The breadth of the social scientists' contributions to the complex systems paradigm has indeed been quite extraordinary, covering both discrete and continuous systems, both the macro- and the micro-perspective, both statistical and process modelling, both analysis and policy-oriented synthesis, and both conceptual and applied research. Building on that wealth of previous efforts, this chapter attempts to glimpse fuzzy visions of the future rather than retread the brilliant past.

### 6.2.1 Complex Time

In a book entitled "The economics of time and ignorance", O'Driscoll and Rizzo (1985) examine the nature of prediction in economics and conclude that under no circumstances can prediction be complete because of the existence of "real" time and "real" ignorance. The authors contrast "real" time with Newtonian time which is simply a framework for ordering events, a reference line against which events can be mapped as either points or intervals. A basic property of time-as-framework is that it does not in itself affect events. In other words, Newtonian time does not bring change; it only serves to register change as it happens. Time is fully analogous to (Newtonian, absolute) space, and has the same three basic properties: homogeneity (all time-points are the same except for their position along the time line); continuous divisibility (implying that neighbouring time points are independent of one another); and causal inertness (time is independent of its contents: in itself it causes nothing). In any model based on Newtonian time, even a fully dynamic one, it is the present as we know it that is sent rolling along the time line. As the great economist F. H. Hahn observed, in such models "*the future is merely the unfolding of a tapestry that exists now*".[1]

"Real" time by contrast is characterized by the properties of dynamic continuity, heterogeneity, and causal efficacy. These properties preclude prediction, hence the notion of "real" ignorance. *Dynamic continuity* is based of the two aspects of memory and expectation. The meaning of each moment depends on its place in the context of what we remember of the past and expect for the future, just as in the experience of music each note can only be appreciated relative to those heard a moment before and those anticipated yet to come. More generally, the timing of an event changes its nature to the extent that the unique context of other events within which it occurs affects its role in the determination of subsequent events. This is the case, for example, with economic agents whose response to events today depends on what they learned yesterday (which includes the responses of other agents to yesterday's events), as well as on what they expect to happen tomorrow (which includes how they expect other agents will act). The property of *heterogeneity* of real time follows from dynamic continuity in that no two instants can be the same, each one relating to a different set of preceding and succeeding moments and their remembered or anticipated contents. This makes events in real time genuinely

[1] Original emphasis, cited in O'Driscoll and Rizzo (1985, p. 52).

non-repeatable. Thus non-repeatability emerges from an event's temporal "place value" – its order in the flow of events. *Causal efficacy* is a further corollary of the above in that dynamic, heterogeneous time is causing actions and events to be different now from what they would have been under the same conditions some time earlier or later. Related examples also from economics are the notions of time inconsistency and of discounting, whereby the utility of a given option (say, of buying insurance or of the government raising interest rates) can vary greatly depending on the time when the choice must be made. In general, the nature of the uncertainty that this conception of time implies is much more profound than the two kinds commonly considered in science: the case where the value of a specific outcome is unknown but the ex ante probability distribution of outcomes is known, and the case where the underlying probability distribution itself is not known (random).[2] Here we are dealing with situations where not just the probabilities, but even some qualitative characteristics of outcomes – all the way to the very nature of the possible outcomes themselves – cannot be determined ex ante because they are not part of "a tapestry that exists now", in Hahn's famous words quoted above. Under the name of "deep uncertainty" this latter notion is prominent in the work of a group of researchers from the RAND corporation advocating a general approach to planning that takes into account the virtual impossibility of prediction (Lempert et al. 2004).

Real time is much closer to the psychological intuition of a dynamic flow of ever-changing experiences than to the traditional scientific view of a directed axis used as a ruler for pegging events. Its significance is obvious for social science problems involving intentional agents. However what this conception of time addresses is not just human cognition and action but more generally historicity, or the claim that the nature of any phenomenon depends to some extent on its place within a process of historical development. A good example from natural science would be the significance of a particular mutation in an organism, which may or may not have an evolutionary value depending on the timing (and placing) of its appearance. The fact that it is impossible to predict future speciation in biology is further evidence that the processes of evolution work in real time. The similarity of real time with the notion of path-dependence in complexity theory is surely not coincidental.

Historians have their own complex models of historical time. According to a group of historians involved in a major digital atlas project, modelling time as a fourth dimension downgrades it into being only a facet of space whereas, in fact, time operates according to very particular principles.[3] This is because at the core of historical understanding is the event rather than the object or the point process, and historical events are not described as discrete entities but as networks of other events linked together by causal and telic relations at different levels of granularity.

---

[2] In economics this distinction was made by Knight in his seminal dissertation where he used the term 'risk' to describe the first case from the perspective of a decision maker, reserving the term 'uncertainty' for the second case. See Knight (1921).

[3] This section on historical time draws on the work of the TimeMap project (www.timemap.net/timelines) by Johnston et al.

To support such a view multiple lines of real time may be needed, and these would be punctuated rather than continuous since knowledge of events is episodic and fragmentary. This historical view of complex time provides the complementary macro-perspective to O'Driskoll and Rizzo's mostly individual-level real time, and in doing so it transposes it to a level that is at least an order of magnitude more complex. Both these approaches reject the simple one-dimensional view used in practically all complex systems research in favour of conceptualizations that emphasize intimate causal and telic linkages among time, events, choices, and their ever-changing contexts.

The need to broaden the notion of time has also been keenly felt within the hard-nosed, empiricist geographic information science community. Several models have been proposed in the context of "temporal GIS" beyond linear time: cyclic time, branching time, totally- and partially-ordered time, valid and transaction time, clock- vs. event-driven time, etc. Each of these brings some useful modification to the simple axis of classical physics but the characteristic Newtonian causal inertness remains: time is still the neutral framework against which independently unfolding events are projected, sorted and measured. None of these models (with the possible exception of some interpretations of branching time) approaches the dynamic, causally efficient conception of real time that O'Driscoll and Rizzo believe to be so important in economics and the social sciences in general, and that historians would like to further develop into a highly complex structure.

### *6.2.2 Complex Space*

From non-Euclidean geometries to relativistic space–time to today's high-dimensional spaces of string theory, physics has been a treasure trove of complex models of space. However, attempts to transfer some of these conceptions to the social science domain have not on the whole been successful. Social scientists have had much better luck with relational and network spaces such as those of graph and network theory or the even more complex multi-dimensional spaces described by Q-analysis, Galois lattices, self-organizing maps (SOM) and other such techniques (see for example Gatrell 1983; Freeman and White 1993; Agarwal and Skupin 2008). The connection of these relational spaces with the space of everyday social life is however somewhat tenuous, since they cannot deal directly with fundamental quantitative properties of physical space such as distance, direction, shape, the elementary Euclidean transformations, or spatial autocorrelation.

Notions of complex space also abound in geography and related disciplines and have often been used to simplify or visually enhance the representation of particular kinds of phenomena. Space transforms are a particularly prominent family of complex spaces, and of these, cartographic projections are the most widely known and used. Other familiar kinds include cartograms, logarithmic spaces, velocity fields, representations of cognitive maps, and parallel coordinate spaces (see for example Angel and Hyman 1976; Borden 1996; Gould and White 1974;

Golledge and Stimson 1997; Inselberg and Dimsdale 1994). Some of these are explicitly designed to do what Simon's ant metaphor suggests, that is, they complexify the space so as to simplify the representation of the phenomenon of interest.[4] For example, representations of cognitive maps produced by eliciting pair-wise distance estimates from subjects are converted through the technique of bi-dimensional regression into heavily distorted, crumpled and stretched transforms of actual maps. These representations may then be used to show how errors in distance perception correlate with sub-optimal spatial choices by individuals or groups. However, all of the complex space representations mentioned here are either formulated for very specific kinds of problems, or they are too general. For polyplexity a middle road would be desirable, whereby classes of social science and policy problems could be handled by the same general approach to complex space.

A couple of my own attempts at setting up models of complex spaces may be relevant to polyplexity. The first of these is the concept of proximal space (Couclelis 1997). Proximal space is formed by the set of all locations that have some functional or other kind of non-explicitly spatial relation with every location of interest at each time. It is a generalization of the notion of neighbourhood as used in cellular automata and other kinds of models, whereby proximity is defined not in terms of physical distance or adjacency but in terms of the special relationship a location has with other locations. For example, the set of all locations of my physical and virtual social contacts form the proximal social space of my home location. Proximal space is thus a network space, but one that is not only rooted in actual geographical space, but also lends itself to simulation modelling: indeed, it supports a formal generalization of cellular automata called geo-algebra (Takeyama and Couclelis 1997). This is one example of how one could simplify the representation of a dynamic process by relegating some of its complexity to the embedding space. It is possible, though this has not yet been explored, that a model analogous to proximal space ("proximal time") may also be developed for historical time as discussed above. Proximal time would represent the set of key moments and intervals relevant to a specific event of interest and its aftermath – say, the times associated with the genesis and subsequent fate of this chapter, from the original invitation by this volume's coeditors through the fallout resulting from its publication. Proximal time as defined here would thus rejoin O'Driscoll and Rizzo's notion of the heterogeneity of real time, whereby no two instants can be the same because each one relates to a different set of preceding and succeeding moments.

Some earlier work considers not one, but a sequence of interrelated spaces, seeking to capture their distinguishing characteristics in a systematic and reproducible manner (Couclelis and Gale 1986). That project explores the meaning of several more or less vague notions of space used in psychology which include, beyond the Euclidean, spaces referred to as physical, sensorimotor, perceptual,

[4]This indeed seems to be the modus operandi of insects (including ants!): "...the insects write their spatial memories in the environment, while the mammalian cognitive map lies inside the brain." See Chialvo and Millonas (1995).

cognitive, and symbolic. In that research we propose a hierarchy of six nested levels corresponding to the above sequence of notions of space and representing, psychologically, a progression of increasingly complex levels of an individual's spatio-temporal awareness. The same empirical experience or phenomenon may be defined against any one (or all) of these spaces, with different implications each time. To represent the linkages between levels the model relies on the notion of selective operators as used in spectral theory, while the first four levels are also differentiated internally by means of the family of algebraic structures that are part of group theory. In this model the operands are spatial 'atoms' the empirical interpretations of which vary from level to level (points, locations, positions, vantage points, or places), and the group-theoretic operators are the links between atoms, called "moves" but again meaning different things at each level.

Group theory focuses on operations and transformations, rather than operands, and involves five axioms known as the closure law (G1), the associative law (G2), the existence of an identity element (G3), the existence of inverses (G4), and the commutative law (G5). An algebraic structure conforming to all five axioms (for example the set of integers) is called an *abelian group*. The other members of the group family are obtained by dropping one or more of these axioms. Thus axioms G1–G4 (but not G5) define a *group*; axioms G1–G3 (but not G4 and G5) define a *monoid*; and axioms G1 and G2 (but not G3–G5) define a *semi-group*. A correspondence between these algebraic structures and the hierarchy of spaces is tentatively set up as shown in Table 6.1, based on certain empirical properties of each space in the sequence. Thus, for example, in the physical space of everyday experience – unlike in pure Euclidean space – the commutative property (G5) does not hold with the force of an axiom because the direction of gravity causes space to be anisotropic in the up/down direction. (Bodies that are "up" can easily go "down" but the reverse is usually not true). Similar considerations result in the elimination of one more, then two more group axioms for sensorimotor and perceptual space, respectively. Thus sensorimotor space, the space in which living organisms (and also robots) move, is like physical space in that it lacks the commutative property, but it also lacks a true inverse (G4) because moves in sensorimotor space can never be completely reversed. Even if an animal or machine returns to the exact same location it started from, its state will no longer be exactly what it was when the move was initiated: it will have become more tired, more hungry, more worn down, or it will have acquired new bodily experiences: it will

**Table 6.1** Concepts of space and corresponding algebraic structures

| Concept of space | Axioms | | | | | Structure |
|---|---|---|---|---|---|---|
| | G1 | G2 | G3 | G4 | G5 | |
| Symbolic space | 0 | 0 | 0 | 0 | 0 | ? |
| Cognitive space | 0 | 0 | 0 | 0 | 0 | ? |
| Perceptual space | 1 | 1 | 0 | 0 | 0 | Semi-group |
| Sensorimotor space | 1 | 1 | 1 | 0 | 0 | Monoid |
| Physical space | 1 | 1 | 1 | 1 | 0 | Group |
| Pure Euclidean space | 1 | 1 | 1 | 1 | 1 | Abelian group |

have "depleted its batteries" or enriched its sensorimotor memories to some extent. One level up, perceptual space is in many ways like sensorimotor space, but lacks the identity element (G3), because even the "stay-as-you-are" move is no guarantee that perceptual identity will be maintained: it is well known that attention filters what we perceive at any particular time. Beyond that level the model breaks down, because it is difficult to give meaning to a space characterized only by the closure law.

Beside group theory, the other mathematical notion underlying the model is that of selective operators. A selective operator may be thought of as a sieve or filter that sorts the entities corresponding to some particular description out of a universe **U** of entities.[5] This method is used in the model to construct the lower four levels out of each other, by selecting out of the universe of group properties first two, then three, then four, then all five group axioms. It is uncertain if the remaining two levels (cognitive and symbolic) really belong in this hierarchy, since they are not subject to the constraints of pure Euclidean or of physical space – though they are most definitely subject to the *experience* of these spaces.

Regardless of its merit (or lack thereof) as a formalized description of the range of individual awareness of space, two aspects of this model are relevant to the notion of polyplexity. First, at the sensorimotor level we find the first intimations of real time (in the form of the irreversibility of physical effort), and this impression is reinforced at the next level up, though the details cannot be discussed here. Second, it hints at the possibility of developing an ordered sequence of mutually consistent models of space, of varying degrees of complexity, for use in the social and behavioural sciences. This last point is significant because hierarchies of complex social spaces keep being proposed in geography and related fields with insistent regularity. There may be something to that idea that is worth pursuing further.

## 6.3 Prior Structure, Determination, and Hierarchical Spatio-Temporal Ontologies

Spatio-temporal ontologies are a hot topic in geographic information science these days. The motivations are mostly practical, such as the need to improve interoperability among different GIS platforms, but some of the questions raised by that work are decidedly theoretical, if not philosophical. Similar though less formalized efforts also originate in geography as researchers attempt to classify and make

[5] This works as follows: If $\mathbf{O_a}$ is the selective operator that selects out of **U** whatever answers to the description of **A**, then $\mathbf{O_aU}$ is a representation for the set of entities **A**. Now, **A** itself may comprise several other kinds of entities, among which those answering to the description of **B** may be of particular interest. In this case, if $\mathbf{O_b}$ is the operator that selects the **B**'s, then $\mathbf{O_bA = O_b(O_aU)}$ is a way of representing **B** as a function of **A** and **U**. This procedure can be iterated for as many steps as necessary, so that if we have a hierarchy of entities **A, B, C, D,**... such that $\mathbf{D \subset C \subset B \subset A \subset U}$, we may represent these as: $\mathbf{O_aU = A}$, $\mathbf{O_bO_aU = B}$, $\mathbf{O_cO_bO_aU = C}$, $\mathbf{O_dO_cO_bO_aU = D}$, and so on (see Larsen 1970).

sense of the unwieldy variety of available conceptual and quantitative models of geographical phenomena. The vast majority of these proposals are hierarchical, involving "tiers" or "levels" or "spaces" of different degrees of complexity and characterized by very different properties. Here are some quick examples, in chronological order: (1) Mathematical space, physical space, socioeconomic space, behavioural space, experiential space (Couclelis 1992); (2) Physical level, functional level, biological level, intentional level, social level (Guarino 1999); (3) Physical reality, observable reality, object world, social reality, cognitive agents (Frank 2003). Or, more specifically regarding the complexity of spatial decision models: (4) Stimulus–response (basic observation), stimulus–response (controlled experiment), rational decision, production system, advanced computational process model (Couclelis 1986). And also: (5) Decision making as a variable, as a probability function without feedback, as a probability function with feedback, by one type of agent, by multiple interacting agent types (Agarwal et al. 2002).

Note that even though all the above examples are spatio-temporal hierarchies, they are not hierarchies of nested spatial and temporal scales, but rather, of semantically different planes on which qualitatively different kinds of spatio-temporal phenomena can be described. These and several other similar efforts all seem to agree that the physical is simple but that the social and mental are complex and hard, but other than that there are few commonalities in approaches and perspectives. The last two examples however – (4) and (5), involving decision making models – do have something interesting in common in that they take an *informational* rather than an empiricist approach to the issue. The first explicitly, and the second implicitly, they both recognize that the same system of interest may be modelled at different levels of complexity, from elementary to extremely complex, depending on how much information one is able or willing to include in the representation. They thus side with the perspective of mathematical computer science reflected in the hierarchical theory of modelling and simulation by Zeigler et al. (2001),[6] which is itself based on the hierarchy of automata theory (finite automata, pushdown automata, linear bounded automata, Turing machines) and the corresponding one of formal language theory (regular, context-free, indexed, recursively enumerable languages; see Hopcroft and Ullman 1979).

Somewhat along similar lines is the notion of *prior structure* in modelling that I briefly explored many years ago (Couclelis 1984). That was part of an attempt to figure out where the predictive power of some simple (and very unrealistic) mathematical urban models comes from.[7] The idea was that in every complex system there are a number of constraints, formal as well as empirical, that can be

[6]Zeigler's hierarchy of system specifications comprises the following four levels: Input–output relation observation, input–output function observation, discrete event system, discrete event network. Couclelis (1986) specifies four models of decision of increasing complexity in term of that hierarchy.

[7]There may be some connection between prior structure as discussed here and Bunge's notion of "determination" as the basis for causality. If so, my idea would stand on fairly respectable philosophical ground! See Bunge (1979).

known a priori to limit the range of observable system states. Empirical constraints (called historical prior information) derive from certain aspects of the system – physical, biological, technological, institutional or social – that can be more or less reliably assumed to remain reasonably constant within the forecasting horizon of the model. For example: the rate of change in the life expectancy of a population, the rate of transformation of raw materials into structures, or the fact that there will still be fewer commuters on the roads on Sundays than on most weekdays. Such considerations have of course been at the basis of numerical forecasting techniques for many years and are expressed in the distinction between "fast" and "slow" variables in dynamic modelling. The notion of prior structure stresses the importance of being able to specify the level of analysis at which these kinds of empirically derived constraints become operative.

Much more intriguing however is the second class of constraints, called structural (or logical) prior information. This derives from the formal invariances that characterize the fundamental logico-mathematical structures (such as set theory, topology, number theory and logic) that underlie mathematical and computational models. As with the case of historical prior information, the nature and amount of logical prior information available depend on the level of model specification. Together, empirical and logical prior information make up the model's prior structure, that is, the envelope of constraints which incorporates all positive knowledge about the system of interest at a specific level. Within that envelope, all allowable microstates are equiprobable, but Wilson's (1970) entropy maximizing approach can be used to identify the most likely system macrostates. Wilson's seminal statistical–mechanical derivation of spatial interaction (formerly "gravity") models rescued these from the prevailing crude planetary analogies, while also providing a philosophically significant insight into the value of an informational – as opposed to empiricist – perspective.

And what about polyplexity? Well – complex time and complex space, described in some appropriate, orderly hierarchical sequence, may constitute a third kind of prior information, along with the historical and logical. Polyplexity would take the idea of prior structure in models one step further. This would not suddenly render predictable what is fundamentally unpredictable in complex social systems (the notion of real time alone settles this issue), but it may tighten the envelope of constraints within which the genuinely surprising can happen, while also helping to clarify the limits of modelling in the social and policy sciences.

## 6.4 Some Concluding Thoughts

An unspoken word behind much of the preceding discussion – a discussion at times quite dry and technical, is *intentionality*. Intentionality, along with the human purposes it drives, is why the notion of real time makes immediate intuitive sense, it is what guides the weaving of disparate locations and moments into places and events meaningful to people, and it is what distinguishes cognition and abstract thought from the mechanical sorts of awareness represented in, say, the hierarchical group-theoretic model discussed earlier. Intentionality and the closely associated notions of purposeful action set limits to what we can model in the social world since, qua telic concepts, they are not compatible with current causal scientific paradigms, including the paradigms of complexity. Indeed, social processes and events involve both "because" and "in order to", and we yet have no tools to deal with the latter. Purpose is a major factor in the evolution, adaptation and learning in social systems, whereas in natural systems that also can evolve, adapt and learn it clearly is not. The role of purpose in the social world is a defining qualitative difference between natural and social complex systems. The more advanced models of artificial cognitive agents are designed to mimic purposeful behaviour; however, to ask where these agents get their purposes from is to promptly end the conversation.

Considering how difficult it is to build reliable models of complex natural systems, what should models of complex social and policy-oriented systems be expected to do?

For years now several researchers have argued for a softer role for models in social science and policy, beyond the traditional triad of description–explanation–prediction. They talk about models as narratives about possible things to come, as plots around which stories of warning or encouragement may be woven. This is not just a nonchalant New Age stance but is informed by multiple evidence that validation of complex system models is not really possible. I sympathize with this view but feel that it goes too far in abdicating all responsibility in trying to anticipate at least some aspects of the future. Polyplexity is meant as an effort to figure out what kinds of things may be known in advance, under what conditions, through what kinds of representational manipulations, and thus perhaps to help restore a modicum of respect in the predictive power of complex social system models.

There are obviously more questions than answers in what I presented here. Does the idea of polyplexity make sense in principle? If yes, could it help simplify the study of the many intractable problems that the social and policy sciences deal with? Could it handle phenomena of the information age that appear to enfold against a hybrid physical/virtual space–time? What may be the role of polyplexity in forecasting and scenario development, especially as used in the policy sciences? What may be, in particular, the contribution of polyplexity to robust adaptive planning as defined by Lempert et al. (2004)? Can we figure out how best to distribute complexity considerations among actor, context, and spatio-temporal background? What are the computational implications of this approach? How may familiar, successful models of complex social science systems be usefully recast in polyplexity terms? Because ideas evolve in real time it is not possible to predict at this point to what extent these speculations about polyplexity may survive scrutiny. But writing this chapter was a complex spatiotemporal event closely linked to a number of other, similar events, all intersecting at the time and place of the meeting out of which this volume was eventually born. Taken together, these intertwined trajectories in time, space and ideas may express an emerging message on complexity and simplicity in social science that no-one could have predicted.

## References


Agarwal P, Skupin A (2008) Self-organising maps: applications in geographic information science. Wiley, New York

Agarwal C, Green GM, Grove JM et al. (2002) A review and assessment of land-use change models: dynamics of space, time, and human choice. NE-297. Department of Agriculture, Forest Service, Northeastern Research Station, and Indiana University, Center for the Study of Institutions, Population, and Environmental Change, USA

Angel S, Hyman GM (1976) Urban fields: a geometry of movement for regional science. Pion, London

Borden D (1996) Cartography thematic map design, 4th edn. C. Brown, Dubuque, IA

Bunge M (1979) Causality and modern science, 3rd edn. Dover, New York

Chialvo DR, Millonas MM (1995) How swarms build cognitive maps. In Steels L (ed) The biology of intelligent autonomous agents, vol 144. NATO ASI series, Belgium, pp 439–450

Copeland BJ (ed) (2004) The essential turing: seminal writings in computing, logic, philosophy, artificial intelligence, and artificial life plus the secrets of enigma. Oxford University Press, Oxford

Couclelis H (1984) The notion of prior structure in urban modelling. Environ Plan A 16:319–338

Couclelis H (1986) A theoretical framework for alternative models of spatial decision and behavior. Ann Assoc Am Geogr 76:95–113

Couclelis H (1992) Location, place, region, and space. In: Abler RF, Marcus MG, Olson JM (eds) Geography's inner worlds. Rutgers University Press, New Brunswick, NJ, pp 215–233

Couclelis H (1997) From cellular automata to urban models: new principles for model development and implementation. Environ Plann B Plann Des 24(2):165–174

Couclelis H, Gale N (1986) Space and spaces. Geografiska Annaler 68(1):1–12

Frank AU (2003) Ontology for spatio-temporal databases. Spatio-temporal databases: the CHOROCHRONOS approach, vol. 2520. Springer, Berlin, pp 9–77

Freeman LC, White DR (1993) Using Galois lattices to represent network data. Sociol Methodol 23:127–146

Gatrell AC (1983) Distance and space: a geographical perspective. Oxford University Press, Oxford

Golledge RG, Stimson RJ (1997) Spatial behavior: a geographic perspective. Guilford, New York

Gould P, White R (1974) Mental maps. Penguin Books, New York

Guarino N (1999) In: Christian Freksa, David M. Mark (eds) The role of identity conditions in ontology design. Spatial information theory: a theoretical basis for GIS. Proceedings, International conference COSIT '99, Stade, Germany. Springer, Berlin, pp 221–234

Haken H (1983) Synergetics, an introduction: nonequilibrium phase transitions and self-organization in physics, chemistry, and biology, 3rd edn. Springer, New York

Hopcroft JE, Ullman JD (1979) Introduction to automata theory, languages, and computation. Addison-Wesley, Reading

Inselberg A, Dimsdale B (1994) Multidimensional lines 1: representation. SIAM J Appl Math 54(2):559–577

Knight FH (1921) Risk, uncertainty and profit. Hart, Shaffner and Marx, Houghton Mifflin, Boston, MA

Larsen MD (1970) Fundamental concepts of modern mathematics. Addison-Wesley, Reading, MA

Lempert RJ, Popper SW, Bankes SC (2004) Shaping the next one hundred years: new methods for quantitative, long-term policy analysis. The RAND Corporation, Santa Monica, CA

O'Driscoll GP, Rizzo MJ (1985) The economics of time and ignorance. Basil Blackwell, Oxford

O'Sullivan D (2004) Complexity science and human geography. Trans Inst Br Geogr 29:282–295

Prigogine I (1980) From being to becoming: time and complexity in the physical sciences. Freeman, San Francisco

Simon HA (1969) The sciences of the artificial. MIT, Cambridge

Takeyama M, Couclelis H (1997) Map dynamics: integrating cellular automata and GIS through Geo-Algebra. Int J Geogr Inform Sci 11(1):73–91
Rene T (1975) Structural stability and morphogenesis. W.A. Benjamin, Reading
Wilson AG (1970) Entropy in urban and regional modelling. Pion, London
Zeigler BP et al (eds) (2001) Methodology in systems modelling and simulation. North-Holland, Amsterdam

# Part B
# Evolutionary Networks in a Socio-Economic Context

# Chapter 7
# Complexity, Evolution and Learning

## Empirical and Experimental Validation of Heterogeneous Expectations

**Cars Hommes**

## 7.1 Introduction

A paradigm shift in economics is taking place. In traditional, neoclassical economics a representative agent who behaves perfectly rational has been the main working hypothesis and mathematical analysis of simple tractable models its main focus. A problem with this approach is that it requires unrealistically strong assumptions about individual behaviour, such as perfect knowledge and information about the economy and extremely high computational abilities to do what is optimal. An advantage of the neoclassical research programme, partly explaining its success, is that rationality imposed through optimizing behaviour and model consistent expectations enforces strong discipline on the modelling framework leaving no room for market psychology and unpredictable, irrational behaviour.

An alternative complexity view is now emerging, based on interaction of many heterogeneous agents, whose behaviour is only *boundedly rational*. In this new behavioural agent-based approach, computer simulation models are the main modelling framework. An advantage is that it becomes possible to describe in detail individual behaviour at the micro level based on realistic assumptions. The Santa Fe conference proceedings Anderson et al. (1988) and Arthur et al. (1997a) contain many contributions within the complexity view. The recent *Handbook of computational economics* (Tesfatsion and Judd 2006) contains many chapters describing the state of the art of agent-based economics. There is however still an important problem with the bounded rationality research programme: it leaves too many degrees of freedom. There is only one way (or perhaps a few ways) one can be right, but there are many ways one can be wrong. To turn the alternative view into a successful research programme, one has to "tame the wilderness of bounded rationality".

---

C. Hommes
School of Economics, University of Amsterdam, Amsterdam, The Netherlands

A. Reggiani and P. Nijkamp (eds.), *Complexity and Spatial Networks*, Advances in Spatial Science,
DOI: 10.1007/978-3-642-01554-0_7, © Springer-Verlag Berlin Heidelberg 2009

A key feature that distinguishes economics from natural sciences is that market realizations depend on future *expectations* and, at the same time, expectations about future developments are based on current and past realizations. An economy is an expectations feedback system in which beliefs and realizations co-evolve. Agents are "smart" and will adapt their behaviour if it benefits them. If all agents are perfectly rational, in equilibrium individual expectations and realizations must coincide on average, leading to the neoclassical representative rational agent model. But if agents are heterogeneous and only boundedly rational, one needs a convincing theory of heterogeneous expectations. In this chapter we discuss, a simple story of *heterogeneous expectations* and some empirical and experimental validation. Agents can choose from a class of simple heuristics disciplined by *adaptive learning* and *evolutionary selection*. An extensive recent survey of this approach including many references to related work can be found in Hommes (2006).

This chapter is organized as follows. Section 7.2 describes a simple example, an asset pricing model with heterogeneous beliefs, and illustrates how the asset price dynamics may become unstable when expectations are driven by reinforcement learning based on past strategy performance. Section 7.3 discusses the empirical validity of a simple version of the model with two types of traders, fundamentalists and technical analysts, and how it explains the "dot com bubble" in stock prices in the late 1990s. Section 7.4 discusses how this approach matches the stylized facts of learning to forecast laboratory experiments with human subjects. Finally, Sect. 7.5 briefly describes a future perspective.

## 7.2 An Asset Pricing Model with Heterogeneous Beliefs

As a simple example of a model with heterogeneous expectations we consider the asset pricing model with heterogeneous beliefs of Brock and Hommes (1998). This model may be viewed as a simple stylized version of the Santa Fe artificial stock market model introduced by Arthur et al. (1997b). Agents can invest in a risk free asset that pays a fixed return $1 + r$ or in a risky asset that pays uncertain dividends $y_t$ in each period. The market clearing pricing equation is given by

$$(1 + r)p_t = \sum_{h=1}^{H} n_{ht} E_{ht}(p_{t+1} + y_{t+1}) + \varepsilon_t, \tag{7.1}$$

where $p_t$ is the price of the risky asset, $n_{ht}$ the (time varying) fraction of trader type $h$, $E_{ht}(\cdot)$ the belief of type $h$ about next period's price plus dividend, and $\varepsilon_t$ a noise term representing, for example a small fraction noise traders. In the special case when *all* agents are rational the asset price will be equal to the *rational, fundamental benchmark* $p_t^*$, given by the discounted sum of expected future dividends

$$p_t^* = \frac{E_t[y_{t+1}]}{1+r} + \frac{E_t[y_{t+2}]}{(1+r)^2} + \cdots.$$

This fundamental benchmark is nested as a special case within the general heterogeneous agent model. In the case of IID dividends with mean $\bar{y}$, the fundamental price becomes constant,

$$p^* = \bar{y}/r.$$

Assuming that the beliefs about future dividends are correct (for example because they can be inferred from past observations of the exogenous dividend process), the model can be rewritten in deviations

$$x_t = p_t - p^*$$

from the fundamental and simplifies to:

$$(1+r)x_t = \sum_{h=1}^{H} n_{ht} E_{ht} x_{t+1} + \varepsilon_t. \tag{7.2}$$

Strategy choice follows an *evolutionary selection* principle, that is, "strategies that have performed better attract more followers". This can be modelled in several ways, but we follow Brock and Hommes (1997) where the fractions of belief type $h$ are determined by the discrete choice model (a random utility model)

$$n_{ht} = \frac{e^{\beta U_{h,t-1}}}{Z_{t-1}}, \tag{7.3}$$

where

$$Z_{t-1} = \sum_j e^{\beta U_{j,t-1}}$$

is normalization factor and $U_{h,t-1}$ measures the *past performance* or *fitness* (for example realized profits, forecasting performance, etc.) of strategy $h$. The parameter $\beta$ is the *intensity of choice* measuring the sensitive of agents to differences in strategy performance. In the extreme case $\beta = 0$, agents behave randomly and all fraction types are fixed with equal weights; at the other extreme, $\beta = \infty$, all agents immediately switch to the best predictor (the "neoclassical limit").

Which ones out of an ocean of possible forecasting rules will agents use? In a real market, it seems unlikely that many agents will coordinate on a very complicated rule. Therefore, we use simple rules, such as linear rules with only one time lag (written in deviations

$$x_{t-1} = p_{t-1} - p^*$$

from the fundamental):

$$f_{ht} = p^* + g_h x_{t-1} + b_h,$$

where $g_h$ is a *trend* parameter and $b_h$ a *bias* parameter. Another simple rule not using any fundamental price information is the *trend extrapolating rule*

$$f_{ht} = p_{t-1} + g_h(p_{t-1} - p_{t-2}),$$

which simply extrapolates the last price change. So far, the parameters in the forecasting rules have been fixed, but one can introduce *adaptive learning* to learn the parameters over time. For example, agents may update forecasting parameters by sample average or by employing a recursive ordinary least squares scheme (OLS-learning), as additional observations become available (see for example Evans and Honkapohja (2001) for an extensive treatment of adaptive learning in macroeconomics and Sargent (2008) for a recent discussion of the importance of learning in macroeconomics and monetary policy).

Figure 7.1 shows simulations of the price fluctuations in an example with four belief types, including fundamentalists and trend followers, and fitness given by last period's realized profits. When the intensity of choice is small, the steady state is typically stable and the asset prices converge to the fundamental benchmark. Intuitively this may be understood by observing that for small intensity of choice, agents are more or less randomly distributed over the different strategies, and as a result the average forecast is close to the fundamental enforcing convergence to the fundamental price. In contrast, when the intensity of choice is large agents typically coordinate on a common strategy and the dynamics destabilizes. In particular, coordination on a trend following strategy may occur, leading to persistent price deviations from fundamental. Indeed the asset pricing dynamics in Fig. 7.1 is characterized by irregular switching between phases of close to the fundamental price fluctuations with fundamentalists dominating the market and phases of temporary bubbles when trend following strategies dominate the market. Excess volatility and temporary bubbles are driven by short run profit opportunities. The noisy simulation illustrates that even in this simple model the start and burst of the temporary bubbles are highly unpredictable.

## 7.3 Empirical Validation

How relevant are these bubble and crash dynamics to real financial market data? We briefly discuss the estimation of a simple version of the model with two types of agents, using yearly S&P 500 data; see Boswijk et al. (2007) for a detailed analysis.

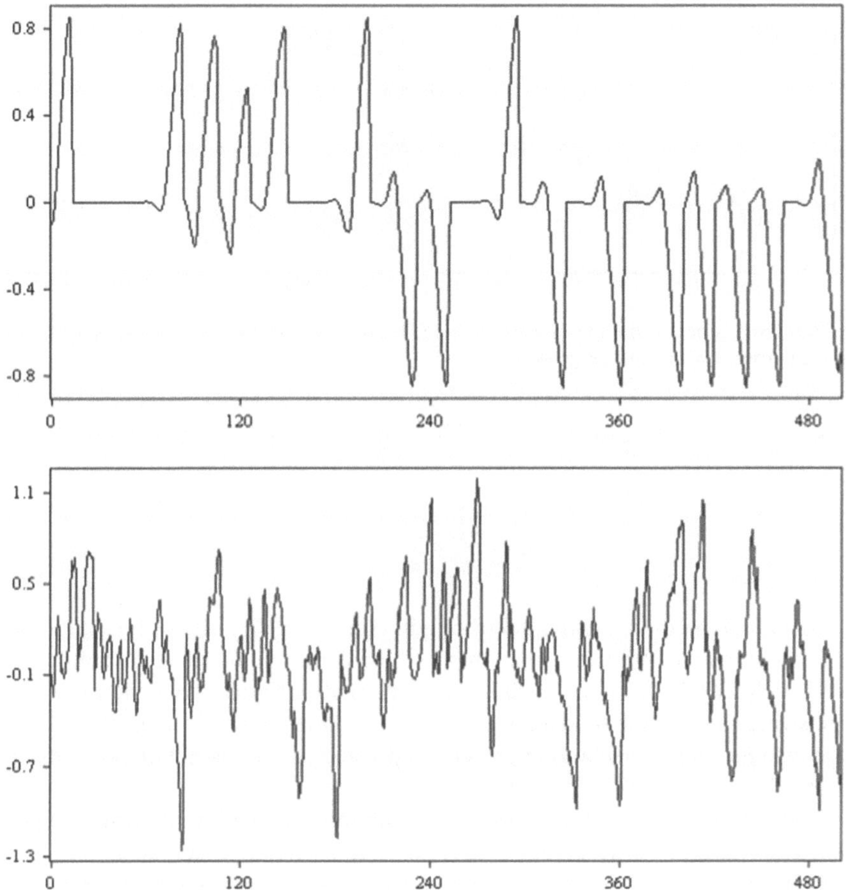


**Fig. 7.1** Chaotic (*top*) and noisy chaotic (*bottom*) time series of asset prices (in deviations from the fundamental price) in an example with four trader types. Prices fluctuate irregularly around the benchmark fundamental price (which corresponds to 0). Parameters are: $g_1 = 0, b_1 = 0$; $g_2 = 1.1$, $b_2 = 0.2$; $g_3 = 0.9$, $b_3 = -0.2$ and $g_4 = 1.21$, $b_4 = 0$, $r = 0.1$ and $\beta = 90$

Figure 7.2 shows the logs of yearly S&P 500, 1871–2003, and a benchmark fundamental price based on dividends with a constant growth rate $g$. This is the standard Gordon model and the fundamental benchmark is given by

$$p_t^* = \frac{1+r}{r-g} y_t,$$

where $g$ is the growth rate of dividends and $r$ is the required rate of return for investors to hold the risky asset (given by the sum of the risk premium to hold

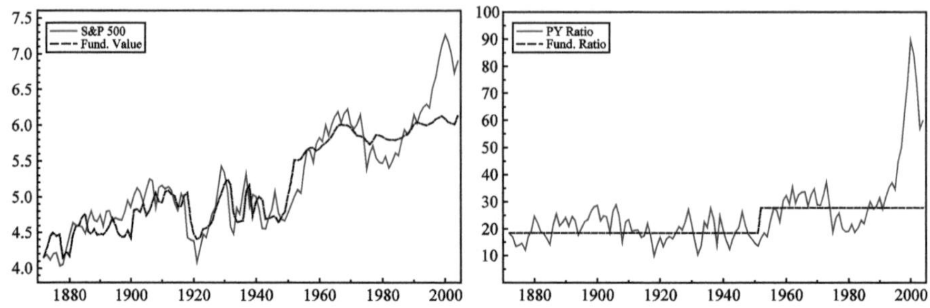


**Fig. 7.2** Time series of the log of S&P 500, 1871–2003 and the benchmark fundamental for a dividend process with constant growth rate

stocks and the risk free interest rate). The corresponding fundamental price to cash flow ratio

$$\delta_t^* = \frac{p_t^*}{y_t} = \frac{1+r}{r-g} = m$$

is constant along the fundamental (the right plot in Fig. 7.2 allows for one jump in the fundamental in 1950, due to a jump in the risk premium; see for example Fama and French (2002)). Figure 7.2 shows that the realized price–dividend ratio shows large swings around the fundamental benchmark, fluctuating between 10 and 30 for more than a century, rising to unprecedented values of almost 90 in the 1990s, and coming down to values below 60 in recent years.

There are two competing views concerning the explanation of swings in price-to-cash flows. Some attribute them to rational responses to macroeconomic fundamentals, while others judge that irrational swings in investor sentiment play a significant role. Shiller (2000) gives a lucid description of both views, stressing the relevance of psychological factors.

Boswijk et al. (2007) estimated a simple two-type model of the form

$$R^* x_t = n_t \phi_1 x_{t-1} + (1 - n_t)\phi_2 x_{t-1} + \varepsilon_t, \tag{7.4}$$

where

$$R^* = (1+r)/(1+g), \ x_t = \delta_t - m$$

is the deviation of the price-to-cash flow from the fundamental, $n_t$ and $(1 - n_t)$ are the fractions of the two types (depending on past realized profits) and $\phi_h x_{t-1}$, $h = 1, 2$, are the forecasts of the two types of next period's deviation from the fundamental (only the first lag was significant). The estimation results yield significant estimates $\phi_1 = 0.76$ and $\phi_2 = 1.14$, implying that type 1 are fundamentalists believing in *mean reversion* of the price towards its fundamental value, while type 2 are trend followers, believing that the price bubble will continue. Figure 7.3

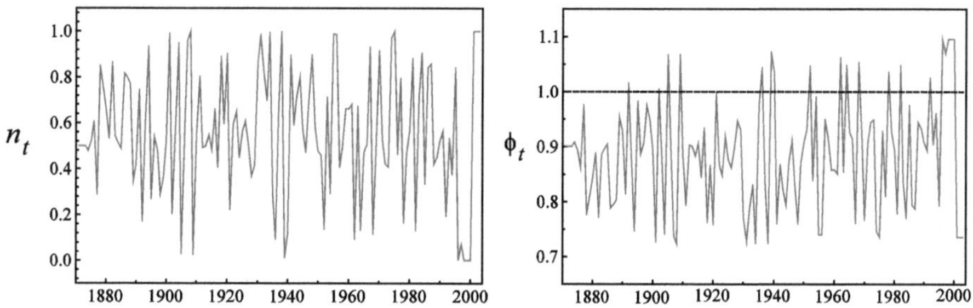


**Fig. 7.3** Time series of the estimated fraction $n_t$ of fundamentalists (*left panel*) and average market sentiment $\phi_t = \{n_t\phi_1 + (1 - n_t)\phi_2\}/R^*$ (*right panel*)

shows the time variation of the estimated fraction $n_t$ of fundamentalists. Significant heterogeneity with strategy switching and large fluctuations in the fractions of both types occur. In particular, one observes a low fraction of fundamentalists for 5 or 6 subsequent years in the late 1990s. The average coefficient

$$\phi_t = \{n_t\phi_1 + (1 - n_t)\phi_2\}/R^*$$

in Fig. 7.3 shows that market sentiment fluctuates considerably over the years, with average traders believing in explosive asset prices in the late 1990s. This simple model explains the "dot com bubble" in the late nineties as being *triggered by fundamentals*, in the form of good news (a new technology) about the economy, and *strongly amplified by trend following strategies* based on reinforcement learning driven by short run profits.

The PD-ratio has come down to values below 60 in recent years, and one may ask the question: *Will the bubble resume?* Figure 7.4 shows prediction of both the nonlinear model with strategy switching and the linear model with a representative fundamentalist, believing in average mean reversion. Clearly the nonlinear switching model predicts much larger swings in price-to-cash flow fluctuations of asset prices than its linear, representative agent counterpart.

## 7.4 Learning to Forecast Experiments

Laboratory experiments with human subjects are well suited to discipline the "wilderness of bounded rationality". In laboratory markets with carefully controlled market fundamentals one can investigate which behavioural rules are more likely to be used by human subjects in different market environments. In this section we briefly discuss laboratory experiments on expectations formation in the asset pricing framework of the previous sections. We address the following questions:

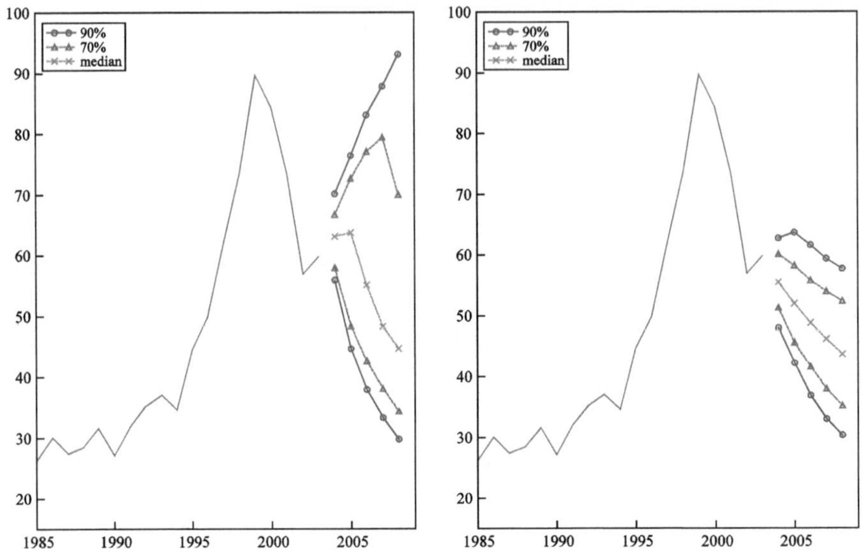


**Fig. 7.4** Quantiles of 2,000 simulated predictions of the PD-ratio for the nonlinear evolutionary switching model (*left*) and the linear, representative agent model (*right*). Both models are estimated using data until 2003 and then predict up to 5 years ahead

- How do *boundedly rational* agents form expectations and how do they learn in a *heterogeneous* world?
- How do individual forecasting rules *interact* and what is the *aggregate outcome* of individual interaction?
- Will *coordination* occur, even when there is limited market information?
- Does *learning* enforce convergence to rationality?

Hommes et al. (2005) performed learning to forecast experiments in an asset pricing framework similar to that used in Sects. 7.2 and 7.3. Six human subjects have to forecast the price of a risky asset for 50 periods, and their payment is inversely related to their forecasting errors. There is expectations feedback, since the realized market price is determined by aggregation of individual forecasts. After all individual make a forecast, the computer computes a market clearing price derived from standard mean–variance maximization demand functions using the individuals forecasts as inputs. Since subjects only forecast and trading is completely computerized, agents may be viewed as rational optimizers, given their individual forecasts. Such a laboratory environment thus produces "clean data" on expectations, and one can test various expectations hypotheses. Except for the six subjects, there is a seventh robot trader in the market, who always predicts the fundamental price and whose weight increases (from 0 to at most 0.2) when prices deviate more from fundamental.

Subjects thus have *limited information* about the market. They are told that they are advisors to a pension fund, which will invest more in the risky asset, when the subject makes a higher forecast. They also know that the asset price is determined by market clearing. From the qualitative market information, subjects should be able to understand that the asset market exhibits *positive feedback*, that is, higher forecasts lead to higher realized market prices. Subjects also know the interest rate $r = 0.05$ and the mean dividend $\bar{y} = 3$, and could use these to compute the fundamental price

$$p^{\mathrm{f}} = \bar{y}/r = 60.$$

Furthermore in forecasting $p_{t+1}$, they know past realized prices (up to $p_{t-1}$), their own past forecasts (up to $p^e_{t,h}$) and their own earnings (up to $e_{t-1,h}$). However, subjects do *not* know market equilibrium equations, the forecasts of others and the number of pension funds in the market. The information in these experiments is therefore similar to what is often assumed in models with boundedly rational traders.

The (unknown) price generating process is given by

$$p_t = \frac{1}{1+r}\left((1-n_t)\frac{p^e_{t+1,1} + \cdots + p^e_{t+1,6}}{6} + n_t p^{\mathrm{f}} + \bar{y} + \varepsilon_t\right), \tag{7.5}$$

where $n_t$ is the share of robot traders given by

$$n_t = 1 - \exp\left(-\frac{1}{200}|p_{t-1} - p^{\mathrm{f}}|\right), \tag{7.6}$$

$p^e_{t+1,h}$, $1 \leq h \leq 6$, are the individual forecasts and $\varepsilon_t$ is a small noise term. If all subjects would forecast rationally and use the fundamental price of 60 as their individual forecast, the realized market prices would be close to 60 with small random fluctuations around it. This is perhaps not what one would expect right from the start in a market with limited information, but an interesting question is whether the market price will at least converge to the fundamental price. It is useful to briefly mention two other homogeneous agents benchmarks. If all subjects would use *naive expectations*, that is, use the last price observation to forecast

$$p^e_{t+1} = p_{t-1},$$

starting say with an initial forecast of 50, then realized market prices will converge monotonically to the fundamental price 60. If on the other hand all subjects use a simple trend extrapolation rule

$$p^e_{t+1} = (p_{t-1} + 60)/2 + (p_{t-1} - p_{t-2}), \tag{7.7}$$

then prices will fluctuate around the fundamental for 50 periods (about six oscillations). One may wonder how individual subjects would arrive at such a rule, but

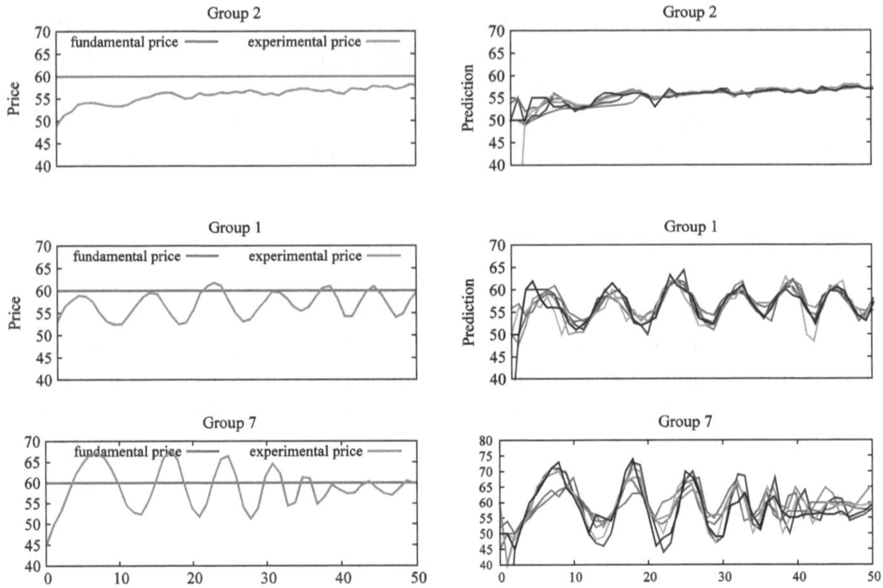


**Fig. 7.5** Three typical outcomes of realized prices (*left panel*) and individual forecasts (*right panel*) of the learning to forecast laboratory experiments: (1) monotonic convergence, (2) permanent oscillations, and (3) dampened oscillations

remarkably estimation of the forecasting rules showed that a number of individuals use a rule very similar to (7.7).

Figure 7.5 shows some typical outcomes of realized prices (left panel) and individual forecasts (right panel). Three qualitatively different patterns are observed: (1) monotonic convergence, (2) permanent oscillations, and (3) dampened oscillations. Monotonic convergence is very similar to what would happen if all subjects use a naive forecast. The permanent oscillations are similar to what would happen if all subjects use a simple linear AR2 rule such as (7.7). In the third case of dampened oscillations a strong price trend emerges in the beginning of the experiment, but the strong trend gets weaker and reverses when prices deviate too much from their fundamental value. Another striking feature of the experiment is that in all cases there is strong *coordination* on a common prediction rule, as illustrated in the right panel of Fig. 7.5. Coordination however is *path dependent*, since different qualitative outcomes are observed in different markets.

Estimation of individual prediction rules shows that for most subjects (more than 90%) forecasting is well explained by a simple linear model with no more than three lags in prices and individual forecasts. In fact, for a majority of subjects (more than 50%) a simple rule with only one or two lags fits the forecasting behaviour very well. Some simple rules that have been estimated include:

- *Adaptive expectations*

$$p^e_{t+1} = wp_{t-1} + (1-w)p^e_t$$

(in converging groups)
- *Linear rules*

$$p^e_{t+1} = \alpha + \beta_1 p_{t-1} + \beta_2 p_{t-2}$$

(in oscillating groups)
- *Trend-extrapolating rules*

$$p^e_{t+1} = p_{t-1} + \gamma(p_{t-1} - p_{t-2})$$

(in oscillating groups)

In order to explain these experiments Anufriev and Hommes (2009) recently developed a *heuristics switching model*. There are a number of *simple heuristics* and in the beginning agents choose heuristics *randomly*. Agents evaluate the *past performance* of these heuristics based on forecasting accuracy, and subsequently tend to *switch* to more successful heuristics. Figure 7.6 shows simulations of the heuristics switching model reproducing the three different patterns observed in the laboratory experiments.

The four forecasting heuristics used in the simulations are adaptive expectations, a weak and a strong trend-extrapolating rule and an *anchoring and adjustment heuristic*

$$p^e_{4,t+1} = 0.5p^{\mathrm{av}}_{t-1} + 0.5p_{t-1} + (p_{t-1} - p_{t-2}), \tag{7.8}$$

where $p^{\mathrm{av}}$ is the sample average of all past prices. This rule uses an anchor (the average of the last observed price and the sample average) and extrapolates a trend from there. Following the terminology of Tversky and Kahneman (1974), it may be viewed as a forecasting anchoring and adjustment heuristic.

The price dynamics in the heuristics switching model is given by

$$\begin{aligned} p_t = \frac{1}{1+r_f}&((n_{1,t}p^e_{1,t+1} + n_{2,t}p^e_{2,t+1} + n_{3,t}p^e_{3,t+1} + n_{4,t}p^e_{4,t+1}) \\ &\times (1-n_t) + p^{\mathrm{f}} n_t + \bar{y} + \varepsilon_t). \end{aligned} \tag{7.9}$$

The fractions $n_{h,t}$, $1 \le h \le 4$, of the four heuristics are determined by a discrete choice model with *asynchronous updating*

$$n_{i,t} = \delta n_{i,t-1} + (1-\delta)\frac{\exp(\beta U_{i,t-1})}{\sum_{i=1}^{4}\exp(\beta U_{i,t-1})}, \tag{7.10}$$

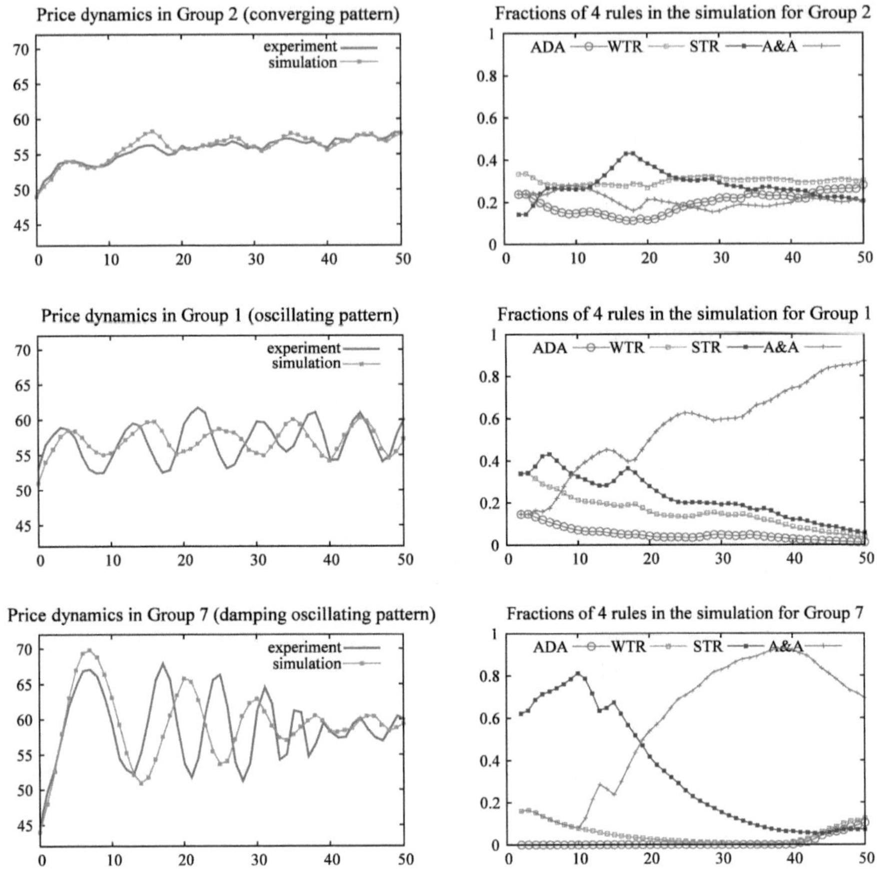


**Fig. 7.6** Simulations of the heuristics switching model. Prices for laboratory experiments and evolutionary model. Fractions of four forecasting heuristics: adaptive expectations (ADA), weak trend followers (WTR), strong trend followers (STR), and anchoring adjustment heuristic (A&A). The simulations only differ in initial price forecasts and initial distribution of strategies

with the fitness measure given by (minus) squared prediction errors, that is,

$$U_{i,t-1} = -(p_{t-1} - p^e_{i,t-1})^2 + \eta U_{i,t-2}. \tag{7.11}$$

The parameter $\eta \in [0,1]$ represents the *memory strength* in the fitness measure, the parameter $\delta \in [0,1]$ represents the inertia of traders' switching behaviour (in each period, only a fraction $1-\delta$ of traders will switch strategy) and the parameter $\beta \geq 0$ is the intensity of choice as before. The fraction $n_t$ of robot traders evolves according to (7.6), as in the experiment.

The only differences in the simulations of Fig. 7.6 are the initial price forecasts and the initial distribution over the four heuristics. Trends in realized market prices are more likely when the initial fractions of the weak and strong trend followers are sufficiently large. Interestingly, the anchoring and adjustment heuristic is important in keeping the fluctuations alive, since in both the permanent and the dampened oscillatory cases their fractions becomes large (more than 80%). Coordination of individual forecasts on simple forecasting heuristics thus explains the three different observed aggregate market outcomes. Oscillations may be triggered by initial prices and small random shocks, are reinforced when the initial fraction of weak and strong trend heuristics is relatively large and may be sustained by the anchoring adjustment heuristic.

## 7.5 Concluding Remarks

We have discussed a simple theory of heterogeneous market expectations, in which bounded rationality is disciplined through simple heuristics, adaptive learning and evolutionary selection. This theory matches important stylized facts in financial market, such as excess volatility and (temporary) bubbles and crashes. In particular, coordination on trend following strategies, driven by experience based reinforcement learning, may strongly amplify a rise or decline in asset prices triggered by fundamental news. As we have seen, the theory matches for example the "dot com" bubble in stock prices in the late 1990s. The theory is also consistent with learning to forecast laboratory experiments with human subjects and explains observed path-dependent stable and unstable outcomes. In particular, laboratory experiments confirm that coordination on simple trend following strategies may occur and lead to persistent deviations from fundamental and fluctuations in asset prices.

In future work the theory should be tested in different market environments. Complexity models in economics are often based on heterogeneous expectations, and a satisfactory theory of heterogeneous expectations is therefore necessary for a successful research programme on bounded rationality, complexity, agent based economics and evolution.

## References


Anderson PW, Arrow KJ, Pines D (eds) (1988) The economy as an evolving complex system II. Addison-Wesley, Reading, MA

Anufriev M, Hommes CH (2009) Evolution of market heuristics. Knowl Eng Rev, forthcoming

Arthur WB, Durlauf SN, Lane DA (eds) (1997a) The economy as an evolving complex system II. Addison-Wesley, Reading, MA.

Arthur WB, Holland JH, LeBaron B et al. (1997b) Asset pricing under endogenous expectations in an artificial stock market. In: Arthur W, Lane D, Durlauf S (eds) The economy as an evolving complex system II. Addison-Wesley, Reading, MA, pp 15–44

Boswijk HP, Hommes CH, Manzan S (2007) Behavioral heterogeneity in stock prices. J Econ Dyn Control 31:1938–1970
Brock WA, Hommes CH (1997) A rational route to randomness. Econometrica 65:1059–1095
Brock WA, Hommes CH (1998) Heterogeneous beliefs and routes to chaos in a simple asset pricing model. J Econ Dyn Control 22:1235–1274
Evans GW, Honkapohja S (2001) Learning and expectations in macroeconomics. Princeton University Press, Princeton
Fama EF, French KR (2002) The equity premium. J Finan 57:637–659
Hommes CH (2006) Heterogeneous agent models in economics and finance. In: Handbook computational economics, vol 2: Agent-based computational economics. Elsevier, Amsterdam, pp 1109–1186
Hommes CH, Sonnemans J, Tuinstra J, et al. (2005) Coordination of expectations in asset pricing experiments. Rev Financial Stud 18:955–980
Sargent TJ (2008) Evolution and intelligent design, Am Econ Rev 98:5–37
Shiller RJ (2000) Irrational exuberance. Princeton University Press, Princeton
Tesfatsion L, Judd KL (eds) (2006) Handbook of computational economics, vol 2: Agent-based computational economics. Elsevier Science, Amsterdam
Tversky A, Kahneman D (1974) Judgment under uncertainty: heuristics and biases. Science 185:1124–1131

# Chapter 8
# Homophily, Conformity, and Noise in the (Co-)Evolution of Complex Social Networks

**George Ehrhardt, Matteo Marsili, and Fernando Vega-Redondo**

## 8.1 Introduction

Social networks constitute the backbone underlying much of the interaction conducted in socioeconomic environments.[1] Therefore, when this interaction attains a *global* reach it must have, as its counterpart, the emergence of a social network with a wide range of overall (typically indirect) connectivity. Naturally, for such a social network to emerge, agents must be able to link profitably. But this in turn demands that they display similar – at least compatible – behaviour. Thus, for example, they must use coherent communication procedures, share key social conventions, or have similar technical abilities. Here, we may quote the influential work of Castells (1996).

> Networks are open structures, able to expand without limits, integrating new nodes as long as they are able to communicate within the network, namely as long as they share the same communication codes (for example, values or performance goals).

Reciprocally, of course, such a convergence of behaviour is facilitated by the range of social interaction being global rather than local or fragmented. This suggests the idea that the buildup of a global social network might be understood as the outcome of twin cross-reinforcing processes: one that facilitates the convergence of norms and behaviour, and another that extends the range of

F. Vega-Redondo (✉)
Department of Economics, European University Institute, Florence 50133, Italy

[1]Some important examples include labor markets (Granovetter 1974; Montgomery 1991; or Calvó-Armengol and Jackson 2004), trade arrangements (Kirman et al. 2000; Kranton and Minehart 2001), or networks of interfirm collaboration in either research and development (Hagedoorn 2002; Powell et al. 2005).

A. Reggiani and P. Nijkamp (eds.), *Complexity and Spatial Networks*, Advances in Spatial Science,
DOI: 10.1007/978-3-642-01554-0_8, 

social connectivity. This mutual reinforcement also suggests that if such a global transition indeed takes place, it should be relatively *fast* and *resilient*.

To explore these ideas, Ehrhardt et al. (2006) – thereafter referred to as EMV – proposed a stylized model in which agents are involved in a local coordination game with their neighbours in a coevolving network. Two features characterize the dynamics. First, we postulated that while links vanish at a constant exogenous rate, new links are created only between agents who randomly meet and happen to be "coordinated", that is, display the same action/strategy in the coordination game. On the other hand, we assumed that, on the same time scale (that is, at a comparable rate), agents adjust their action towards that which maximizes the extent of coordination with their current neighbours. For brevity, we referred to the first feature as *homophily* and the second as *conformity*.

The way in which EMV conceive both homophily and conformity is particularly stark but also restrictive. In particular, both dynamic forces are implemented through noiseless mechanisms, which implies that, in the long run, the only links existing in the network are those connecting agents displaying the same action – that is, network components are homogeneous. This seems too extreme a setup, which make one wonder whether the analysis is robust to the introduction of some noise. Our aim here will be to conduct such a "robustness check" by studying a model where some persistent noise may perturb both the establishment of links and the adjustment of actions.

In a nutshell, our conclusion is that neither of these generalizations affects the main predictions of the model. As in the original benchmark model, we continue to observe:

1. Sharp qualitative transitions in network connectivity and coordination, as a discontinuous (upward and downward) response to slight changes in the underlying parameters beyond corresponding thresholds.
2. The transitions mentioned in (1) display hysteresis, that is, they are locally irreversible in the long run, even if the environmental parameters revert to their original values.
3. As a consequence of (2), there is a sizable range of parameter values for which the system exhibits long-run multiplicity, which is resolved depending on history or/and the initial conditions.

EMV observed the phenomenology (1)–(3) as the rate of volatility, the only significant parameter of the model, was varied gradually along its full range. We now confirm that, in the generalized model, the same qualitative behaviour concerns changes in the noise parameters now introduced, both concerning homophily and conformity. In fact, since their implications are fully parallel, we shall focus our present discussion on the parameter modulating the noise of action adjustment, as will be explained in detail below.

The remainder of the chapter is organized as follows. Section 8.2 presents the model, Sect. 8.3 carries out the analysis and motivates it, while Sect. 8.4 concludes with a recapitulation.

## 8.2 The Model

Let there be a certain population of agents, $P = \{1, 2, \ldots, N\}$, who interact bilaterally over time as specified by the evolving social network. Time is modelled continuously, with $t \in 0, \infty)$. At any $t$, the state of the system $\omega(t)$ consists of two items: (1) the social network *g(t)* that specifies the set of undirected links *ij* (=*ji*) prevailing at *t*; (2) the action profile $\alpha(t) \in A^M$, where $A = \{a_1, a_2, \ldots, a_q\}$ is the set of $q$ possible actions.

Players adjust both actions and links over time. The dynamics is described by a continuous Markov process for the state $\omega(t)$, and is therefore completely determined by the rates governing all possible transitions $\omega \to \omega'$. These transitions pertain to adjustments that involve (1) link creation, (2) link destruction, (3) action revision. We now describe each of these in turn.

**Link creation:** We posit that at a certain positive rate $\eta$ each agent $i$ receives a link creation opportunity. When such an opportunity arrives at some $t$, another agent $j$ is randomly chosen in the population (all with the same probability). When no link exists between $i$ and $j$ (that is, $ij \notin g(t)$), the link $ij$ is formed with probability one if

$$\alpha_i(t) = \alpha_j(t).$$

Otherwise, the link is formed with probability $\varepsilon$, conceived small.

**Link destruction:** It is assumed that existing links decay at a rate $\lambda$. This component of the process may be provided with different (non-exclusive) interpretations. For example, it may be conceived as a reflection of unmodelled environmental *volatility* that affects the value or feasibility – and thus the persistence – of some of the existing links.

**Action revision:** At every $t$, each agent $i$ independently receives at a rate $v$ the opportunity to revise her current action. If this revision opportunity materializes, then she chooses every possible action $a_r \in A$ with probability

$$\Pr(\alpha_i(t) = a_r) = \frac{1}{H} \exp\left[\beta\left|\{\alpha_j(t) = a_r : ij \in g(t)\}\right|\right], \tag{8.1}$$

where

$$H \equiv \sum\nolimits_{r'=1}^{q} \exp\left[\beta\left|\{\alpha_j(t) = a_{r'} : ij \in g(t)\}\right|\right]$$

is a normalization factor and $\beta \geq 0$ is a parameter that modulates the sensitivity of agents' adjustment to conforming (or coordinating) with the local environment. It can be understood as embodying a desire of the agents to play optimally in a local coordination game, where the *instantaneous* payoffs that may be obtained from each action are linear in the number of neighbours *currently* displaying that same choice.

The above formulation yields the model studied in EMV as a particular case when

$$\varepsilon = 1/\beta = 0,$$

that is, when the noise associated to link creation and action revision are both zero. As advanced, since the two sources of noise lead to totally analogous implications, we abstract from the former by still making $\varepsilon = 0$ while we focus our attention on the latter by assuming that $\beta$ is finite. Note that, in this case, (8.1) becomes the specific exponential form that has been amply used in modern evolutionary literature to model gradual adjustment and learning in games (see, for example, Blume (1993), Durlauf (1997), or Young (1998)). It is in the spirit of the well-known formulation of *logistic quantal response equilibrium* proposed by McKelvey and Palfrey (1995), which has been provided with a natural bounded-rationality interpretation by Chen et al. (1997). The parameter $\beta$ modulates the noise impinging on agents' adjustment. If $\beta = 0$, noise is the overwhelming force and all actions are chosen with the same probability, irrespectively of local conditions and payoffs. In contrast, in the polar case where $\beta$ is very large, only if a particular action is a genuine best response is it chosen with a sizable probability.

## 8.3 Analysis

For finite $\beta$, the model is substantially more complex than the degenerate version studied in MEV. Consequently, we are unable to obtain an exact characterization of its stable equilibria and thus have to base our analysis on some simplifying assumptions. Naturally, this implies that the solution we arrive at can no longer be regarded as a fully accurate description of the long-run behaviour of the model. The entailed approximation, however, turns out to be quite effective since, as we shall explain, it matches very closely the results obtained from numerical simulations for large populations.

The adjustment rule given by (8.1) not only has the precedents in economics summarized in Sect. 8.2 but is also formally identical to the "spin dynamics" postulated by the so-called Potts model in statistical physics, itself a generalization of the canonical Ising model. In this context, $1/\beta$ plays the role of the temperature at which the particle interaction takes place and the $q$ different actions are the possible spins. (See, for example, Vega-Redondo (2007) for a detailed explanation of these models and their relationship to the economic and evolutionary literature.) The Potts model has been studied in detail by physicists and exact solutions for it exist for low-dimensional lattices as well as trees (cf. Baxter (1982)). Recently, the analysis has been extended to random networks by Dorogovtsev et al. (2004) and Ehrhardt and Marsili (2005). We crucially rely on the latter in our present analysis.

One of the simplifying assumptions we make is that the network prevailing at any given point in time is a random network suitably characterized by a degree distribution

$$\boldsymbol{p} \equiv \{p(k)\}_{k=0}^{\infty}$$

that specifies the fraction of nodes *p(k)* that display each possible degree $k$. The defining property of a *random network* is the absence of statistical correlations. Thus, in particular, it is presumed that the degree of a node is stochastically independent of any of its neighbours. Such a property does not strictly hold in our present generalized context – only approximately so.[2] This is why the random-network postulate must be viewed, in this case, as a convenient, but not fully accurate, description of the system at any point in time.

Another simplifying assumption we shall make concerns the relative speed of action and link adjustment. For technical tractability (and in particular, to rely on the solutions of the Potts model available in the literature), it is convenient to posit that the network adjusts much slower than agents' actions. Formally, this is captured by making $v \to \infty$. It amounts to assuming that actions adjust at a much brisker pace than links, so that the current underlying network may be taken as fixed while the action distribution reaches a stable configuration.

Under these assumptions, the analysis of the model can be decomposed into the following steps. First, we need to derive the law of motion for the degree distribution **p**, which is governed by the subprocesses of link creation and link destruction. As explained, the latter simply has every existing link disappear at a constant rate $\lambda$. Link creation, on the other hand, depends on the probability that any two agents who meet and have the potential of creating a new link happen to display the same action. For any given agent/node $i$, this ex ante probability must generally depend on its degree $z_i = k$, so we denote it by $\pi(k)$. In essence, this probability results from the combination of the following three constituent probabilities:

1. The (unconditional) probability $\zeta$ that, when node $i$ selects another node at random, the latter belongs to the (unique)[3] giant component of the network.
2. The conditional probability $\gamma(k)$ that node $i$ of degree $k$ belongs to the giant component.

[2]By way of illustration, one of the features of the process that introduces internode correlations can be explained as follows. First note that the postulated action dynamics leads high-degree nodes to exhibit, on average, stronger social "conformity" than lower-degree nodes. That is, they have a higher probability of choosing the action that is in the majority in the population. This in turn implies that links between high degree nodes will be formed with higher probability (that is, at a higher rate) than between lower-degree nodes. In the end, therefore, positive degree correlations will tend to arise, high-degree nodes being more likely to be connected to other high-degree nodes than what is prescribed by the *unconditional* average.

[3]As is well known in the theory of random networks, if a giant component exists in this context, it is unique.

3. The conditional probability $\mu(k)$ that, if node $i$ of degree $k$ does belong to the giant component, its action coincides with that of a randomly selected node in that component.

The first probabilities, $\zeta$ and $y(k)$ for each $k$, only depend on the underlying degree distribution $\mathbf{p}$. The probabilities $\mu(k)$, on the other hand, depend both on $\mathbf{p}$ and the value of $\beta$ in (8.1) that modulates the pressure towards local conformity induced by the action dynamics. To compute the probabilities in (1)–(2), one can directly use the standard techniques of the modern theory of random networks, as explained, for example, in Vega-Redondo (2007). And concerning the probability in (3), we may rely on the aforementioned solution of the Potts model in random networks that has been developed by Ehrhardt and Marsili (2005).

Thus let $\zeta$, $\gamma(k)$ and $\mu(k)$ for each $k$ be the probabilities prevailing at some point in time when the underlying network is modelled as a random network with degree distribution $\mathbf{p}$. Then, the probability $\pi(k)$ that a randomly chosen node of degree $k$ meets a node that displays its own action is simply given by:

$$\pi(k) = (1 - \zeta\gamma(k))\frac{1}{q} + \zeta\gamma(k)\mu(k). \tag{8.2}$$

The second term in the right-hand side of (8.2) corresponds to the event that some arbitrary node $i$ of degree $k$ happens to be part of the giant component and it meets another node $j$ in that same component. In that case, the link between $i$ and $j$ is formed with probability $\mu(k)$, that is, the probability that both display the same action.[4] The first term, on the other hand, contemplates what happens when either node $i$ or/and the node $j$ it meets are *not* in the giant component. Then, with probability essentially one in a large random network, both nodes are in different components. This implies that they will only display the same action (and thus form a link) "by chance", that is, with probability $1/q$ since there are $q$ possible actions.

Given the probabilities $\pi(k)$ specified in (8.2), the evolution of the degree distribution in time can be modelled through the following differential equation:

$$\dot{p}(k) = (k+1)\lambda p(k+1) + 2\eta p(k-1)\pi(k-1) - k\lambda p(k) - 2\eta p(k)\pi(k), \tag{8.3}$$

where we dispense with the time index for notational simplicity. The first two terms in the right-hand side of (8.3) reflect the inflow into the frequency of nodes of degree $k$. This inflow consists of those nodes that had degree $k+1$ and lost one of its links (which happens at a rate $\lambda$ per link), combined with the rate at which nodes with degree $k-1$ form a new link. (In the latter respect, note that the factor of 2 accounts for the fact that a link is created if either the node in question receives the initiating opportunity or some other node does.) On the other hand, the two last terms embody the opposite flow that decreases the frequency of nodes of degree $k$ when these nodes either loose or create a link.

[4]Of course, this presumes that the link between $i$ and $j$ is not already in place, which is an event that can be essentially ignored in large populations.

We are interested in characterizing the pair

$$\mathbf{p}^* = \{p^*(k)\}_{k=0}^{\infty}, \pi^* = \{\pi^*(k)\}_{k=0}^{\infty}$$

that defines a stationary point of the dynamical system. Such a stationarity embodies a twin requirement. First, given $\mathbf{p}^*$, the corresponding $\pi^*$ induced by (8.2) must suitably characterize the long-run coordination probabilities induced by the fast dynamics governed by (8.1). Second, given $\pi^*$, the degree distribution $\mathbf{p}^*$ must define a stationary point of (8.3). This latter requirement simply amounts to stating that, for all $k = 1, 2, \ldots$,

$$(k+1)\lambda p^*(k+1) + 2\eta p^*(k-1)\pi^*(k-1) = k\lambda p^*(k) + 2\eta p^*(k)\pi^*(k).$$

This defines a system of difference equations that can be solved recursively as follows:

$$p^*(1) = 2\eta p^*(0)\pi^*(0), \tag{8.4}$$

along with

$$p^*(k+1) = \frac{2\eta p^*(k)\pi^*(k) + k\lambda p^*(k) - 2\eta p^*(k-1)\pi^*(k-1)}{\lambda(k+1)} \quad (k = 1, 2, \ldots), \tag{8.5}$$

once we impose the normalization

$$\Sigma_{k=0}^{\infty} p^*(k) = 1.$$

To understand the essential gist of the argument, it is useful to make the assumption that the underlying degree distribution $\mathbf{p}$ is Poisson. (This assumption is not exactly true, but happens to be quite a good approximation for most parameter values in the interesting range.) Then, we can define the average probability $\pi(c, \beta)$ that two randomly selected nodes happen to be well coordinated (and thus will establish a link), as a function of $\beta$ (the noise parameter) and the average degree $c$ (the only parameter characterizing a Poisson degree distribution). The desired stationarity of the situation then requires that average link destruction be equal to average link creation, that is,

$$\lambda c = \pi(c, \beta). \tag{8.6}$$

The situation is described in Fig. 8.1 for different values of $\lambda$, while we normalize the rate of link creation $\eta$ to unity. It shows that when $\lambda > \lambda_2$ there is a single solution, representing a sparse network. At $\lambda_2$ other two solutions arise, one of which is unstable. At a further point $\lambda_1 < \lambda_2$ the sparse-network solution merges

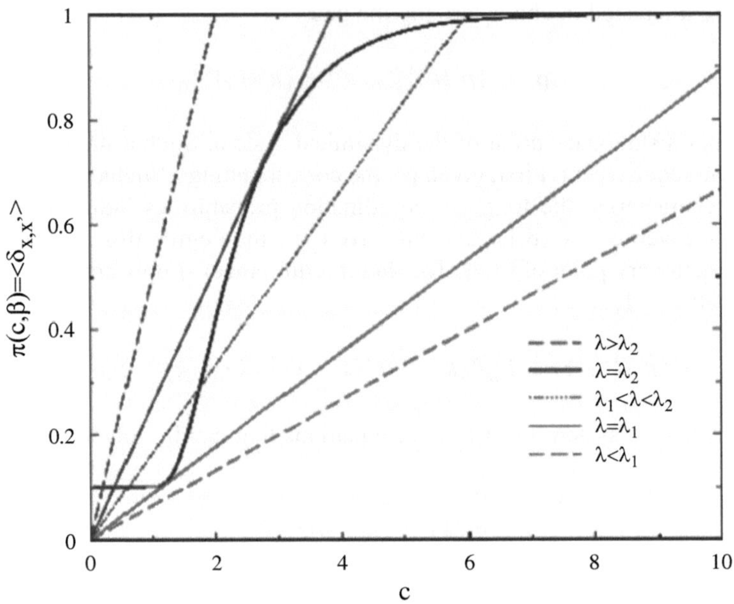


**Fig. 8.1** Graphical illustration of the solutions for the stationarity condition (8.6) for $q=10$ and $\beta=4$, under the assumption that the underlying random network is Poisson. The solutions are given by intersections of the curve representing the function $\pi(c;\beta)$ with rays $\lambda c$ for different values of $\lambda$. For $\lambda=\lambda_1$ and $\lambda=\lambda_2$ the rays are tangent to the curve, thus marking the values that bound the region $\lambda \in [\lambda_1, \lambda_2]$ where multiple intersections (that is, solutions) exist. Outside this region, (8.6) displays a unique solution

with the unstable one and both disappear for $\lambda < \lambda_1$, leaving only a solution with a stable and dense network. This reproduces the phenomenology summarized in (1)–(3) in Sect. 8.1 and, as explained, coincides fully with that obtained in EMV for the noiseless degenerate model.[5]

To conclude, we focus our discussion on the role played by the new parameter $\beta$ that marks the only difference with the EMV model. Interestingly, we find that changes in $\beta$ induce the same qualitative pattern of long-run behaviour as observed before for changes in $\eta$. The conclusions – both theoretical and simulations – are depicted in Fig. 8.2 for two different values of $\eta$. (In this case, we find it convenient to scale time by normalizing the rate $\lambda$ to unity.)

Figure 8.2 shows that, as one implements gradual changes in the noise impinging on action adjustment (which can be suitably parameterized by $1/\beta$), the long-run

[5]Mathematically, the behavior displayed by the model reflects the onset of a bifurcation towards instability and equilibrium multiplicity as the parameter $\lambda$ enters the region $[\lambda_1, \lambda_2]$. Such a bifurcation is analogous to that found by Brock and Hommes (1997) as the intensity of choice (here captured by either a change in the volatility rate $\lambda$ or the choice-sensitivity parameter $\beta$) varies in a suitable range. See also Hommes (2006) for an extensive discussion of the issue.

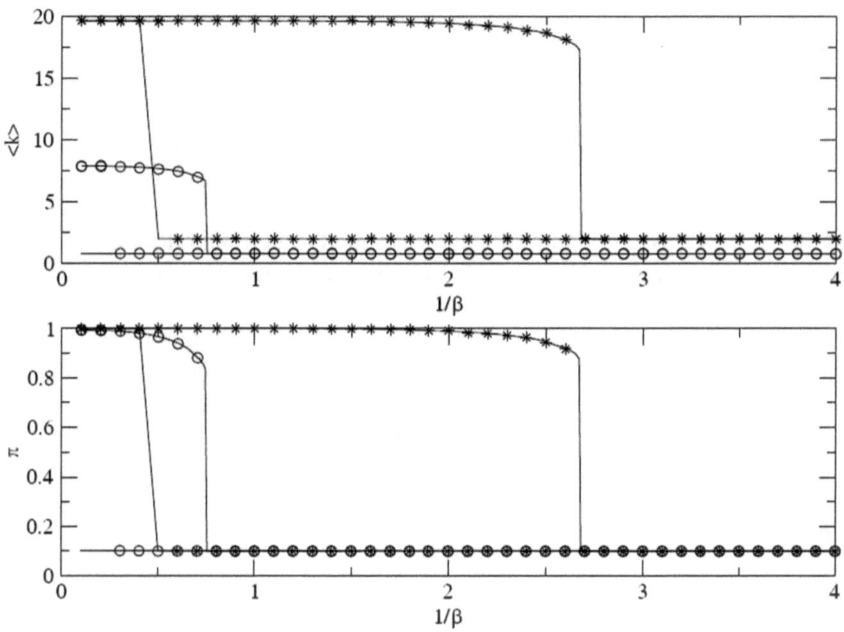


**Fig. 8.2** The *upper panel* plots the average connectivity $\langle k \rangle$ predicted by the model against the noise level $1/\beta$, for $q = 10$ and two different values of $\eta$, that is, $\eta = 4$ (*lower curves*), $\eta = 10$ (*higher curves*). The *solid lines* trace the theoretical prediction while the points represent simulation results for $n = 1{,}000$.The *lower panel* displays analogous results for the average probability $\pi$ that two randomly chosen nodes display the same action

behaviour of the system displays the same three features that are obtained for changes in $\eta$ That is, both connectivity as well as social conformity exhibit sharp and resilient transitions across multiple equilibria as the noise level changes "slightly" around certain thresholds. It is worth stressing that the theoretical predictions are well supported by numerical simulations, even though, as we have explained, the analytically solved model can only be conceived as an approximate description of the system dynamics.

## 8.4 Conclusion

We conclude, therefore, that the introduction of noise into the model studied by MEV maintains the essential phenomenology that was encountered there for the degenerate noiseless version of the model, that is, sharp transitions, hysteresis, and equilibrium multiplicity are robust features of the long-run dynamics of the process when the link-destruction or link-creation rates change in a relevant range. In fact, we have found that analogous behaviour arises as well concerning changes in the parameter controlling for the action-revision noise on which we have focused our analysis here. This suggests that such a phenomenology may well represent a solid (and, in a sense, universal) pattern to be expected in network formation processes reflecting the forces of homophily and conformity.

Building upon those insights, there are two different avenues we want to explore in future research. First, we would like to understand how the model fares when the mechanism of link creation is endowed with a natural local dimension – for example, if new linking opportunities are assumed to be found through the "intermediation" of current neighbours, as in Marsili et al. (2004). Second, we would like to enrich the present abstract modelling approach with features (say, payoffs and purposeful decision making) that would allow one to study important and concrete economic problems, for example, the role of R&D collaboration on firm innovation in oligopolistic industries.

**Acknowledgments** We thank an anonymous referee for very useful comments. Vega-Redondo also acknowledges financial support from the Spanish Ministry of Education under grant SEJ2007–62656.

## References

Baxter RJ (1982) Exactly solved models in statistical mechanics. Academic Press, London
Blume L (1993) The statistical mechanics of strategic interaction. Games Econ Behav 4:387–424
Brock WA, Hommes CH (1997) A rational route to randomness. Econometrica 65:1059–1095
Calvó-Armengol A, Jackson MO (2004) The effects of social networks on employment and inequality. Am Econ Rev 94:426–454
Castells M (1996) The information age: economy, society, and culture, volume I: The rise of the network society. Blackwell, Massachusetts
Chen H-C, Friedman JW, Thisse J-F (1997) Boundedly rational Nash Equilibrium: a probabilistic choice approach. Games Econ Behav 18:32–54
Dorogovtsev SN, Goltsev AV, Mendes JFF (2004) Potts model on complex networks. Eur Phys J B 38:177–182
Durlauf S (1997) Statistical mechanics approaches to socioeconomic behavior. In: Arthur WB, Durlauf SN, Lane DA (eds) The economy as an evolving complex system II. Addison-Wesley, Reading, MA, pp 81–104
Ehrhardt GCMA, Marsili M (2005) Potts model on random trees. J Stat Mech P02006
Ehrhardt GCMA, Marsili M, Vega-Redondo F (2006) Networks emerging in a volatile world. Preprint Abdus Salam Center for Theoretical Physics and Universidad de Alicante
Granovetter M (1974) Getting a job: a study on contacts and careers. Chicago University Press, Chicago
Hagedoorn J (2002) Inter-firm R&D partnerships: an overview of major trends and patterns since 1960. Res Pol 31:477–492
Hommes CH (2006) Heterogeneous agent models in economics and finance. In: Tesfatsion L, Judd KJ (eds) Handbook of computational economics, vol 2: Agent-based computational economics. Elsevier, Amsterdam, pp 1109–1186
Kirman A, Herreiner DK, Weisbuch G (2000) Market organization and trading relationships. Econ J 110:411–436
Kranton R, Minehart D (2001) A theory of buyer-seller networks. Am Econ Rev 91:485–508

Marsili M, Vega-Redondo F, Slanina F (2004) The rise and fall of a networked society: a formal model. Proc Natl Acad Sci USA 101:1439–1442
McKelvey RD, Palfrey TR (1995) Quantal response equilibria for normal form games. Games Econ Behav 10:6–38
Montgomery J (1991) Social networks and labor market outcomes: toward an economic analysis. Am Econ Rev 81:1408–1418
Powell WW, White DR, Koput KW et al. (2005) Network dynamics and field evolution: the growth of inter-organizational collaboration in the life sciences. Am J Sociol 110:1132–1205
Vega-Redondo F (2007) Complex social networks, econometric society monograph series. Cambridge University Press
Young P (1998) Individual strategy and social structure: an evolutionary theory of institutions. Princeton University Press, Princeton, NJ

# Chapter 9
# Complex Evolution and Learning

## The Role of Constraints

**Massimo Ricottilli**

## 9.1 A Short Introduction

Searching and learning are key processes in determining the complex evolution of technological capabilities but take place according to idiosyncratic rules that constrain the way they unfold. In this respect, constraints are often understood as barriers hampering efficient functioning. They are, indeed, seen as orienting activity in specified directions generating performance patterns the upshot of which is likely to be suboptimal. This view may be thought as applying to the domain of social undertakings as well as to activities of a biological nature. Constraints are thus perceived as fetters to efficiency, bounding the required freedom often deemed as crucial to achieve, if not optimal, at least improving solutions. In a recent paper (Ricottilli 2008), it has been argued that constraints set to restrain full freedom of choice, but the argument may well apply to across-the-board functioning of biological entities, act as focusing devises quite often resulting in better performance. The problem at hand takes full contours when dealing with the issue of searching and learning, that is when problem-solving is set in an evolutionary, hence dynamic, context. More particularly, this issue is of utmost relevance when this process is meant to lead to innovation, be it technological or organizational or, in fact, both. It is a well established fact that a firm's strategy to survive and thrive in a market environment does require innovative activity and investment. Although markets rarely function according to the classical competitive paradigm, they become contestable precisely thanks to the likelihood of product and process innovation. It is therefore important that a model of searching and learning take fully into account the role that constraints are likely to play in these processes.

---

M. Ricottilli
Department of Economics and Centro Interdipartimentale 'Luigi Galvani' (CIG) per la Complessità, University of Bologna, Bologna, Italy

A. Reggiani and P. Nijkamp (eds.), *Complexity and Spatial Networks*, Advances in Spatial Science,
DOI: 10.1007/978-3-642-01554-0_9, 

In this chapter I wish to report the work that has recently been done on these matters by a small research group at the University of Bologna.[1] It is now an accepted fact-of-life that searching and learning are cognitive processes of knowledge acquisition that are radically uncertain as to their final outcome. It is, accordingly, a warranted assumption to postulate that, in this context, bounded rationality holds. They are radically uncertain because paucity of information rules, computing capabilities are limited and forecasting is therefore severely hindered. This assumption then implies that agents, more precisely firms in an economic framework, although in theory facing a normally complex space of possibilities when searching can effectively explore only a small fraction. The notion of space of possibilities is, of course, difficult to define. It may nevertheless be taken to encompass all the conceivable new characteristics that a present state may likely take given the current stock of knowledge: a clearly partial and ill defined concept since radical innovations may well lie outside such space precisely since they cannot be foreseen and presently envisaged. Yet, even this rudimentary conceptual device may be of help in understanding the complexity that lies at the root of a searching process. An interesting way to portray it is to consider that any extant frontier technology is made up by a large number of elements each of which can lend itself to change and thus give rise to an useful innovation. Assume, for instance, that there be a number $a$ of conceivable ways of changing each of the elements that describe such a technology. Let the latter make up a countable set of cardinality $N$. Even if, for simplicity's sake, $N$ is taken to remain constant the theoretical space of possibilities is: $a^N$, a magnitude that scales exponentially with $N$ and linearly with $a$. Exploring and testing for performance each of these likely configurations in quest of the optimum is clearly an unaffordable task even for the best equipped and resource-provided firm or organization for $N$ even moderately large. To further complicate this matter it is important to retain that, historically, processes of production have lengthened in terms of the components technically required to obtain a given output; they have, in other words, become more technically roundabout: $N$ is bound to increase in time. Yet, more complications enhancing the level of complexity are generated by the interdependence existing between technologically complementary elements, a fact entailing a thorough reshaping and fine-tuning of the whole process. What is implied is a complex coevolutionary pattern of change.

## 9.2 The Evolution of Technological Capabilities

Recent literature has dealt at length with the evolutionary process leading to the shaping of technological capabilities. In particular, the approach dating back to the contributions of Nelson and Winter (1982), Dosi (1988) for an extensive review,

[1] See, for instance (Andergassen et al. 2005, 2006; Castellani et al. 2007).

has argued that searching in a space of opportunities is a process that is bounded both by limited rationality and by the confines of a neighbourhood that be reachable either spatially or cognitively. More recently, theories that have investigated the features of networks have forcefully shown that information diffusion and sharing are necessarily supported by the properties of network architecture (Albert and Barabási 2002; Jackson 2008).

### *9.2.1 Search for Information and Knowledge Building*

The search for innovations is therefore a knowledge-building process that is necessarily local and that requires gradual information collection. It is gradual since information is likely to come from a variety of sources and because it comes in packets that are first sought then evaluated and finally acquired as part of a consistent design that upon completion develops into a full fledged innovation. In order to account for the basic characteristics of this process it is expedient to view it as taking place according to two distinct but clearly overlapping activities. The first concerns an *in-house*, direct effort to autonomously generate new technological knowledge by single-handed investigation. This is the activity normally carried out by R&D departments but also by informal, piecemeal trial-and-error searching. The second is an activity directed at collecting information from spillovers originating from firms that have innovated and developed cutting-edge technologies. This is a somewhat artificial distinction but it is useful to highlight the fact that success at bringing to the fore new methods of production and products much depends on firms' interaction while at the same time recognizing that if there were no independent effort to conjure up something entirely new mere diffusion and circulation of ideas would produce no progress. The first activity can usefully be viewed as ruled by a Poisson stochastic process whilst the second can appropriately be modelled by information diffusion that occurs with a given strength of interaction.

Because of the assumption of bounded rationality, it is straightforward to retain that firms cannot search the whole economy in their quest for information but that only a small section can actually be explored. The following paragraphs report on the finding of a simulation model that emphasizes this crucial feature of the innovation process by highlighting the importance of interaction. Firms are accordingly seen as acting out their innovative activity placed in a network of other firms within which they can draw information although from a normally small neighbourhood. In this context, a network is to be understood as a cognitive environment taking shape as firms assess and adjust their information-contributing neighbourhood. The upshot of this activity is, first of all, an increase in firms' technological capability, a quality that is rendered by a quantity measured by a performance index that cumulates the impact of both independent and interactive searching. The latter is the result of the transmission of capabilities from neighbouring firms whose number is assumed, for the purpose of retaining simplicity, as being constant: $\overline{k}_{\text{in}}$. The neighbourhood in question is defined as *inward* since it is the medium gathering performance-enhancing information for any given firm and it is distinguished from the *outward* neighbourhood that emerges as other firms include it in their own *inward* one. A firm's *outward* neighbourhood is accordingly defined as made up of all firms that have chosen it as an information source and evolves as they choose whom to receive information from. For simplicity's sake, the model that has been implemented to simulate the involved dynamics is described by the following equations:

$$V_i(t) = \sum_{j=1}^{J} a_{ij} b_{ij}(t) V_j(t) + C_i(t), \tag{9.1}$$

solving for

$$V(t) = [I - M(t)]^{-1} C(t). \tag{9.2}$$

$V_i(t)$ is firm $i$ technological capability extant at time $t$, $C_i(t)$ is its autonomous part while $a_{ij}$ is the broadcasting strength enabling a fraction of $V_j(t)$ to be passed on to $i$. While the latter are assumed to be given parameters, $b_{ij}(t)$ is the element of the adjacency matrix determining which $j$ belongs to $i$'s neighbourhood. $M(t)$ is the resulting matrix[2]

$$M(t) = (a_{ij} b_{ij}(t)).$$

Since $C_i(t)$ is a purely stochastic element, it is simply taken to be a value drawn from the uniform distribution (0,1): after a waiting time $\mu$ a firm is randomly chosen for this purpose.

It is clear that the crux of the matter lies with the updating process that leads firms to change their *inward* neighbourhood as they search for better contributors to their technological capability. While it is rather straightforward to assume that every now and then, in fact when randomly chosen for the purpose, firms check for the worst contributing neighbour and then attempt to change it, it is quite clear that the real task is to device an appropriate protocol to substitute the low performer. Several procedures can be examined that are consistent with the likely searching attitudes of firms that are limited by bounded rationality. More specifically, the relative efficiency of alternative *routines* can be tried: in this framework a *routine* is a procedure that firms are assumed to follow in order to assess, first, their neighbours' performance in contributing information and, second, to choose a new neighbour when the low performer is evicted. Evaluating the performance contribution of existing neighbours may simply be rendered by the following criterion:

[2] It is interesting to note that $M(t)$ can be defined as a cognitive interaction matrix. Its equivalent in the framework of spatial theory is the *flow matrix*. Since it is of a cognitive nature, it is clearly an artificial construction although it can be likened to a formal representation establishing the linkages within a virtual space of interacting, information-exchanging nodes.

$$\gamma_i(t-1) = \arg\min_{j \in \Gamma_i(t)} [a_{ij} b_{ij}(t-1) V_j(t-1)] \quad (9.3)$$

$\gamma_i$ being the identified neighbour. A satisfactory replacement choice can be rendered by a simple improvement rule:

$$V_i(t) > V_i(t-1), \quad (9.4)$$

a procedure that redefines at each time step $M(t)$ thus generating a new set of solutions. The actual choice of a new neighbour, however, can be made in a variety of ways. Two broad procedures are distinguished. The first defines what can be called a strong but bounded rationality principle: firms consider the whole economy for a replacement and randomly draw from the whole set of non-neighbouring firms. The second is a more constrained protocol: it is recognized that only firms placed close by, in a cognitive sense, can be understood and their spillovers fruitfully incorporated. Thus, choice is a random draw from the excluded neighbour's own neighbourhood. The rewiring process, in this case, obeys a specific constraint imposed to the searching process by cognitive limits restricting the virtual space of likely candidates for substitution. This second procedure, however, can be designed to give rise to mixed *routines* in which choice is either always local in the sense just mentioned above or it is a composite rule of both local and global search. More precisely, calling $\pi$ the probability of exploring across the entire firm space at each attempt to change a randomly chosen neighbour, define $\pi \equiv \frac{1}{\tau}$. When $\tau=1$, ($\pi=1$) the search routine is such that firms always look over the whole economy for a new neighbourhood member but $\tau$ tends to infinity, $\tau \to \infty$, when searching is entirely restricted within their neighbours' neighbourhood ($\pi \to 0$, they never look over the whole economy). Intermediate values state the number of times searching occurs inside the firm's proximate neighbourhood and just *once* outside. Choosing a specific routine is of great relevance to the way the network evolves and finally becomes structured. In turn, the emergent network topology is expected to have a crucial impact on the economy's average technological capability and on the role some firms are likely to play in broadcasting information and passing on capability spillovers. Thus, constraints act to shape the network architecture and ultimately the economy's technological performance. What follows is a short report on results obtained from simulating the model.

### 9.2.2 Spillovers Through an Average Parameter

Simulations[3] that are illustrated below have been performed under a very restrictive assumption, namely that all the $a_{ij} = a$ a specified parameter that applies to all

[3] A detailed description of the simulations that have been carried out appears in Andergassen et al. (2005) and Andergassen et al. (2006).

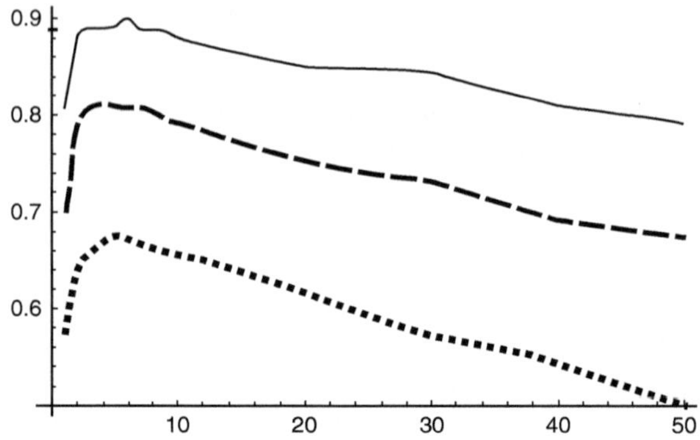


**Fig. 9.1** Efficiency index $\phi(T)$, on the $y$-axis, as a function of $\tau$, on the $x$-axis, for mean-waiting time. $\mu = 8$ (*dots*), $\mu = 16$ (*dashed line*) and $\mu = 32$ (*continuous line*)

firms. This assumption can be justified by taking a mean-field approach measuring, in other words, the average strength of capability interaction. A further paper, Andergassen et al. (2005), attempts to investigate a more knowledge-wise heterogeneous environment.

Figure 9.1 illustrates the behaviour of the efficiency index $\phi(T)$[4] at the terminal date $T$ plotted against $\tau$, the type of routine used. This index is obtained by dividing the system's average gain resulting from the actual choice as shown in (9.1) and (9.2) over the potential maximum one. The three lines correspond to as many waiting times. It is very interesting to note that the efficiency index rises as the implemented routine varies from one that targets the global economy to a progressively more local one to reach a maximum at a magnitude roughly set between four and five. The best searching strategy, therefore, is not the one that adopts a routine targeting the whole economy, that chooses a new and more performing neighbour by randomly drawing from the whole set of firms, but one in which local and broad targeting mingle. A $\tau = 4$ or 5 means that it is best to search locally four or five times and once globally. As searching becomes more local and as it tends to an exclusively local routine, efficiency falls. The implication is that it pays to constrain the searching routine to a partially local one but drawing widely from the whole economy once every a given number of times. The reason behind this result owes to the fact that firms, as they are randomly drawn to adjust their neighbourhood, tend to incorporate better capability-contributing firms. Thus, if a firm encounters high performers when it gets the chance to expel the relatively lowest one of its given neighbours, sampling randomly from the latter's neighbours implies a high probability of meeting an even higher contributor. It follows that searching in one's own neighbour's neighbourhood stands a chance of finding increasingly better spillovers. Yet, given some waiting time, neighbouring firms may be shocked by a random draw from the uniform distribution; they stand a chance of worsening their own independently achieved capability $C_j$. Hence, getting locked in an exclusively

[4]The formal definition of this index can be found in Andergassen et al. (2006).

local search has an increasing probability of being stuck with performers that are likely to decrease their technological capability and, thus, it pays to look further for better substitutes. In the economy as a whole, given a large enough number of firms, meeting better performers has roughly the same probability of meeting worse ones. There is then a trade-off between local and global search: shortsightedness and longsightedness have an important role to play in the quest for higher technological capability.

It is now important to verify the relative firms' weight in spreading relevant information. As it has been seen the *routine* that is adopted to implement searching and that constrains it to a specific narrow neighbourhood is very important as to the results that firms, on average, are likely to achieve. It is, likewise, expected that *routines* also play a significant role in the emergence of firms as crucial providers of information.

Figure 9.2 reports interesting findings. It depicts the distribution of the size of *outward* neighbourhoods. The *x-axis* plots 32 quantiles, from the 0–2 to the 62–64 quantile. A quantile corresponds to how many firms are likely to look at and retrieve information from any specific one; the *y-axis* plots the corresponding frequency. While data for several waiting times have been collected, overall results do not vary substantially with the latter; thus, an average $\mu = 32$ has been chosen to represent these findings. The important result that is highlighted by data shown in Fig. 9.2 is that for routines that blend local and global search, that is for $\tau > 1$, there is a positive probability that the last quantile be filled. The implication is that there exist one or two firms that have a positive probability of being in the neighbourhood of the remaining ones and that, therefore, broadcast their technological knowledge to the rest of the economy. Because of this property, they are dubbed technological *paradigm setters*. The emergence of these very special firms is barred in the case of the across-the-board routine basically for the reason that has already been mentioned above. Given the fact that good and bad performers have a roughly equal probability of being encountered by random searching over the whole economy, it is unlikely that any particular firm gets locked-in into a neighbourhood of very high performers in which every one else will finally also get into.

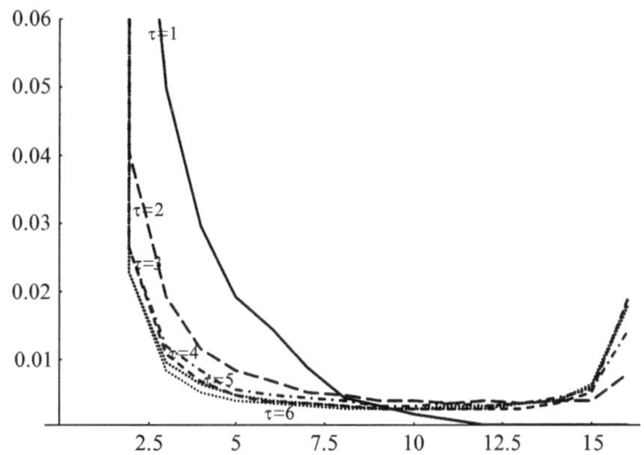


**Fig. 9.2** Distribution of outward links according to the chosen routine. The $x$-axis plots the quantiles, whilst the $y$-axis their frequency

It is quite apparent that the emergence of paradigm setters and average performance be related. Data indicate that the emergence of the former precedes the achievement of the maximum efficiency for $\tau = 4$ or 5 . Indeed, paradigm setters begin to appear as soon as some locality is introduced in searching ($\tau = 2$) highlighting the importance of constraints, in this case of *routines* that confine the choice of new contributors of technological spillovers to a relatively small neighbourhood. Thus, the probable existence of firms that drive the technological capability of almost every other firm seems to be a prerequisite for high average performance. By contrast, while for very local routines paradigm setters are also very likely to appear, average performance is probably poor on account of lock-in into badly performing neighbourhoods. This occurrence implies that, in these circumstances, paradigm setters need not be the highest performers within the economy: better firms are likely to exist but will never be discovered through a very local search. If this is the case, technological leadership when searching *routines* are very local turns out to be quite inefficient.

### 9.2.3 Heterogeneous Knowledge

An attempt at testing the behaviour of the model when spillovers do not spread according to an average parameter $a$ has been made. In following figures data from simulations have been collected by assuming that the economy is split in blocks of firms belonging to different cognitive areas. The assumption is that within each area the broadcasting strength between firms is highest but weakens when transmission involves firms belonging to different ones. Three cases are actually studied. The first replicates the experiment reported above in which a single interaction parameter $a$ is assumed. The second considers two blocks in which $a_1 > a_2$ defining a parameter

$$\delta = \frac{a_2}{a_1} = 0.8$$

whilst the third resets $\delta = 0.6$ , the latter being the case of greatest heterogeneity.[5]

The major difference in respect to the economy in which an unique, average, spillover broadcasting parameter is considered does not lie so much with the behaviour of the average technological performance, measured by an efficiency index, as in the network topology as seen from the outward neighbourhood perspective. Thus while the data shown in Fig. 9.1 basically hold, Figs. 9.3–9.5 illustrate the emergent pattern of the outward neighbourhood distribution. It is, clearly, within the

[5] In the framework of spatial theory, this procedure is equivalent to adding a cost factor to links between nodes belonging to areas of heterogeneous knowledge. More precisely, matrix *M(t)* can be re-arranged such that coefficients belonging to areas of homogeneity be multiplied by $\delta = 1$ whilst those belonging to areas of heterogeneity by $\delta < 1$. Cost shows up in lower spillover strength.

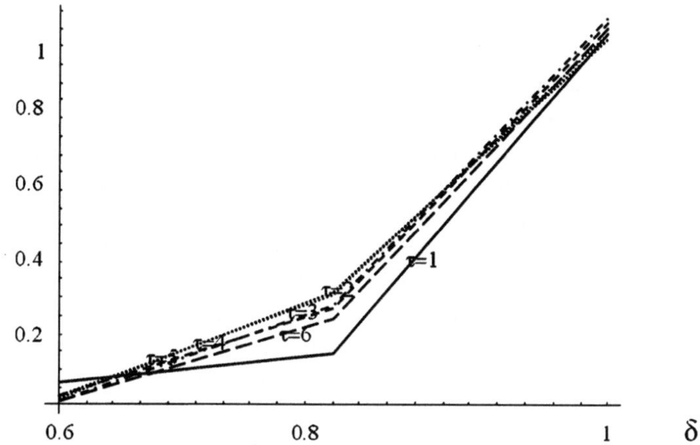


**Fig. 9.3** Share of connected firms plotted, on the $y$-axis, against the degree of heterogeneity $\delta$, on the $x$-axis, and according to the chosen routine $\tau$

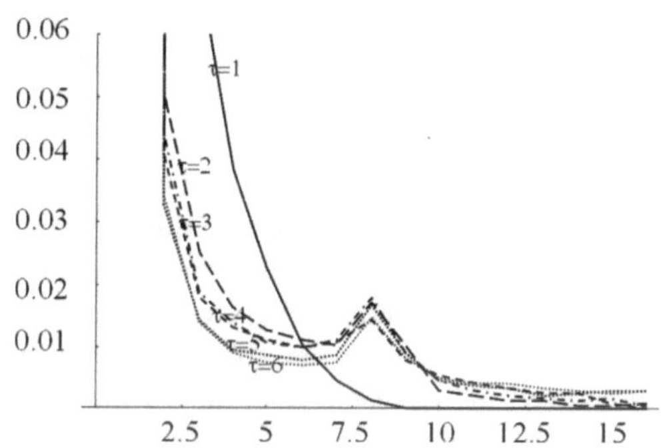


**Fig. 9.4** Distribution of outward links according to the chosen routines when the heterogeneity degree is 0.8. Quantiles are on the $x$-axis, frequencies on the $y$-axis

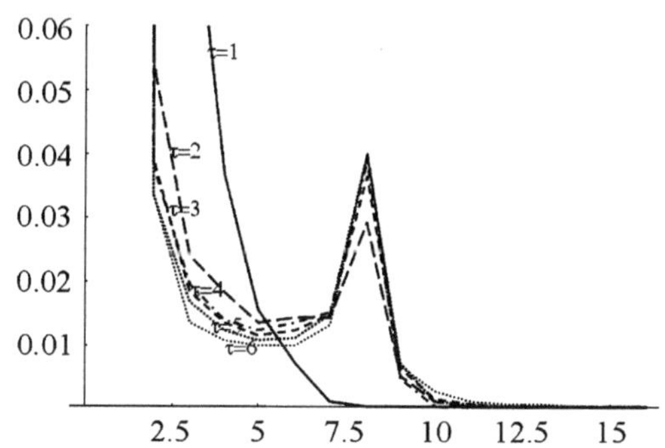


**Fig. 9.5** Distribution of outward links according to the chosen routine when the heterogeneity degree is 0.6. Frequencies on the $y$-axis and quantiles on the $x$-axis

model logic that connectivity between blocks declines as the cognitive distance increases. Figure 9.3 plots the share of firms that are connected against parameter $\delta$.

It is immediately seen that connectivity is lowest when the *routine* followed to adjust the inward neighbourhood is global but rises as it becomes increasingly local.

Evidence is clearest for $\delta = 0.8$: when $\delta = 1$ connectivity is, obviously, not an issue whilst it becomes negligible for $\delta = 0.6$ . The latter case indicates that the economy is split in two heterogeneous areas while inter-area connectivity practically disappears. Yet, for moderate heterogeneity ($\delta = 0.8$) routines do seem to make a difference. Global search quite understandably generates low connectivity. Since blocks possess the same membership cardinality, the probability of drawing an heterogeneous candidate for replacement according to (9.4) is about 1/2 and given also that that high and low performers are randomly distributed, an attempt to include a cognitively distant firm in one's own neighbourhood will most likely be frustrated. As *routines* get to be more local and searching becomes more constrained in neighbours' neighbourhoods, having discovered heterogeneous but high contributors leads to yet better performers, albeit heterogeneous, a fact that enhances connectivity.

The connectivity problem is effectively exposed when observing the outward neighbourhood distributions. Figure 9.4 reports them according to the chosen *routine* ($\tau$) and for $\delta = 0.8$ (moderate heterogeneity).

It is immediate to notice that when $\tau = 1$ no paradigm setters emerge whilst for $\tau > 1$ there is a positive but scant probability that they do. But the main feature that emerges from the data is a relative probability peak in the approximately median quantile: the one which includes about one half of all firms, that is those placed in the homogeneous area. The implication is that while there is weak evidence of global paradigm setters emerging, sectional ones do, that is, paradigm setters within each of the two blocks of firms. The evidence described above is much more apparent for $\delta = 0.6$, the case of high heterogeneity as shown in Fig. 9.5. The conclusion is therefore warranted that, quite generally, when technological capabilities are heterogeneous the economy tends, other things being equal, to split into separated cognitive areas, each with its own paradigm setters. This tendency towards separation is greater for global *routines*. The expression "other things being equal" is to be understood as meaning that only statistical properties have been observed for given cognitive interaction parameters. The latter are weaker when heterogeneity is present. Thus, even high, potential contributors may quite likely be discarded simply because the capability that they can pass on is small owing to heterogeneity. Interesting results would be obtained if this heterogeneity draw-back were offset by a risk-taking factor. It is conceivable that firms ready to take a technological risk by venturing into alien knowledge areas would most likely increase connectivity. While a formal investigation of this problem is yet to be made, it is expected that patterns similar to the homogeneity case would emerge.

## 9.3 Shaping the Cardinality of Inward Neighbourhoods

The experiments that have been carried out above have taken each firm's *inward* neighbourhood cardinality as given. The number of other firms that each is allowed to observe is set to a conventionally fixed number $\overline{k}_{\text{in}}$ (in the simulations discussed above the cardinality is three). Clearly this is a procedure that preempts the topology of both inward and outward networks. An improving extension of the analysis discussed in foregoing sections is to deal with the specific problem of generating the initial number of inward and, consistently, outward neighbours for each firm present in the economy. Once the network structure has taken shape, it is then fitting to study how it adjusts when it is finally able to perform the task for which it is designed, namely to allow for efficient cognitive interaction through which spillovers travel. It is, clearly, a somewhat artificial procedure since linking and rewiring are often concomitant processes but it is expedient to separate the two for analytical clarity. An experiment has been made following an approach according to which edges are established as the outcome of a process which gradually tests and assigns them some measure of strength until a link is finally set up. The issue of strength building implies a task of weighing prospective partners that ends with eventual rejection or positive selection. In this sense, it is a competitive process leading to some firms finally being chosen and others shunned. The model that has been considered draws inspiration from models that have been adopted in biology (Castellani et al. 1999; Intrator and Cooper 1992). The mathematical backbone of this network-building procedure consists in assuming a loss function that cumulates an index of strength, let it be called $u_{ij}$, that is nevertheless subject to a constant loss if not brought to its final result: $u_{ij} = 1$, for an effectively active link, $u_{ij} = 0$, for a definite rejection. Let $\alpha$ be the rate at which the cumulated index loses strength. Without loss of generality, this function can then be rendered as:

$$L(u_{ij}) = -\alpha \int_0^{u_{ij}} \Phi(s, \theta_{ij}) \mathrm{d}s, \tag{9.5}$$

in which $\Phi(s, \theta_{ij})$ depends on the distance of the link strength from a threshold $\theta_{ij}$. $\Phi(s, \theta_{ij})$ is the general function that sets up the rule according to which the $ij$ edges gain or lose strength. In this formulation the threshold $\theta_{ij}$ that is applied in the assessment of link $ij$ plays a crucial role since it acts both as a constraint and as a filter in the partner choice process. It is, accordingly, a determining factor in shaping the network topology. It can be shown that loss is at a minimum either when the strength is close to zero or when it is well above the threshold. The function in question can then be rendered as

$$\Phi(s, \theta_{ij}) = s(s - \theta_{ij});$$

more precisely the strength-building process can follow the dynamics generated by the following Lotka–Volterra type equation:

$$\dot{u}_{ij} = u_{ij}(u_{ij} - \theta_{ij}). \tag{9.6}$$

At this stage, the crucial question lies with the method to be used in order to define the all-important threshold. It seems reasonable, in the light of the previous discussion, that when faced with heterogeneous agents that carry idiosyncratic knowledge each firm places them in neighbourhoods defined as having similar technological characteristics. It is equally expedient to assume that firms will not attempt to link up with those whose knowledge is quite similar and thus whose spillovers are likely to bear a small impact. The latter firms can be defined as cognitive neighbours. A rather practical way of dealing with this approach is to assume firms as vertices of a regular polygon having as their immediate cognitive neighbours those placed in both their right and left as well as their neighbours' neighbours up to order $p$, a magnitude that can be used as a tuning parameter. Thus each firm turns out having $2p$ cognitive neighbours. Once the neighbourhood, $U_i$, defined, the link-building process can be dealt with by imposing that each firm $i$ attempts to connect with firm $j$ by setting it against its own cognitive neighbourhood, $U_j$, that is, with firms carrying a similar technological knowledge. The threshold $\theta_{ij}$ can then be defined as

$$\theta_{ij} = u_{ij}^2 + \sum_{k \in U_j - U_i} (u_{ik}^2 + u_{ki}^2). \tag{9.7}$$

The constraint that here operates is one that limits the eligibility of an edge to nodes that are not immediate neighbours and that, therefore, are likely to carry novel technological information. As it is apparent, index $k$ spans the $j$th neighbourhood, $U_j$, that is to say the nodes in this network that are set in competition for a link with $i$, save for the part of $j$ which overlaps with it and for which the latter has no interest. $\theta_{ij}$ can then be read as the average strength within $(U_j - U_i)$ over which the $ij$ link must jump if a gain has to materialize. Varying $p$ yields quite different outward link distributions. Figure 9.6 compiles data concerning a small sample of firms in which $p = 2$.[6] There seems to be a prima facie evidence that the distribution is a quasi-random one: a result that is explained by the fact that there is little neighbourhood overlapping and thus each edge gets established quite independently. Yet, as overlapping rises on account of a higher $p$, the distribution begins to be more skewed. Following Figs. 9.6–9.8, are, in fact, obtained by increasing the tuning parameter $p$. The last figure shows that when overlapping is very high, hubs emerge: most firms tending to look up at just two: the initial leaders of the technological network. The case of high overlapping is interesting since, given (9.6) and (9.7), it implies considerable competition amongst firms in who is going to link up with whom. At the same time, it is also a case of greater co-operation since it also implies greater homogeneity.

[6] The simulation procedure is detailed in Castellani et al. (2007).

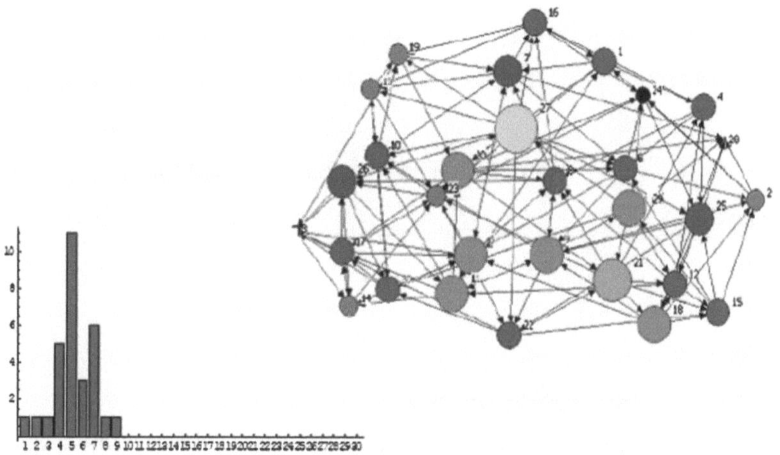

**Fig. 9.6** Distribution of outward links and network given 30 firms and $2p = 4$ cognitive neighbours. The histogram shows the distribution while nodes and edges the network architecture

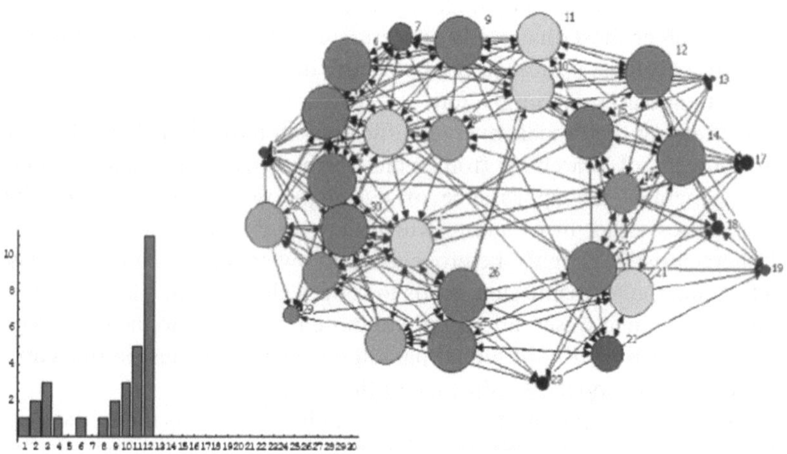

**Fig. 9.7** Distribution of outward links and network given 30 firms and $2p = 18$ neighbours

## 9.4 Some Short Conclusions

This research work has been aimed at highlighting the role played by constraints and *routines*. It has been shown that interaction among firms is crucial in transmitting technological capabilities through knowledge spillovers. A process that is local, that is, taking place within limited neighbourhoods, as well as routinized, that is constrained by searching protocols of various content. It is a quite robust

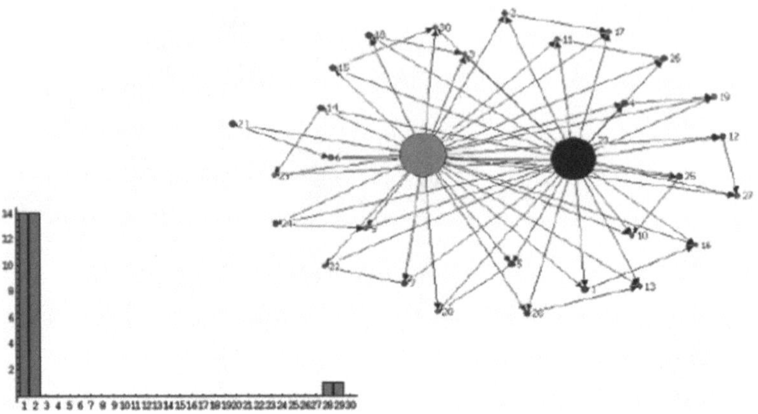

**Fig. 9.8** Distribution of outward links and network given 30 firms and $2p = 28$ neighbours

finding that *routines* that perform better in conjuring up average technological capabilities are not those that randomly explore the whole set of firms in the economy but rather those that partly concentrate in exploring a limited neighbourhood of firms. The best searching protocols are, in fact, those that merge localized with across-the-board searching: a four or five to one ratio appears from simulations as the one yielding the highest average efficiency in terms of achieved capability. Yet, this process of searching from which firms attempt to collect knowledge spillovers generates the appearance of *paradigm setters,* a few firms that are observed by most other firms and to which they necessarily transfer their technological characteristics. Highly heterogeneous knowledge, by hampering easy understanding and thus transmission, tends to form technological islands in the sense that firms confine their observations only to those that belong to the same cognitive area. Whilst the probability of global paradigm setters becomes, in this case, very small, local ones do appear within each of these areas.

An important extension of this model of searching concerns the initial construction of a network topology. Shaping a network implies establishing who links up with whom: a procedure in which edges gradually form until stable links appear: Lotka–Volterra-like equations preside over the mathematical framework. The chapter shows that the definition of a "natural" neighbourhood, namely one in which technological capabilities are similar and thus in which interaction is of little impact, bears considerable consequences. In particular, it is shown that when large overlapping between "natural" neighbours occurs high node competition leads to a distribution of links in which a few hubs appear: a few nodes that are prone to provide technological information to most others.

## References


Albert R, Barabási AL (2002) Statistical mechanics of complex networks. Rev Mod Phys 74: 47–97

Andergassen R, Nardini F, Ricottilli M (2005) 'Firms' network formation through the transmission of heterogeneous knowledge. Working Paper of the Department of Economics, University of Bologna, n. 543

Andergassen R, Nardini F, Ricottilli M (2006) The emergence of paradigm setters through firms' interaction and network formation. In: Namatame A, Kaizouji T, Aruka Y (eds) The complex networks of economic interactions. Lecture notes in economics and mathematical systems, vol 567. Springer, Berlin

Castellani GC, Intrator N, Shouval H, et al. (1999) Solutions to the BCM learning rule in a network of lateral interacting nonlinear neurons. Network 10:111–121

Castellani GC, Nardini F, Ricottilli M (2007) Node evolution: from randomness to full connectedness. Paper Presented at Net 2007 Conference, Urbino

Dosi G (1988) Sources, procedures and macroeconomic effects of innovations. J Econ Lit 26:1120–1171

Intrator N, Cooper LN (1992) Objective function formulation of the BCM theory of cortical plasticity: statistical connections, stability conditions. Neural Netw 5:3–17

Jackson OM (2008) Network formation. In: The new Palgrave dictionary of economics and the law, MacMillan Press, Basingstoke

Latora V, Marchiori M (2001) Efficient behaviour of small world networks. Phys Rev Lett 87:198701

Nelson RR, Winter GS (1982) An evolutionary theory of economic change. The Belknap Press, Cambridge, MA

Ricottilli M (2008) Constraints and freedom of action: a fitness trade-off. In: Castellani GC, Fortunati V, Franceschi C, Lamberti E (eds) Biocomplexity at the cutting edge of physics, systems biology and humanities. Bononia University Press, Bologna

Watts DJ, Strogatz SH (1998) Collective dynamics of "small worlds" networks. Nature 393: 440–442

# Chapter 10
# Proximity, Social Capital and the Simon Model of Stochastic Growth

**Koen Frenken**

## 10.1 Introduction

It is customary to define economic geography as a discipline that deals with the uneven distribution of economic activity across space. From a historical perspective, stochastic growth models are of particular use (Simon 1955). Such models explain the current distribution of activities from the dynamics of the long historical process that has produced these patterns. This approach might also be labelled "evolutionary economic geography" (Boschma and Frenken 2006), referring to the evolutionary economics tradition, since stochastic growth models account for *path dependence* in which each event changes the probability of a next event to occur (Arthur 1989; David 1985).

In geography, stochastic models of urban growth have a long intellectual history. In particular, The Simon model of the Zipf's rank-size rule is still regarded as one of the canonical models of urban size distribution (Batty 2005). A shortcoming of these urban growth models is that they take spatial entities as the unit of analysis. Since spatial entities are not behavioural entities, the explanation of urban growth in such models is not grounded in the micro-behaviour of agents. What is more, the delineation of spatial entities is a notoriously difficult exercise. Organizational units are less problematic, because these are the agents of change and relatively easy, though by no means trivial, to delineate. More recently, some of the urban growth models have explicit micro-foundations, including neoclassical models (Duranton and Puga 2004) and agent-based models (Batty 2005).

It is for these reasons that scholars have turned to micro-founded theories. One such attempt has been to take product divisions as the unit of analysis and define growth as stemming from product innovations leading to new product divisions within an existing firm or a new firm, and within an existing city or a new city.

---

K. Frenken
Urban and Regional Research Centre Utrecht (URU), Faculty of Geosciences, Utrecht University, Utrecht, The Netherlands

A. Reggiani and P. Nijkamp (eds.), *Complexity and Spatial Networks*, Advances in Spatial Science,
DOI: 10.1007/978-3-642-01554-0_10, © Springer-Verlag Berlin Heidlberg 2009

Reasoning from product divisions allows us to model the firm size distribution and the city size distribution simultaneously (Frenken and Boschma 2007). It also allows to model social networks emerging from the mobility decisions of entrepreneurs moving between product divisions within or between firms, and within or between cities. The structure of these networks can be analysed using a spatial interaction equation as to analyse what types of "proximities" affect the interdependencies between cities and between firms. In this way, we can understand the formation of higher order entities, particular business groups and city-regions, as a logical "multilevel" outcome of stochastic growth models.

## 10.2 Industrial Dynamics and Urban Growth

City size distributions are well approximated by Zipf's law, which states that the size of the $n$th ranked city is $1/n$ times the size of the largest city (Zipf 1949). To understand this distribution as the result of a historical growth process, Simon (1955) modelled city growth by discrete increments (lumps). The probability that a city receives this lump is proportional to its size and with some small probability the lump can create a new city. Having the latter probability approach to zero, the resulting distribution will be Zipf distributed. Notwithstanding the limitations, Simon's model provides a useful analytical starting point in thinking about uneven distributions in geographical space for two reasons. First, the model performs remarkably well empirically and is extendable as to account for more specific empirical outcomes. Second, the model is, in essence, an evolutionary model in that the probability of a particular event to occur is affected by the events that have taken place in the past (*path dependence*).

The model, however, lacks micro-foundations as urban growth is modelled as stemming from exogenous lumps rather than from agents' decisions. From an evolutionary perspective, reasoning from spatial units makes it difficult to introduce explicit firm dynamics into a theoretical framework (Boschma 2004). Yet, firm dynamics ultimately drive economic growth through the diffusion of routines in the economic system. An evolutionary approach to economic geography can thus build on a demographic perspective, which focuses on changing spatial patterns resulting from entry, growth and exit of firms, or as in the discussion below, on birth processes alone.

Reasoning from firms, we take product divisions as the unit of analysis, where each division belongs to a particular firm and is located in a particular city. One can then derive the firm size distribution by aggregating product divisions into firms and the city size distribution by aggregating product divisions into cities. In this framework, firm growth and urban growth occur simultaneously through the establishment of new product divisions, where the size of a firm or a city is simply defined as the number of product divisions belonging to a firm or a city, respectively. In terms of Simon's model, the lumps that drive growth can be considered as product innovations that are exploited by entrepreneurs by establishing a new product division. By reformulating Simon's stochastic model as a growth process fuelled by new product divisions, and by assigning each new product division simultaneously to a firm and a city, the firm size distribution and the city size distribution can be derived from one single growth process.

Following Frenken and Boschma (2007), one can introduce two *organizational parameters* (*p and p**) and two *locational parameters* (*q and q**). With probability $p$ the employee will commercialize the innovation in-house leading to a new product division within the parent firm. With probability $p^*$ the employee will commercialize the product innovation in another firm by changing jobs. The remaining probability $(1-p-p^*)$ is the probability that the employee creates a spinoff firm (which, following the Simon model, should be very small). And, with probability $q$ the innovation will be commercialized in the city of origin. With probability $q^*$ the innovation will be commercialized in another city. And with the remaining probability $1-q-q^*$ the innovation will be commercialized in a new city (which, again, should be very small). The probabilities can be multiplied since organizational and locational events are orthogonal to each other. For example, a firm can grow internally $(p)$ but at a different location than the division from which the idea originated $(q^*)$ or even at a new location $(1-q-q^*)$.

This reformulation of Simon's model incorporates nine possible events resulting from a product innovation (see Table 10.1). As such, the framework provides a rich repertoire for formal modelling approaches with only four parameters ($p$, $p^*$, $q$ and $q^*$). Firms and cities being the aggregates of product divisions, the model will produce the Zipf law for both the firm size distribution and the city size distributions in a single model as long as (Frenken and Boschma 2007):

1. $(1-p-p^*)$ and $(1-q-q^*)$ are close to zero and
2. In case of inter-firm or inter-city mobility, the probability that an employee chooses a firm or city, respectively, is proportional to its size (otherwise growth ceases to be proportional to size).

It is the latter assumption that we loosen in the following to account for biased mobility patterns. The bias we assume comes from the social networks of inventors who have previously worked together in a product division. As a result, we obtain correlated growth rates between "networked" firms and cities, which we elaborate in the form of hypotheses for future empirical research.

**Table 10.1** Possible events resulting from a product innovation

| | |
|---|---|
| $(p)(q)$ | Internal firm growth in city of origin |
| $(p)(q^*)$ | Internal firm growth in another city |
| $(p)(1-q-q^*)$ | Internal firm growth in a new city |
| $(p^*)(q)$ | Firm growth though labour mobility in city of origin |
| $(p^*)(q^*)$ | Firm growth though labour mobility in another city |
| $(p^*)(1-q-q^*)$ | Firm growth though labour mobility in a new city |
| $(1-p-p^*)(q)$ | Spinoff in city of origin |
| $(1-p-p^*)(q^*)$ | Spinoff in another city |
| $(1-p-p^*)(1-q-q^*)$ | Spinoff in a new city |

*Source*: Frenken and Boschma (2007)

## 10.3 Proximity

Mobility patterns of employees setting up their own product division create links between divisions. Since the Simon logic prescribes that each existing division has the same probability to give birth to an entrepreneur as any other division, the resulting network structure between divisions in the Simon model will be a perfect tree. Starting from the first product division in the economy, a new entrepreneur is added to the network at random. Each new division is connected through the parent division by a link. And, since division give birth to new division randomly, the resulting structure is a simple random tree where the degree of each node will be proportional to its age.

The tree-like network structure between divisions can be aggregated at the level of firms and at the level of cities. Note that this aggregation yields a network structure including intra- and inter-organizational links and intra- and inter-city links, respectively. The aggregated network structure at the level of firms or cities will be different from the random tree obtained at the level of product divisions as long as $(1-p-p^*) < 1$ and $(1-q-q^*) < 1$. If not, all new product divisions would lead to new firms and new cities, and the aggregated network at the level of firms and at the level of cities would yield the exact same network structure.

In the model, we have mobile agents between firms in case $p^* > 0$ and mobile agents between cities in case $q^* > 0$. In these cases, one has to specify the choice behaviour of entrepreneurs. The assumption proposed by Frenken and Boschma (2007) is to assume that the probability that an employee chooses a particular firm and city is proportional to their size as to ensure that the proportionate growth feature of the Simon-model is replicated. This assumption is equivalent to random interaction models, which state that migration flows between entities are proportional to their size[1] (Pumain 2006, p. 202).

The assumption of random interaction is obviously too crude, as mobility patterns are influenced by macroscopic structures. Labour mobility does not take place randomly; rather, most people move within the firm, or between firms whose activities are organizationally coordinated. One can measure the extent to which two organizations are tied by coordination, also called *organizational proximity* (Boschma 2005), for example, in terms of co-ownership or interlocking corporate boards. Thus, one can expect the probability of someone setting up his or her new product division in a particular firm to be dependent on the organizational proximity between the sending firm and the receiving firm.

Similarly, mobility between cities does not take place randomly; rather, most people move within the same city or to neighbouring cities within the region. Locational inertia stems from many sources (family, friends, local knowledge,

[1] See Pumain (2006, p. 202): "The proportionality between resident population and inward and outward migratory flows which is derived from the multiplication of the population at the origin by the population at destination in the numerator of the model can be seen as merely an application of a random interaction process".

identity). Thus, one can expect the probability of someone moving between two cities to be dependent on the *geographical proximity* between two cities.

To include a proximity structure in a simulation model based on the Simon logic of stochastic growth, one should also include intra-firm and intra-city mobility (the diagonal of a flow matrix). Obviously, these flows are characterised by the highest degree of organizational and geographical proximity, respectively. Entrepreneurs who stay within the firm or within the city also move, yet at the shortest distance possible. Once two proximity dimensions are introduced in the model, one can dispense with parameters $q$ and $q^*$ and $p$ and $p^*$ and replace these by a single organizational proximity parameter controlling the bias to move to a firm at a particular organizational distance, and a single geographical parameter controlling the bias to move to a location at a particular geographical distance.

Empirically, proximity can be introduced in empirical estimations of people flows between firms and between cities. One way to model such flows is by using spatial interaction models (Tinbergen 1962; Wilson 1970). In such models, the strength of flows between two entities is determined by the size of two entities (MASS), and a vector of proximities. Following our two proximity dimensions (organizational and geographical) defined above, organizational and geographical proximity can be introduced in the model to specify the probability that an entrepreneur who leaves a firm chooses for a particular firm and a particular city (possibly the same firm and the same city (s)he already worked in). Organizational proximity (OP) between firms can be expected to increase the probability of labour mobility between two firms (IFIRMS), and geographical proximity (GP) can be expected to increase the probability of labour migration between two cities (ICITIES). The equations to be estimated are:

$$\mathrm{IFIRMS}_{ij} = \mathrm{MASS}_i^{\alpha_1}\mathrm{MASS}_j^{\alpha_2}\mathrm{OP}_{ij}^{\alpha_3} + \varepsilon_{ij}, \tag{10.1}$$

$$\mathrm{ICITIES}_{ij} = \mathrm{MASS}_i^{\beta_1}\mathrm{MASS}_j^{\beta_2}\mathrm{GP}_{ij}^{\beta_3} + \varepsilon_{ij}. \tag{10.2}$$

Such a model could thus, in a simple manner, replicate the existence of "business groups" that rotate key personnel and "city-regions" as a set of cities within a region with strong labour market linkages.

## 10.4 Social Capital

To view economic growth as an ongoing process of product innovations introduced by entrepreneurs who set up their own product division, allows one to introduce an explicit evolutionary dynamic of social tie formation. With each creation of a new product division, a social tie is created between the parent division and the entrepreneur's new division (Breschi and Lissoni 2003, 2006). This social tie stems from the shared history of employer and former employee. A shared history here is defined as having worked within the same product division.

One can now understand the set of social ties of a product division as the "social capital" of a division. The amount of social capital is thus a function of the number of previous employees who have set up a new product division. As social capital gives a division access to previous employees, social capital can be thought of as a channel for knowledge spillovers. In a stochastic growth framework, differences in social capital among product divisions implies that firm growth and urban growth are no longer random, but positively dependent on the social capital of its divisions. If a firm or city harbours divisions with a high amount of social capital, the probability that these divisions will generate new entrepreneurs is consequently higher as well.[2] Because entrepreneurs are biased to set up their product divisions is the firm and/or city of origin, firms and cities with more social capital will grow faster. As the amount of social capital depends on previous growth events, firm growth and city growth are self-reinforcing.

A theory as outlined above, would provide an alternative explanation for endogenous growth and increasing returns to scale different from the one proposed by Krugman (1991) based on economies in production. Such an explanation is compatible with the argument by Boschma and Frenken (2006) that endogenous growth stems from recombination if one assumes that social capital supports the recombination of knowledge residing in different people.

So far, we have only assumed that organizational and geographical proximity matter in that it biases mobility decisions towards more proximate organizations. However, one can also assume that the "quality" of social capital is higher when social ties between parent firm and previous employees are organizational and geographically proximate, since social ties are easier to maintain within organizations and between similar organizations, and within cities or between proximate cities. This means that the probability of a product division generating an entrepreneur becomes not only dependent on the number of its social ties but also its quality.

Importantly, the joint effect of organizational and geographical proximity of the quality of social ties is expected to reflect substitutability (Boschma 2005; Ponds et al. 2007). When the entrepreneur sets up a business within the parent firm, organizational proximity is maximum. As a consequence, the social ties with the previous product division can be easily maintained even at large distance. This allows multi-locational firms to exploit their innovations at the best location without losing too much "social capital" involved in the relation between entrepreneur and his/her previous division. When the entrepreneur decides to remain in the city of origin, (s)he can easily maintain the social ties between the parent division even if it leaves the firm. This allows cities to exploit innovations in different organizational formats while profiting from the social capital involved in the

[2] Put differently, the probability of a product division producing a new entrepreneur who sets up his or her own product division is proportional to the number of previous employees who set up an own product division, an example of preferential attachment where the probability that a node acquires a new link is proportional to the node's degree (Barabási and Albert 1999).

relation between entrepreneur and his/her previous division. It thus provides us with a simple logic of why cities and multi-locational firms can be advantageous loci of growth.

Similarly, such substitution effects can exist when entrepreneurs move to similar organizations in "business groups" (allowing them to migrate over larger distance) and to nearby cities in "city-regions" (allowing them to move to different firms). As for multi-locational firms, business groups allow for the exploitation of innovations at different locations without losing too much social capital. And, city-regions allows, as do cities, for the exploitation of innovation in different organizational format while still providing sufficient geographical proximity to support social capital. Thus, both forms of proximity are expected to contribute to the probability of innovation, yet their combined effect will be less than the sum of their effect separately reflecting proximities are substitutes.

## 10.5 Concluding Remarks

The framework proposed here builds on the general framework laid down by Frenken and Boschma (2007). Our perspective combines industrial dynamics and urban growth in a "proximity" perspective. Empirically, this approach can rely on traditional spatial interaction equations/methodologies. Theoretically, it would allow a further extension of the role of multi-locational firms (facilitating the maintenance of social capital between previous colleagues over large distance) and multi-organizational cities (facilitating the maintenance of social capital between previous colleagues between different firms). By doing so, the now popular concepts of "network society" (Castells 1996), "global city regions" (Scott 2001) and "global pipelines and local buzz" (Bathelt et al. 2004) can be further elaborated analytically and analysed systematically using data on labour mobility and labour migration.

## References


Arthur WB (1989) Competing technologies, increasing returns, and lock-in by historical events. Econ J 99:116–131

Barabási A-L, Albert R (1999) Emergence of scaling in random networks. Science 286:509–512

Bathelt H, Malmberg A, Maskell P (2004) Clusters and knowledge: Local buzz, global pipelines and the process of knowledge creation. Prog Hum Geogr 28:31–56

Batty M (2005) Cities and complexity. Understanding cities with cellular atomata, agent-based models, and fractals MIT Press, Harvard, MA

Boschma RA (2004) The competitiveness of regions from an evolutionary perspective. Reg Stud 38(9):1001–1014

Boschma RA (2005) Proximity and innovation. A critical assessment. Reg Stud 39(1):61–74

Boschma RA, Frenken K (2006) Why is economic geography not an evolutionary science? Towards an evolutionary economic geography. J Econ Geogr 6(3):273–302

Breschi S, Lissoni F (2003) Mobility and social networks: Localised knowledge spillovers revisited. CESPRI Working Paper 142. http://www.cespri.unibocconi.it/, forthcoming in the Annales d'Economie et de Statistique

Breschi S, Lissoni F (2006) Mobility of inventors and the geography of knowledge spillovers. New evidence on US data. CESPRI Working Paper 184. http://www.cespri.unibocconi.it/

Castells M (1996) The rise of the network society. Blackwell, Oxford

David PA (1985) The economics of QWERTY. Am Econ Rev (Papers and Proceedings) 75:332–337

Duranton G, Puga D (2004) Micro-foundations of urban agglomeration economies. In: Henderson JV, Thisse JF (eds) Handbook of regional and urban economics, 1st ed., vol 4. Elsevier: Amsterdam, pp 2063–2117

Frenken K, Boschma RA (2007) A theoretical framework for evolutionary economic geography: industrial dynamics and urban growth as a branching process. J Econ Geogr 7(5):635–649

Krugman PR (1991) Increasing returns and economic geography. J Polit Econ 99(3):483–499

Ponds R, van Oort FG, Frenken K (2007) The geographical and institutional proximity of research collaboration. Pap Reg Sci 86:423–443

Pumain D (2006) Alternative explanations of hierarchical differentiation in urban systems. In: Pumain D (ed.) Hierarchy in natural and social sciences. Springer, Dordrecht, pp 169–222

Scott AJ (2001) Global city-regions. Trends, theory, policy. Oxford University Press, Oxford

Simon HA (1955) On a class of skew distribution functions. Biometrika 42(3–4):425–440

Tinbergen J (1962) The world economy. Suggestions for an international economic policy. Twentieth Century Fund, New York, NY

Wilson AG (1970) Entropy in urban and regional modelling. Pion, London

Zipf G (1949) Human behavior and the principle of least effort. Addison-Wesley, Cambridge MA

# Chapter 11
# Evolutionary and Preferential Attachment Models of Demand Growth

**Terry L. Friesz, Changhyun Kwon, and David Bernstein**

## 11.1 Introduction

It is widely acknowledged that, to create models for transportation planning that recognize the essential dynamic character of passenger network flows, one must consider two time scales: the so-called within-day time scale and the day-to-day time scale. Substantial progress has been made in modelling within-day dynamic flows for fixed trip matrices; one of the most widely acknowledged models for this purpose is the dynamic user equilibrium model proposed by Friesz et al. (1993) and studied by Xu et al. (1999), Wu et al. (1998), Friesz et al. (2001), Bliemer and Bovy (2003), and Friesz and Mookherjee (2006). In this chapter we propose two day-to-day models of demand growth compatible with a differential variational inequality formulation of the Friesz et al. (1993) model. The first of these employs dynamics inspired by evolutionary game theory, while the second uses the perspective of preferential attachment familiar from the network science and social network literature to create a model of demand growth. Additionally, numerical experiments to compare and contrast the two proposed theories of demand growth are described, along with hypotheses that one might address via such experiments.

## 11.2 Dynamic User Equilibrium

First, however, we need to make a few comments about modelling and computing dynamic flow patterns on traffic networks. *Dynamic traffic assignment* is the name given to the determination of time varying traffic flows for road networks.

T.L. Friesz (✉)
Department of Industrial and Manufacturing Engineering, The Pennsylvania State University, University Park, PA 16802, USA

A. Reggiani and P. Nijkamp (eds.), *Complexity and Spatial Networks*, Advances in Spatial Science,
DOI: 10.1007/978-3-642-01554-0_11, © Springer-Verlag Berlin Heidelberg 2009

When those flows obey a differential Nash-like equilibrium relative to departure rates and route choice, we say we have a dynamic user equilibrium flow pattern. To define a dynamic user equilibrium, we introduce the notion of an effective path delay operator $\Psi_{\mathrm{p}}(t,h)$, which expresses the unit path delay for departure time $t$ and traffic conditions $h$. The vector $h$ is time dependent and its $p$th component is $h_{\mathrm{p}}(t)$, the departure rate from the origin of path $p$ at time $t$. A dynamic user equilibrium flow pattern has the property that

$$h_{\mathrm{p}}^{*} > 0, p \in P_{ij} \Rightarrow \Psi_{\mathrm{p}}(t,h^{*}) = v_{ij}, \tag{11.1}$$

where $P_{ij}$ is the set of paths that connect origin–destination pair $(i,j) \in W$, while $W$ is the set of all origin–destination pairs. Furthermore, $v_{ij}$ is the minimum travel delay that can be experienced for $(i,j) \in W$. Embedded within each effective path delay operator $\Psi_{\mathrm{p}}(t,h)$ is a notion of arc delay (congestion) for the arcs comprising a given path and a penalty for early/late arrival. In fact the path delay operators are really a shorthand for a separate model, frequently called a network loading model, which determines the propagation of flows through a given network, as well as the path delays experienced, in response to a given vector of departure rates.

Additionally all path-specific departure rates are non-negative so we write

$$h = \left(h_{\mathrm{p}} : p \in P\right) \geq 0, \tag{11.2}$$

where $P$ is the set of all network paths. As a consequence

$$\Psi_{\mathrm{p}}(t,h^{*}) > v_{ij},\ p \in P_{ij} \Rightarrow h_{\mathrm{p}}^{*} = 0.$$

as can easily be proven from (11.1) by contradiction. We next comment that the relevant notion of flow conservation is

$$\sum_{p \in P_{ij}} \int_0^T h_{\mathrm{p}}(t)\mathrm{d}t = Q_{ij} \quad \forall (i,j) \in W,$$

where $Q_{ij}$ is the fixed travel demand (expressed as a traffic volume) for $(i,j) \in W$. Thus, the set of feasible solutions is

$$\Lambda = \left\{ h > 0 : \sum_{p \in P_{ij}} \int_0^T h_{\mathrm{p}}(t)\mathrm{d}t = Q_{ij} \quad \forall (i,j) \in W \right\}. \tag{11.3}$$

Friesz et al. (1993) show that a dynamic user equilibrium is equivalent to the following variational inequality:

$$\left.\begin{array}{l}\text{find } h^* \in \Lambda \text{ such that} \\ \sum_{p \in P} \int_0^T \Psi_p(t, h^*)\left(h_p - h_p^*\right)dt \geq 0 \;\; \forall h \in \Lambda\end{array}\right\}. \tag{11.4}$$

Suffice it to say that algorithms exist for solving (11.4); these include the fixed point algorithm presented in this chapter.

## 11.3 Demand Dynamics

Consider a transportation network for which a set $W$ of origin–destination pairs $(i, j)$ have been defined. Let

$$\tau \in \Upsilon \equiv \{1, 2, ..., L\}$$

be one typical discrete day, and take the length of each day to be $\Delta$, while the continuous clock time $t$ within each day is

$$t \in [(\tau - 1)\Delta, \tau\Delta]$$

for all

$$\tau \in \{1, 2, ..., L\}.$$

The entire planning horizon spans $L$ consecutive days. As noted above, we assume the travel demand for each day changes based on the previous day.

### 11.3.1 The Dual Time Scale Model

Let us suppose we have a demand growth model of the abstract form

$$Q_{ij}^{\tau+1} = F_{ij}\left(Q_{ij}^{\tau}, h^{\tau}, \theta\right) \quad \forall (i,j) \in W, \tau \in \{0, 1, 2, ..., L-1\}$$

$$Q_{ij}^{\tau} \geq 0 \quad \forall (i,j) \in W, \tau \in \{0, 1, 2, ..., L\}$$

$$Q_{ij}^{0} = K_{ij}^{0} \quad \forall (i,j) \in W,$$

where $Q_{ij}^{\tau}$ is the travel demand between origin-destination pair $(i, j) \in W$ during day $\tau$. Then a dual time scale model of dynamic user equilibrium with endogenous demand growth is

$$\left.\begin{array}{c} \text{find } Q \geq 0 \text{ and } h^* \in \Lambda(Q) \text{ such that} \\ \sum\limits_{p\in P} \int_0^T \Psi_p(t,h^{\tau*})(h_p^\tau - h_p^{\tau*})dt \geq 0 \ \ \forall \tau \in \Upsilon, h^\tau \in \Lambda_\tau(Q^\tau) \\ Q_{ij}^{\tau+1} = F_{ij}\left(Q_{ij}^\tau, h, \theta\right) \quad \forall (i,j) \in W \\ Q_{ij}^0 = K_{ij}^0 \quad \forall (i,j) \in W \end{array}\right\}. \tag{11.5}$$

This model may be solved by time stepping, so that exactly one variational inequality is faced for each value of $\tau$.

### 11.3.2 An Ad Hoc Model of Demand Growth

We postulate that the travel demands $Q_{ij}^\tau$ for day $\tau$ between a given OD pair $(i,j) \in W$ are determined by the following system of difference equations:

$$Q_{ij}^{\tau+1} = \left[ Q_{ij}^\tau - s_{ij}^\tau \left\{ \frac{\sum\limits_{p\in P_{ij}} \sum\limits_{j=0}^{\tau-1} \int_{j\cdot\Delta}^{(j+1)\cdot\Delta} \Psi_p(t,h^*)]dt}{|P_{ij}| \cdot \tau \cdot \Delta} - \chi_{ij} \right\} \right]^+ \tag{11.6}$$
$$\forall \tau \in \{0,1,2,...,L-1\}$$

with boundary condition

$$Q_{ij}^0 = \tilde{Q}_{ij}, \tag{11.7}$$

where $\tilde{Q}_{ij} \in \Re_+^1$ is the fixed travel demand for the OD pair $(i,j) \in W$ for the first day and $\chi_{ij}$ is the so-called fitness level. The operator $[x]^+$ is shorthand from $max[0,x]$. The parameter $s_{ij}^\tau$ is related to the rate of change of inter-day travel demand. The above system of difference equations assumes that the moving average of effective travel delay plus any imposed toll is the principal signal that influences demand learning.

### 11.3.3 Replicator Dynamics for Demand Growth

The model (11.6) does not precisely capture the structure proposed by Hofbauer and Sigmund (1998) for the fundamental dynamics of evolutionary game theory, namely replicator dynamics. For a state variable $Q$, replicator dynamics have the structural form

$$\frac{\dot{Q}}{Q} = \alpha\{\text{fitness} - \text{averagefitness}\}, \tag{11.8}$$

where $\alpha$ is a constant of proportionality and the notion of fitness of a given system is given a broad interpretation. We may modify the story behind (11.6) to more closely correspond with (11.8) by writing

$$\frac{Q_{ij}^{\tau+1} - Q_{ij}^{\tau}}{Q_{ij}^{\tau}} = \alpha_{ij}^{\tau} \left\{ \chi_{ij} - \frac{\sum\limits_{p \in P_{ij}} \sum\limits_{j=0}^{\tau-1} \int_{j \cdot \Delta}^{(j+1) \cdot \Delta} \Psi_p[t, h^*(t)] \mathrm{d}t}{|P_{ij}| \cdot \tau \cdot \Delta} \right\} \quad \forall \tau \in \{0, 1, 2, ..., L-1\} \tag{11.9}$$

with the same boundary condition (11.7). We now introduce a specific definition of instantaneous fitness. In particular, we assume instantaneous fitness for a given origin–destination pair $(i,j) \in W$ is

$$\chi_{ij} \equiv v_{ij} = \min_{p \in P_{ij}} \Psi_p[\tau, h^*(\tau)]. \tag{11.10}$$

In words, instantaneous fitness is least travel delay achieved at the end of the previous discrete time period (yesterday) and hence known at the start of the current discrete time period (today). Obviously (11.9) may be manipulated to give

$$Q_{ij}^{\tau+1} = Q_{ij}^{\tau} - \alpha_{ij}^{\tau} \left\{ \frac{\sum\limits_{p \in P_{ij}} \sum\limits_{j=0}^{\tau-1} \int_{j \cdot \Delta}^{(j+1) \cdot \Delta} \Psi_p[t, h^*(t)] \mathrm{d}t}{|P_{ij}| \cdot \tau \cdot \Delta} - v_{ij} \right\} Q_{ij}^{\tau} \quad \forall \tau \in \{0, 1, 2, ..., L-1\}. \tag{11.11}$$

If information technology is increasing the speed of access to data about least travel delay, then the length of each "day" can be shortened, as the notion of day used herein is arbitrary. If one wishes to assure demand does not become negative, then (11.11) is replaced by

$$Q_{ij}^{\tau+1} = \left[ Q_{ij}^{\tau} - \alpha_{ij}^{\tau} \left\{ \frac{\sum\limits_{p \in P_{ij}} \sum\limits_{j=0}^{\tau-1} \int_{j \cdot \Delta}^{(j+1) \cdot \Delta} \Psi_p[t, h^*(t)] \mathrm{d}t}{|P_{ij}| \cdot \tau \cdot \Delta} - v_{ij} \right\} Q_{ij}^{\tau} \right]^{+} \quad \forall \tau \in \{0, 1, 2, ..., L-1\}. \tag{11.12}$$

## 11.4 Demand Dynamics Based on Preferential Attachment

In network science affinity networks are widely thought to evolve according to the notion of preferential attachment. Bianconi and Barabási (2001) suggest an improved form of preferential attachment they call *quenched noise*. In that model they denote the connectivity of node $i$ by $k_i(t)$ and postulate an associated fitness parameter $\eta_i$ that accounts for differences among nodes with regard to their potential to attract and sustain attachments. They view network growth as a process whereby a new node with its distinct fitness is added randomly to a network during each period of time considered. The probability that a new node will connect to node $i$ already present in the network is taken to be

$$\pi_i = \frac{\eta_i k_i}{\sum\limits_j \eta_j k_j}. \tag{11.13}$$

Accordingly node $i$ will increase its connectivity at the rate

$$\frac{\partial k_i}{\partial t} = m \frac{\eta_i k_i}{\sum\limits_j \eta_j k_j}, \tag{11.14}$$

where $m$ is the number of new arcs added upon introduction of a new node. An initial condition must be associated with each (11.14) in order for the system of partial differential equations created in this fashion to be numerically solved.

Our interest in the above version of the preferential attachment model lies in the fact that it suggests a relationship between an underlying social network and the formation of travel demand. In particular one may partition, without loss of generality, the nodes of an affinity network into spatially related subsets of nodes that correspond to origins or destinations; when a given social network node is both an origin and a destination, a copy of it can be made and included as a member of both categories. Thus, arcs added to the social network, in light of the partition just described, join origin–destination pairs. As each arc of the affinity network represents a "travel relationship", it also represents an increment to the corresponding origin–destination travel demand. In this way, the Bianconi–Barabási network growth model, when applied to a social network, becomes a model of travel demand growth.

The above observations not withstanding, it is not really possible to directly employ the mathematical analysis surrounding the quenched noise model within a dynamic traffic assignment or congestion pricing model because the Bianconi–Barabási model lacks the spatial and agent detail needed for transportation network modelling. As a consequence, we need to provide a separate articulation of travel demand induced by affinity network growth, based on preferential attachment, that involves the variables and concepts introduced in previous sections and includes randomness. To that end we propose the following model:

$$Q_{ij}^{\tau+1} = Q_{ij}^{\tau} + s_{ij}^{\tau} \frac{\eta_{ij} Q_{ij}^{\tau}}{\sum\limits_{k \in N_0} \eta_{kj} Q_{kj}^{\tau}} \quad \forall \tau \in \{0, 1, 2, ..., L-1\}, \tag{11.15}$$

where all notation is as before but now fitness is doubly subscripted and appears as $\eta_{ij}$ for each $(i,j) \in W$ and we use $N_0$ to denote nodes that are origins. Importantly each $s_{ij}^{\tau}$ is a random variable that is naturally suited for treatment by a learning process. Note also that (11.15) is a discrete time version of preferential attachment, in that demand growth is greatest for origin–destination pairs with the largest current demand. Variations of (11.15) are easily constructed. For instance

$$Q_{ij}^{\tau+1} = Q_{ij}^{\tau} + s_{ij}^{\tau} \frac{\sum\limits_{k \in N_0} \eta_{kj} Q_{kj}^{\tau}}{\sum\limits_{(k,\ell) \in W} \eta_{k\ell} Q_{k\ell}^{\tau}} \quad \forall \tau \in \{0, 1, 2, ..., L-1\} \tag{11.16}$$

considers all origin–destination pairs in assessing the probability of a new increment in demand for a given pair.

Note that both model (11.15) and model (11.16) have the property that demand grows monotonically, which cannot be deemed realistic for all time. Thus, a needed modification is the introduction of a term that corresponds to the retirement of individuals from the underlying social network; if the rate of such retirements is $\rho_{ij}^{\tau}$, then (11.15) and (11.16) may be re-stated, respectively, as

$$Q_{ij}^{\tau+1} = \left[ Q_{ij}^{\tau} + s_{ij}^{\tau} \frac{\eta_{ij} Q_{ij}^{\tau}}{\sum\limits_{k \in N_0} \eta_{kj} Q_{kj}^{\tau}} - \rho_{ij}^{\tau} Q_{ij}^{\tau} \right]^{+} \quad \forall \tau \in \{0, 1, 2, ..., L-1\}, \tag{11.17}$$

$$Q_{ij}^{\tau+1} = \left[ Q_{ij}^{\tau} + s_{ij}^{\tau} \frac{\sum\limits_{k \in N_0} \eta_{kj} Q_{kj}^{\tau}}{\sum\limits_{(k,\ell) \in W} \eta_{k\ell} Q_{k\ell}^{\tau}} - \rho_{ij}^{\tau} Q_{ij}^{\tau} \right]^{+} \quad \forall \tau \in \{0, 1, 2, ..., L-1\}, \tag{11.18}$$

where we have introduced the $[\cdot]^{+}$ operator to assure demand does not become negative. The retirement rates $\rho_{ij}^{\tau}$ may be determined empirically, by a separate model or randomly. The random approach seems more in keeping with notion of preferential attachment we have borrowed from network science to describe the addition of new demand.

## 11.5 Demand Learning via the Kalman Filter

In this section we will base our remarks on the ad hoc model (11.6). However, it should be clear that a completely analogous discussion of demand learning may be crafted for each of the demand models suggested above. The model parameters $s_{ij}^{\tau}$

will typically be unknown to the modeller and follow stochastic distributions. Assuming that the modelling error and observation error follow normal distributions, in this section, we adapt a well-known forecasting method, so-called Kalman filtering. Recall the ad hoc day-to-day dynamics for travel demand:

$$Q_{ij}^{\tau+1} = \left[ Q_{ij}^{\tau} - s_{ij}^{\tau} \left\{ \frac{\sum_{p \in P_{ij}} \sum_{j=0}^{\tau-1} \int_{j \cdot \Delta}^{(j+1)\cdot\Delta} \Psi_{\mathrm{p}}[t, x(h^*, g^*)]\mathrm{d}t}{|P_{ij}| \cdot \tau \cdot \Delta} - \chi_{ij} \right\} \right]^{+} \tag{11.19}$$

$$\forall \tau \in \{0, 1, 2, \ldots L-1\}.$$

Each parameter $s_{ij}^{\tau}$ is treated as fixed during the solution process, but it is stochastic and its real value is unknown. After one day is completed, we want to update the model parameter $s_{ij}^{\tau}$ to obtain a better estimate of demand for the next planning horizon. The dynamics of $s_{ij}^{\tau}$ are assumed to be

$$s_{ij}^{\tau+1} = s_{ij}^{\tau} + \xi_{ij}^{\tau},$$

where $\xi_{ij}^{\tau}$ is a random noise from a normal distribution $N(0, B_{ij})$. The matrix $B_{ij}$ is known and called the process-noise covariance matrix.

The value of the parameter $s_{ij}^{\tau}$ cannot be observed directly but only through the change of realized travel demand, which can be defined as

$$z_{ij}^{\tau} \equiv \Delta Q_{ij}^{\tau} = Q_{ij}^{\tau+1} - Q_{ij}^{\tau}.$$

Note that

$$\Delta Q_{ij}^{\tau} = -s_{ij}^{\tau} \left\{ \frac{\sum_{p \in P_{ij}} \sum_{j=0}^{\tau-1} \int_{j \cdot \Delta}^{(j+1)\cdot\Delta} \Psi_{p}[t, x(h^*, g^*)]\mathrm{d}t}{|P_{ij}| \cdot \tau \cdot \Delta} - \chi_{ij} \right\} + \omega_{ij}^{\tau}, \tag{11.20}$$

and $\omega_{ij}^{\tau}$ is a random noise of observation from a normal distribution $N(0, R_{ij})$. The matrix $R_{ij}$ is known and called the measurement noise covariance matrix. Referring to Sect. 12.6 Bryson and Ho (1975), we obtain the Kalman filter dynamics

$$\begin{aligned} \bar{s}_{ij}^{\tau+1} &= \hat{s}_{ij}^{\tau} = \bar{s}_{ij}^{\tau} + V_{ij}^{\tau}\left[z_{ij}^{\tau} - H_{ij}^{\tau}\bar{s}_{ij}^{\tau}\right] \\ P_{ij}^{\tau} &= \left[\left(M_{ij}^{\tau}\right)^{-1} + \left(H_{ij}^{\tau}\right)^{T}\left(R_{ij}^{\tau}\right)^{-1} H_{ij}^{\tau}\right]^{-1} \\ M_{ij}^{\tau+1} &= P_{ij}^{\tau} + B_{ij}^{\tau}, \end{aligned}$$

where

$$V_{ij}^{\tau} \equiv P_{ij}^{\tau} H_{ij}^{\tau} \left( R_{ij}^{\tau} \right)^{-1}$$

$$H_{ij}^{\tau} \equiv - \left\{ \frac{\sum\limits_{p \in P_{ij}} \sum\limits_{j=0}^{\tau-1} \int_{j \cdot \Delta}^{(j+1) \cdot \Delta} \Psi_{p}[t, x(h^*, g^*)] \mathrm{d}t}{\left| P_{ij} \right| \cdot \tau \cdot \Delta} - \chi_{ij} \right\},$$

and $\bar{s}_{ij}^{\tau}$ is the a priori estimate of $s_{ij}^{\tau}$ (before observation) and $\hat{s}_{ij}^{\tau}$ is the a posteriori estimate (after observation). When estimation process based on the above dynamics is completed, we have $\bar{s}_{ij}^{\tau+1}$, which is the value of $s_{ij}$ used in the next discrete time interval.

## 11.6 Conclusions

We have proposed some dynamics for the growth of travel demand in vehicular traffic networks. Clearly, these ideas are preliminary; they are meant to promote discussion and to motivate future research. A great deal of work still needs to be done.

### *11.6.1 Numerical Experiments*

The models proposed above were constructed to conform with evolutionary game theory and the dynamics of preferential attachment in social networks. However, we do not know what spatial and temporal patterns of traffic flows and network usage at the link level will result from these models. In particular, we do not know if individual demand growth models, drawn from the family of models we have described, will display statistical tendencies to promote or diminish stability, resiliency, sustainability, congestion, the price of anarchy, the Braess paradox, and social cohesion.

For example if the rate of retirements in model (11.17) is described as a feedback mechanism driven by link-level congestion occurring on the transportation network, can preferential attachment dynamics maintain the level of connectivity of a community (origin) with other communities (destinations) sufficient to assure adequate employment and/or other means of sustainability? If the answer to this question were “no” based on numerous simulations, then empirical studies to ascertain whether demand does in fact grow by preferential attachment are needed. If such growth mechanisms are found to occur in the real world, then policies that deter demand growth by preferential attachment are warranted. Many other questions and hypotheses may be proposed and considered using the dual time scale model (11.5) together with one of the demand growth models of Sects. 11.3.2, 11.3.3 and 11.4.

### *11.6.2 Other Network Growth Processes*

The Bianconi–Barabási network growth model is only one of many that have been discussed in the network science literature. Several others are also worth considering in the current context. Erdos and Renyi (1959) start with a set of nodes and simply assume that each pair of nodes is connected by an arc with probability $p$. In the current setting, this corresponds to a situation in which the demand between an origin–destination pair increases by a fixed amount with probability $p$. One can complicate this model by specifying the number of destinations associated with each origin (that is, by specifying the degree of each origin). The properties of the ensemble of graphs that have a given degree distribution have been studied by Molloy and Reed (1998), Newman et al. (2001), Chung and Lu (2004) and others. Finally, one might want to construct the network based on attributes of the network. For example, one might assume that realizations with lower cost are more likely to occur [as in the cost efficiency theory of Smith (1983)] or one might make assumptions about the topological properties. These kinds of ensembles have been studied by Strauss (1986).

## References

Bianconi G, Barabási A (2001) Competition and multiscaling in evolving networks. Europhys Lett 54(4):436–442

Bliemer M, Bovy P (2003) Quasi-variational inequality formulation of the multiclass dynamic traffic assignment problem. Transp Res Part B 37(6):501–519

Bryson AE, Ho YC (1975) Applied optimal control. Hemisphere Publishing Company

Chung F, Lu L (2004) The average distance in a random graph with given expected degrees. Internet Math 1(1):91–113

Erdos P, Renyi A (1959) On random graphs. Publicationes Mathematicae Debrecen 6 (290)

Friesz TL, Mookherjee R (2006) Solving the dynamic network user equilibrium problem with state-dependent time shifts. Transp Res Part B 40:207–229

Friesz TL, Bernstein D, Smith T et al. (1993) A variational inequality formulation of the dynamic network user equilibrium problem. Oper Res 41:80–91

Friesz T, Bernstein D, Suo Z et al. (2001) Dynamic network user equilibrium with state-dependent time lags. Netw Spatial Econ 1:319–347

Hofbauer J, Sigmund K (1998) Evolutionary games and replicator dynamics. Cambridge University Press

Molloy M, Reed B (1998) The size of the giant component of a random graph with a given degree sequence. Comb Probab Comput 7(03):295–305

Newman M, Strogatz S, Watts D (2001) Random graphs with arbitrary degree distributions and their applications. Phys Rev E 64(2):26118

Smith T (1983) A cost-efficiency approach to the analysis of congested spatial-interaction behavior. Environ Plann A 15:435–464

Strauss D (1986) On a general class of models for interaction. SIAM Rev 28(4):513–527

Wu J, Chen Y, Florian M (1998) The continuous dynamic network loading problem: a mathematical formulation and solution method. Transp Res Part B 32(3):173–187

Xu Y, Wu J, Florian M et al. (1999) Advances in the continuous dynamic network loading problem. Transport Sci 33(4):341–353

# Chapter 12
# Modelling the Economy as an Evolving Space of Flows

## Methodological Challenges

**Kieran P. Donaghy**

## 12.1 Introduction

The spatial economy has increasingly come to be viewed, in the felicitous phrase of Manuel Castells (2000), as a *space of flows*. The mental picture we have of this economy is a motion picture, not a still shot. Moving along the links of various networks are ever greater quantities of people, goods, material, money, and information. Settlements, in turn, appear as increasingly interdependent nodes through which these vast quantities pass. The acceleration of flows through space can be accounted for largely by technological advances in communication and transportation and the emergence of far-flung value chains, which are driven by economizing behaviour, and abetted by increasingly liberal trade agreements and industrial deregulation (Wolf 2004).

Many authors have commented on how the spatial economy would seem to manifest characteristics of complex systems – and there are indeed similarities. Steven Durlauf, who has written extensively on economic complexity (both theoretical and empirical), defines complex systems as "those [systems composed] of a set of heterogeneous agents whose behaviour is interdependent and may be described as a stochastic process" (Durlauf 2005, p. 226). Durlauf sees the following four properties as distinguishing complex systems from other systems characterized by stochastic processes and interdependencies.

- *Nonergodicity* (also known as path dependence) or the property that conditional probability statements describing the system do not uniquely characterize the average or long-run behaviour of the system;
- *Phase transition*, or the property that small changes in parameters bring about qualitative changes in aggregate properties;[1]

K.P. Donaghy
Department of City and Regional Planning, Cornell University, Ithaca, NY, USA

[1] See Anderson (1972).

A. Reggiani and P. Nijkamp (eds.), *Complexity and Spatial Networks*, Advances in Spatial Science,
DOI: 10.1007/978-3-642-01554-0_12, © Springer-Verlag Berlin Heidelberg 2009

- *Emergent properties*, or properties that exist at a higher level of aggregation than the original description of the system; and
- *Universality*, or the property that the presence of a system characteristic is robust to alternative specifications of the system's microstructure (see Durlauf 2001, 2005).[2]

While these properties are potentially helpful in explaining and understanding spatial economic systems, their presence in such systems does not imply with necessity that these systems are complex.[3] In fact, it is an open question as to whether or not complexity in social systems has been established. Durlauf remarks that empirical evaluations of complexity are fraught with identification problems, although he sees the social interaction literature as containing the strongest evidence that forces giving rise to complexity are present. Ostrom's (2000) work on the resolution of collective action problems, Pettit's (1996) on the "Common Mind" and Schelling's (1960, 1971) on spatial segregation and the avoidance of calamitous international conflagrations all suggest compelling candidate examples. Still, the four properties Durlauf has identified do provide a useful benchmark for evaluating empirical work on complexity. And if the systems we are modelling are *potentially* complex, these properties ought to be realizable within our models.

Whether or not a "complexity sighting" has been positively confirmed, there are considerable methodological challenges to modelling an economy as a spatial system inclined toward complexity. In this chapter I will discuss a number of these and then illustrate how we might begin to take on some of these challenges in the case of modelling the evolution of commodity flows in the Midwest United States.

## 12.2 Methodological Challenges

If, for the sake of argument, we accept Durlauf's definition of a complex system and view the aggregate behaviours of (possibly, interdependent) networks involving many heterogeneous interdependent actors as our *explanandum*, or "that which is to be explained", we immediately encounter a number of challenges. Perhaps the first is to

[2]There are, of course, other lists of properties characterizing complex systems. David K. Campbell's is as follows: nonlinearity, interaction, irreducibility (behaviour is lost if the system is broken up into parts), hierarchies (multiple scales in space–time), emergent /self-organizing behaviour (more is different), many nearly equivalent configurations, adaptation, life-like behaviour (learning), intelligent agents using if/then rules (Campbell 2000).

[3]Durlauf observes 'The disparate empirical strategies that have been employed to provide evidence on economic complexity have yet to integrate theoretical models of complexity with data analysis in such a way as to show how a given aggregate property is associated with interactions between agents in a way that allows for a plausible finding that a given environment is in fact complex' (2005, p. 240).

1. *Agree on what stylized facts or empirical regularities are to be explained and what would count as an acceptable explanation*

If our investigations are motivated by policy concerns – say, for example, building and maintaining infrastructure to accommodate burgeoning flows of freight – we must concur on which of possibly many systems properties will be taken as indicators of system performance and identify which causal mechanisms need to be modelled. A second but related challenge is to

2. *Identify stable relationships or the "deep structure" of the system*

To be identifiable as a system, even a stochastic path-dependent evolutionary system must have some aspects which confer an enduring integrity to it through however many permutations it may pass. Without such a structure in mind, a modeller is just tracking passing phenomena. We may need to sharpen and reexamine the definitions of stability we apply to the study of dynamical systems (see Rotmans 2006). A third challenge facing modellers of an economy viewed as a space of flows involving interdependent heterogeneous actors is to

3. *Integrate within the same framework different conceptualizations of networks by the agents involved*

For example, firms involved in far-flung production networks and firms involved in freight logistics will view transportation infrastructure networks very differently – the former, focusing on nodes, will see potential sites for the disaggregated operations of sourcing, assembly, and distribution, whereas the latter, focusing on links, will see the means by which freight can be routed between distant nodes. Both perspectives are essential to a well formulated model of a dynamic game between shippers and carriers.

As Durlauf (2005) points out, any discussion of economic complexity entails identifying how system effects (possibly externalities of various sorts or emergent phenomena) result from the purpose-driven economizing behaviour of interacting agents.[4] Hence a fourth challenge our modellers face is to

4. *Relate micro-behavioural decision making and interaction by different types of agents to system effects*

Assuming we can overcome problems of identification and representation alluded to above, we know that determining whether or not a causal explanation is acceptable minimally entails making a fair comparison with other models based on competing explanations – that is, validating individual models of complex systems is important but does not constitute proving a theory, even provisionally (see Miller 1987). Hence we are challenged to

[4] Some theorists of complexity would seem to argue against the possibility of doing just this. See Markose (2005) after Hayek (1945).

5. *Formulate models that enable fair comparisons of competing explanations*

Two related challenges follow logically. The first is to

6. *Determine which of the available competing explanations is better supported by the data*

While obvious, the sixth challenge is very demanding, since we often lack sufficient spatial time-series observations to estimate together the parameters of a complex economic systems model. At this point in time we cannot test empirically the logical implications of our most advanced theoretical formulations – spatial computable general equilibrium models; we can only view them as sources of interesting and suggestive information that is complementary to the outputs of other models. Work by Brock and Durlauf (2001) on discrete choice with social interactions between agents in well circumscribed neighbourhoods is an exception. This sixth challenge, then, calls for particularly creative responses. The second challenge that logically follows from the fifth is to

7. *Determine, when we encounter any of the benchmark indicators of system complexity, whether or not such markers of complexity as "lock-in", "path dependence", "power laws", or "red queen effects" are really occurring or if something else is going on*

There may be several observationally equivalent but logically incompatible explanations. As dynamic systems modellers who operate with a practical interest in understanding how systems operate – so that we can anticipate how policy interventions might steer system behaviour – our primary objective should probably not be to identify genuine manifestations of complexity, *unless* positively identifying these manifestations contributes to our ability to manage systems. With policy interventions in mind, an eighth challenge to modellers of the economy as a space of flows and path dependent development is to be able to

8. *Identify the extent to which there is local autonomy in interdependent networks*

What capacity remains for local or regional policy to make a difference – say, regarding networks for freight movement? Alternatively, if not at the local level, at what level of spatial resolution might policy interventions make a difference?

In the balance of this short chapter, I shall discuss an example of recent work, which illustrates an attempt to meet some of the challenges set out above.

## 12.3 Modelling the Evolution of Commodity-Flow Patterns in the Midwest United States

Developments in transportation and communications technologies have enabled firms to exploit *economies of scale* and *scope* by fragmenting production processes and dispersing activities to least-cost locations (Jones and Kierzkowski 2001).

Consequently, the production of most goods worldwide now takes place in a distributed pattern over many locations in which semi-finished goods are shipped from one *specialized* establishment to another. *What* activities are carried out and *where* they agglomerate appear to be path dependent – initial advantages are reinforced due to scale effects (Venables 2006). And with the increased use of just-in-time inventory management methods, all production is becoming more transport intensive. The obverse of this development is that most freight shipments are now between establishments of firms operating in the same industry. As a consequence, the industrial cores of many regional economies have become hollowed out and regional economies, both near-by and far-flung, have become increasingly interdependent through global supply chains (Munroe et al. 2007).

While the stylized facts of this story of the evolution of goods movement are generally acknowledged to be accurate, this story is a difficult one to model formally. Why should we be concerned to do so? Some familiar reasons are to

- Test theories (causal explanations)
- Forecast further evolution of goods movement for transportation infrastructure planning purposes and
- Conduct thought experiments of possible policy interventions

But also, a broad-based community of politicians, planners, municipal administrators, environmental groups, port facility managers, shippers, carriers, freight handlers, and labour unions is concerned about these developments, in large part because they lack a clear sense of how these interdependent developments are related and what they portend. Moreover, the design of effective policies to accommodate anticipated increases in freight movement and to promote public/private partnerships that can abate and mitigate deleterious externalities, requires a better understanding of how cost and incentive structures affect the form and functioning of supply chains. While empirically supported theoretical explanations of fragmentation at the firm and industry levels, public/private partnerships at urban and regional levels, and network externalities at the systems level are available, we still lack theories and models that explicitly link micro-behavioural decision making of producers (or shippers) and carriers with impacts on nodes as well as links in transportation networks (that is, with aggregate flows).

The principal objective of a recent research project, undertaken by the author jointly with Geoffrey Hewings, Gianfranco Piras, and Jürgen Scheffren at the University of Illinois – see Donaghy et al. (2006) – is to elaborate an empirically oriented framework that can characterize in large the evolution of goods movement, in which the current state of affairs, or a stylized version thereof, can arise. In so doing, we have drawn on contributions to the literatures on fragmentation (Jones and Kierzkowski 2001), the new economic geography (Krugman and Venables 1995), dynamic networks (Nagurney and Dong 2002), and commodity flow modelling (Wilson 1970; Batten and Boyce 1986; Friesz et al. 1998; Boyce 2002; Ham et al. 2005). A prototype model is sketched in the appendix to this chapter. In particular, we specify a non-cooperative dynamic game between shippers and carriers. The specification is such that both economies of scale and scope can be captured (if present) and competing explanations involving interdependent actors – for example, those of the new economic geographers (aggregationists), theorists of fragmentation (disaggregationists), and other theorists can be confronted with data.

Data availability remains the biggest concern checking modelling ambitions, especially with respect to calibration of parameters (but see Donaghy et al. 2006 for a discussion of possible solutions to the challenges encountered). Numerical solutions to the dynamic games framed by the model also will not be trivial but may be obtained in the case of large-scale models by employing a dynamic variational-inequality approach (see Nagurney and Dong 2002 for details). One may also be able to solve the explicit set of first-order necessary conditions for smaller-scale versions of the dynamic optimization problems as in Donaghy and Schintler (1998), using a custom package, such as Wymer's (2004) continuous-time systems modelling tools, WYSEA. Operationalization of the framework elaborated above is presently proceeding with a small proof-of-concept model and with a larger model of the Midwest United States.

## 12.4 Conclusions

In the foregoing we have identified a number of challenges that face modellers attempting to characterize an economy in terms of spatial-dynamic networks. We have also discussed a model that would enable us to meet some of these challenges – including:

- Integrating different conceptualizations of networks by the agents involved (in a dynamic game)
- Relating micro-behavioural decision making and interaction by different types of agents to system effects
- Formulating models that enable fair comparisons of competing explanations, hence
- Determining which of the available competing explanations is better supported by available data

The other four challenges identified

- Agreeing on what empirical regularities are to be explained and what would count as an acceptable explanation
- Identifying stable relationships or the "deep structure" of the system
- Determining if benchmark indicators of complexity – when encountered – indicate real complexity and
- Identifying the extent to which there is local autonomy in interdependent networks

are less easy to meet and will require further discussion within the scientific community, such as is promoted in this volume.[5]

# Appendix. A
# Dynamic Commodity Flow Model of Donaghy et al. (2006)

We adopt the following notation to characterize network flows. Nodes of the network through which goods are shipped are indexed by $l$ and $m$. Links joining such nodes are indexed by $a$ and routes comprising contiguous links are indexed by $r$. The length of some link $a$ connecting two nodes is denoted by $d_{\mathrm{a}}$. If link $a$ is part of route $r$ connecting nodes $l$ and $m$, an indicator variable $\delta^a_{lmr}$ assumes the value 1.0. It is 0 otherwise. The length of a given route from some node $l$ to another node $m$, $D_{\mathrm{tmr}}$, is given by the sum of link distances along the route:

$$D_{lmr} \equiv \sum_a d_a \delta^a_{lmr}. \tag{12.1}$$

Turning to quantities shipped through the network, we index sectors engaged in production in the spatial economy by $i$ and $j$. Types of final demand will be indexed by $k$. Let $X^i_l$ denote the total output (in dollars) of sector $i$ produced at node $l$, $x^{ij}_{lm}$ denote interindustry sales from sector $i$ at location $l$ to sector $j$ at location $m$, and $\mathrm{FD}^{ik}_{lm}$ denote final demand of type $k$ at location $m$ for sector $i$'s product at location $l$. The physical flow of sector $i$'s product from $l$ to $m$ along route $r$ is $h^i_{lmr}$. This quantity is obtained by converting the value flow along route $r$ from dollars to tons by means of the ratio of total annual interregional economic flow to total annual physical flow, $q^i_x$. The total physical flow of all commodities shipped on a link $a$ via all routes using the link is given by

$$f_a \equiv \sum_i \sum_{lmr} h^i_{lmr} \delta^x_{lmr}, \tag{12.2}$$

and the periodic flow capacity of link $a$ is denoted by $k_a$. Conditions that the network must satisfy at any point in time are as follows.
*Material balance constraint*

$$X^i_l = \sum_m \sum_j x^{ij}_{lm} + \sum_m \sum_k FD^{ik}_{lm}, \forall i, \forall l. \tag{12.3}$$

[5]But see Donaghy and Richard (2006) on identifying the deep structure of an evolving system of demand for international currencies, and Piras et al. (2007) on explicitly testing for types of evolutionary dynamics.

*Conservation of flows constraint*

$$\sum_r h^i_{lmr} = \sum_j x^{ij}_{lm}/q^i_x + \sum_k \mathrm{FD}^{ik}_{lm}/q^i_x, \forall i, \forall l, \forall m.. \tag{12.4}$$

*Link capacity constraint*

$$\sum_i \sum_{lmr} h^i_{lmr} \delta^a_{lmr} = f_a \le k_a, \forall a. \tag{12.5}$$

*Non-negativity and feasibility conditions*

$$fa \ge 0, \forall a; h^i_{lmr} \ge 0, \forall i, \forall l, \forall m, \forall r; x^{ij}_{lm} > 0, \forall i, \forall j, \forall l, \forall m. \tag{12.6}$$

Equation (12.3) ensures that shipments from industry $i$ in location $l$ do not exceed production by the industry in that location, while (12.4) reconciles physical and value flows. Inequality (12.5) ensures that flows along links do not exceed capacities and the conditions given in (12.6) ensure that the distribution of goods throughout the network is feasible.[6]

In the sequel we shall assume that at each location $l$ the behavior of all establishments engaged in production in a given industrial sector can be characterized by a *representative establishment*.[7] Following Dixit and Stiglitz (1977), we further assume that firms operating the establishments act as monopolistic competitors of the Chamberlinian sort: they are output-level and input-price takers and they set output prices by a mark-up over marginal cost (which equals average cost in equilibrium). For a firm with an establishment producing in sector $i$ at location $l$, the mark-up, $\pi^i_l$, is given in terms of the price-elasticity of demand for $X^i_l$, $\sigma^i_l$, as

$$\pi^i_l = [\sigma^i_l/(\sigma^i_l - 1)].$$

Under the assumption of Chamberlinian monopolistic competition, the spatial markets in which firms compete are sufficiently competitive – barriers to entry are sufficiently low – so as to drive to a very low margin, if not zero, profits earned by firms from production of commodities at all locations.

Each local representative establishment is assumed to produce its output according to a two-level C.E.S. – constant elasticity of substitution – technology (Sato 1967). This fungible output can be used in production of other commodities or absorbed in final demand (in the forms of household and government consumption, investment,

[6]The assumption that all $x^{ij}_{lm}$ are positive is an assumption of convenience to ensure that marginal products, specified in (12.9) below, are defined. But given the level of sectoral aggregation of available commodity-flow data, this should be of no consequence.

[7]Hence we are allowing for the possibility that firms may have multiple establishments located in different areas.

and export). At the first level, inputs of each industrial type procured locally and non-locally are aggregated into input bundles:

$$c_m^{ij} = \gamma_m^{ij}\left[\sum_l \theta_{lm}^{ij}(x_{lm}^{ij})^{-\varepsilon_m^{ij}}\right]^{-1/\varepsilon_m^{ij}}, \forall i, \forall j, \forall m. \tag{12.7}$$

In (12.7), $c_m^{ij}$ is a bundle of inputs produced by representative establishments operating in industry $i$ at various locations $l$ used by the representative establishment in industry $j$ in its production activities at location $m$. The parameters $\gamma_m^{ij}, \theta_{lm}^{ij}$, and $\varepsilon_m^{ij}$ have standard interpretations as *scale*, *factor-intensity* and *substitution* parameters (see Ferguson 1969).

At the second level of the production function, total output by a representative establishment in a given industry in a given location is produced from the commodity bundle aggregates at the first level and labor and capital services, $L_m^j$ and $K_m^j$. At the second level, we allow explicitly for the possibility of increasing returns to scale in production at the establishment, regardless of the number of varieties aggregated in the commodity bundles, by employing a generalized C.E.S. function in which $\kappa_m^j \geq 1.0$ is the scale parameter (see Henderson and Quandt 1980).

$$X_m^j = \beta_m^j\left[\sum_i \alpha_m^{ij}(c_m^{ij})^{-\rho_m^j} + \alpha_m^{Lj}(L_m^j)^{-\rho_m^j} + \alpha_m^{Kj}(K_m^j)^{-\rho_m^j}\right]^{-\kappa_m^j/\rho_m^j}. \tag{12.8}$$

Again, the parameters of this function have their standard interpretations. The marginal product (in terms of good $j$) at location $m$ of a unit of good $i$ produced at and shipped from location $l$ is

$$\frac{\partial X_m^j}{\partial x_{lm}^{ij}} = \frac{\partial X_m^j}{\partial c_m^{ij}}\frac{\partial c_m^{ij}}{\partial x_{lm}^{ij}} = \frac{\kappa_m^j \alpha_m^i}{(\beta_m^j)^{\rho_m^j/\kappa_m^j}}\frac{(X_m^j)^{(\kappa_m^j+\rho_m^j)/\kappa_m^j}}{(c_m^{ij})^{(\rho_m^j+1)}}\frac{\theta_{lm}^{ij}}{(\gamma_m^{ij})^{\varepsilon_m^{ij}}}\left(\frac{c_m^{ij}}{x_{lm}^{ij}}\right)^{\varepsilon_m^{ij}+1}. \tag{12.9}$$

To make further progress with an explanation of economic behavior, we need to introduce prices as well as technology. Let $p_m^j$ denote the f.o.b. (or mill) price of a unit of industry $j$s output at location $m$ and $p_{lm}^i$ the delivered price of a unit of intermediate good $i$ at $m$. Then, defining $w_m^j$ and $ucc_m^j$ as the wage rate and user cost of capital in industry $j$ at location $m$, the mill price of this good under Chamberlinian monopolistic competition is given by

$$p_m^j = \pi_m^j\left[\sum_i\sum_l p_{lm}^i \cdot x_{lm}^{ij} + w_m^j \cdot L_m^j + ucc_m^j \cdot K_m^j\right] / X_m^j, \forall j, \forall m. \tag{12.10}$$

The *delivered price* at location $m$ of a good $i$ produced at location $l$, $p_{lm}^i$, includes the *unit cost of transport* by a carrier from location $l$ to location $m$, $\vartheta_{lm}^{ti}$, which is set by the carrier. Collecting these various price components, the delivered price of a unit of good $i$ at location $m$ will be

$$p^i_{lm} = p^i_l + \vartheta^{ti}_{lm}, \forall l, \forall m, \forall i. \tag{12.11}$$

Defining several new variables for the time rates of change in installed capacity (net of depreciation), in interindustry and interregional commodity flows, in employment, and the f.o.b. goods price that is,

$$I^j_m = \dot{K}^j_m,\ a^{xij}_{lm} = \dot{x}^{ij}_{lm},\ a^{Lj}_m = \dot{L}^j_m,\ \text{and}\ a^{pj}_m = \dot{p}^j_m,$$

the intertemporal optimization decision of a representative establishment in sector $j$ at location $m$ is to choose $I^j_m, a^{xij}_{lm}, a^{Lj}_m$ and $a^{pj}_m$ so as to minimize the present value of costs of operation at *and adjustment to* equilibrium levels of capital, intermediate goods, and labor:[8]

$$\int_{t_0}^{t_1} \mathrm{e}^{-\lambda^{sj}_m t} \Bigg\{ \sum_i \sum_l p^i_{lm} \cdot x^{ij}_{lm} + w^j_m L^j_m + \mathrm{ucc}^j_m K^j_m + q^j_m I^j_m + \frac{\omega^{Kj}_m}{2} (I - \upsilon^{Kj}_m (K^j_m * - K^j_m))^2$$
$$+ \sum_i \sum_l \frac{\omega^{xij}_{lm}}{2} (a^{xij}_{lm} - \upsilon^{xij}_{lm} (x^{ij}_{lm} * - x^{ij}_{lm}))^2 + \frac{\omega^{Lj}_m}{2} (a^{Lj}_m - \upsilon^{Lj}_m (L^j_m * - L^j_m))^2$$
$$+ \frac{\omega^{pj}_m}{2} (a^{pj}_m - \upsilon^{pj}_m (p^j_m * - p^j_m))^2 \, \mathrm{d}t, \tag{12.12}$$

subject to the following identities

$$\dot{K}^j_m = I^j_m, \tag{12.13}$$

$$\dot{x}^{ij}_{lm} = a^{xij}_{lm}, \forall i, \forall l, \tag{12.14}$$

$$\dot{L}^j_m = a^{Lj}_m, \tag{12.15}$$

$$\dot{p}^j_m = a^{pj}_m,^{9} \tag{12.16}$$

and (12.3) and the non-negativity condition on $x^{ij}_{lm}$ in (12.6). In objective functional (12.12), $\lambda^{sj}_m$ denotes the temporal discount rate of representative establishment $j$ in location $m$, the equilibrium price level is given by (12.10), and the (atemporal)

[8] We assume adjustment for a variable $y(t)$ towards a target value $y^*(t)$ according to $\mathrm{d}y(t)/\mathrm{d}t = a\,(y^* - y)$. For description of such an adaptive approach based on decision rules see Marcellino and Salmon (2002) and Scheffran (2001).

[9] Note that there are now four state equations (12.13)–(12.16). Note also that the objective functional, which involves derivatives of what would be logical control variables for the shippers, introduces *integral action*.

equilibrium (cost-minimizing) levels of capital, intermediate goods, and labor are given by

$$x_{lm}^{ij}* = \left[ \frac{\theta_m^{ij}}{(\gamma_m^{ij})^{\varepsilon_m^{ij}}} \frac{\kappa_m^j}{\pi_m^j} \frac{\alpha_m^{ij}}{(\beta_m^j)^{\rho_m^j/\kappa_m^j}} \frac{p_m^j}{p_{lm}^i} \frac{(X_m^j)^{(\kappa_m^j+\rho_m^j)/\kappa_m^j}}{(c_m^{ij})^{(\rho_m^j+1)}} \right]^{1/(1+\varepsilon_m^{ij})} c_m^{ij}, \tag{12.17}$$

$$L_m^j* = \left[ \frac{\kappa_m^j}{\pi_m^j} \frac{\alpha_m^{Lj}}{(\beta_m^j)^{\rho_m^j/\kappa_m^j}} \frac{p_m^j}{w_m^i} \right]^{1/(1+\rho_m^j)} (X_m^j)^{(\kappa_m^j+\rho_m^j)/(\kappa_m^j+\kappa_m^j\rho_m^j)}, \tag{12.18}$$

$$K_m^j* = \left[ \frac{\kappa_m^j}{\pi_m^j} \frac{\alpha_m^{Kj}}{(\beta_m^j)^{\rho_m^j/\kappa_m^j}} \frac{p_m^j}{\mathrm{ucc}_m^i} \right]^{1/(1+\rho_m^j)} (X_m^j)^{(\kappa_m^j+\rho_m^j)/(\kappa_m^j+\kappa_m^j\rho_m^j)}. \text{[10]} \tag{12.19}$$

We now make an assumption analogous to that made above concerning representative establishments: we assume that at each location $l$ there is a *representative carrier* which (1) takes as given quantities of goods to be transported from $l$ to other locations $m$ and the prevailing cost structure of goods movement, and (2) sets prices of carriage by commodity, origin, and destination and determines the routing pattern. The intertemporal optimization decision of a representative carrier at location $l$ is, then, to determine a time-varying schedule of prices, $\vartheta_{lm}^{ti}$, for shipping commodities from its respective location $l$ to establishments and sources of final demand (households, government agencies, etc.) at all other locations $m$, $x_{lm}^{ij}$ and $\mathrm{FD}_{lm}^{ik}$, and time-varying flows of commodities along available routes $r$, $h_{lmr}^i$, so as to maximize the present value of its anticipated stream of net revenues over the time horizon $t_0$ to $t_1$,

$$\int_{t_0}^{t_1} \mathrm{e}^{-\lambda_l^c t} \left\{ \sum_i \sum_m \vartheta_{lm}^{ti} \left(\sum_j x_{lm}^{ij} + \sum_k \mathrm{FD}_{lm}^{ik}\right) - \sum_i \sum_m \sum_r h_{lmr}^i D_{lmr} p_{lmr}^{ti} \right\} \mathrm{d}t, \tag{12.20}$$

subject to (12.4) and inequalities (12.5) and (12.6). In (12.20), $\lambda_l^c$ is the temporal discount rate of the representative carrier at location $l$. Also in (12.20), $p_{lmr}^{ti}$ denotes the cost to the carrier of delivering a ton of commodity $i$ from location $l$ to location $m$ via route $r$, and is assumed to be determined by the following cost relationship,

---

[10] Note that, with the generalized C.E.S. technology with increasing returns to scale at the second level, the rate of technological substitution between input bundles remains the same as in the case of constant returns to scale, as does the expansion path. Consequently, the cost function dual to the technology manifests all the usual regularity properties of a well-behaved cost function. These properties include the cost function being *non-negative* in input prices and output, *non-decreasing* in input prices and output, *concave* and *continuous* in input prices, *positively linear homogeneous* in input prices (so only relative prices matter), and supportive of *Shephard's lemma* (see Chambers 1988).

$$p_{lmr}^{ti} = p^t \cdot D_{lmr}^{\xi_{1i}-1} h_{lmr}^{i\xi_{2i}-1}, \quad \text{where} \quad \xi_{1i}, \xi_{2i} < 1.0, \forall i, \forall l, \forall m, \forall r, \tag{12.21}$$

where $p^t$ denotes the industry average ton-mile price of shipping a commodity. Cost relationship (12.21) implies that unit transport costs decline with distance and with total weight of shipment.

We shall further assume that volumes of final demand for goods at various locations are affected by the prices carriers set (through the delivered price) and that carriers are aware of this dynamic. The implied feedback relationship can be captured by defining final demand of type $k$ at location $m$ for good $i$ produced at location $l$ as

$$\mathrm{FD}_{lm}^{ik} = \widetilde{\mathrm{FD}}_{lm}^{ik} (p_{lm}^{i} / \bar{p}_{lm}^{i})^{-b_{lm}^{ik}}, \tag{12.22}$$

in which $\widetilde{\mathrm{FD}}_{lm}^{ik}$ is the volume of exogenously given final demand of type $k$ at location $m$ for good $i$ produced at location $l$ when the (normalized) delivered price is constant and $\bar{p}_{lm}^{i}$ is a period-average or reference delivered price.[11]

When taken over all producers and carriers, the first-order necessary conditions for the solution to the above joint intertemporal optimization problem – including the network constraints (12.3)–(12.6) – correspond to a non-cooperative (Nash) game in which each player takes all others' strategic behaviors as given (the first-order conditions are provided in the appendix to Donaghy et al. 2006). Given the curvature properties of the functional forms employed, a solution to the non-cooperative game should exist and should be unique (questions about the stability of the solution remain). Variations on the game set out above can also, and will be, investigated.

Note that the present set-up differs from the usual commodity-flow model formulation in that producers are minimizing transportation costs of inputs used in production along with other input costs, instead of minimizing shipping costs of supplying the market (cf. Boyce 2002). Carriers seek maximal profits through optimal route selection. The present set-up also differs from other formulations of dynamic games of shippers and carriers in that considerations of transportation costs influence production decisions (cf. Friesz and Holguin-Veras 2005).

Realism would dictate that in applied research on the evolution of goods movement and associated systems effects, transportation modes such as rail, air, or water should also be explicitly introduced, as Ham et al. (2005) have done for a static model of interregional commodity shipments and transportation network flows. This should not present great difficulties and would enable the basic model to support simulation and dynamic gaming exercises whose intent is to examine infrastructure policies.

---

[11] The definition of $\mathrm{FD}_{lm}^{ik}$ given in (12.22) should be substituted for all occurrences of the variable in other relationships of the model.

A more satisfying and more complete modeling framework would account for the evolution of final demand components (including exports) and the evolution of labor markets. An expenditure system for a representative household could be introduced along the lines of a modified almost ideal demand system (MAIDS) (see Cooper and McLaren 1992). Capacity expansion of establishments should also be related to purchases of capital goods from other producers. Changes along these lines would bring the model within the ambit of spatial computable general equilibrium frameworks.

## References

Anderson P (1972) More is different. Science 177:393–396

Batten DF, Boyce DE (1986) Spatial interaction, transportation, and interregional commodity shipment models. In: Nijkamp P (ed) Handbook of regional and urban economics, vol 1. North-Holland, Amsterdam, pp 357–406

Boyce D (2002) Combined model of interregional commodity flows on a transportation network. In: Hewings GJD, Sonis M, Boyce D (eds) Trade, networks, and hierarchies: modeling regional and interregional economies. Springer-Verlag, Heidelberg, pp. 29–40

Brock W, Durlauf S (2001) Discrete choice with social interactions. Rev Econ Stud 68:235–260

Campbell DK (2000) Chaos, complexity, and all that: one physicists perspective, power-point presentation at the Rand workshop on complexity and public policy. Complex systems and policy analysis: new tools for a new millennium, Arlington, VA. http://www.rand.org/scitech/stpi/Complexity/index.html

Castells M (2000) The rise of the network society, 2nd ed. Blackwell, Oxford

Chambers RG (1988) Applied production analysis: a dual approach. Cambridge University Press, New York

Cooper RJ, McLaren KR (1992) An empirically oriented demand system with improved regularity properties. Can J Econ 25:652–668

Dixit AK, Stiglitz JE (1977) Monopolistic competition and optimum product diversity. Am Econ Rev 67:297–308

Donaghy KP, Richard DM (2006) Estimating a regular continuous-time system of demand for world monies with divisia data. In: Belongia MT, Binner JM (eds) Money measurement, and computation. Palgrave Macmillan, Basingstoke, pp 76–103

Donaghy KP, Scheffran J, Piras G, Hewings GJD (2006) A micro-foundations approach to modeling the evolution of commodity flows in the mid-west United States. Paper presented at the North American Meetings of the Regional Science Association International, November, Toronto, Canada

Donaghy KP, Schintler LA (1998) Managing congestion, pollution, and pavement conditions in a dynamic transportation network model. Transport Res D 3(2):59–80

Donaghy KP, Hewings GJD, Piras G et al (2006) A micro-foundations approach to modeling the evolution of commodity flows in the midwest United States, paper presented at the North American meetings of the Regional Science Association International. November, Toronto

Durlauf S (2001) A framework for the study of individual behavior and social interactions. Sociol Methodol 31:47–87

Durlauf S (2005) Complexity and empirical economics. Econ J 115:F225–F243

Ethier WJ (1982) National and international returns to scale in the modern theory of international trade. Am Econ Rev 72(3):389–405

Ferguson CE (1969) The neoclassical theory of production and distribution. Cambridge University Press, Cambridge

Friesz TL, Holguin-Veras J (2005) Dynamic game-theoretic models of urban freight: formulation and solution approach. In: Reggiani A, Schintler LA (eds) Methods and models in transport and telecommunications: cross Atlantic perspectives. Springer-Verlag, Heidelberg, pp 143–162
Friesz TL, Suo Z-G, Bernstien DH (1998) A dynamic disequilibrium interregional commodity flow model. Transport Res B 32:467–483
Ham H, Kim TJ, Boyce D (2005) Implementation and estimation of a combined model of interregional multimodal commodity shipments and transportation network flows. Transport Res B 39:65–79
Hayek FA (1945) The use of knowledge in society. Am Econ Rev 35:519–530
Henderson JM, Quandt RE (1980) Microeconomic theory: a mathematical approach (third edition). McGraw-Hill, New York
Jones RW, Kierzkowski H (2001) A framework for fragmentation. In: Arndt SW, Kierzkowski H (eds) Fragmentation: new production patterns in the world economy. Oxford University Press, New York
Krugman P, Venables AJ (1995) Globalization and the inequality of nations. Quart J Econ 110 (4):857–880
Marcellino M, Salmon M (2002) Robust decision theory and the Lucas Critique. Macroecon Dyn 6:167–185
Markose SM (2005) Computability and evolutionary complexity: markets as complex adaptive systems (CAS). Econ J 115:F159–F192
Miller R (1987) Fact and method. Princeton University Press, Princeton
Munroe D, Hewings GJD, Guo D (2007) The role of intraindustry trade in interregional trade in the MidWest of the US. In: Cooper RJ, Donaghy KP, Hewings GJD (eds) Globalization and regional economic modeling. Springer-Verlag, Heidelberg, pp 87–105
Nagurney A, Dong J (2002) Supernetworks: decision-making for the information age. Edward Elgar, Cheltenham
Ostrom E (2000) Collective action and the evolution of social norms. J Econ Perspect 14(3): 137–158
Pettit P (1996) The common mind. Oxford University Press, Oxford
Piras G, Donaghy KP, Arbia G (2007) Nonlinear regional dynamics: continuous-time specification, estimation, and stability analysis. J Geograph Syst 9:311–344
Rotmans J (2006) A complex systems approach for sustainable cities. In: Ruth M (ed) Smart growth and climate change: regional development and adaptation. Edward Elgar, Cheltenham, pp 155–180
Sato K (1967) A two-level constant-elasticity-of-substitution production function. Rev Econ Stud 36:201–218
Scheffran J (2001) Stability and control of value-cost dynamic games. CEJOR 9(7):197–225
Schelling T (1960) The strategy of conflict. Harvard University Press, Cambridge
Schelling T (1971) Dynamic models of segregation. J Math Sociol 1:143–186
Venables AJ (2006) Shifts in economic geography and their causes. Paper prepared for the 2006 Jackson Hole Symposium of the Federal Reserve Bank of Kansas City
Wilson AG (1970) Interregional commodity flows: entropy maximizing procedures. Geogr Anal 2:255–282
Wolf M (2004) Why globalization works. Yale University Press, New Haven
Wymer CR (2004) WYSEA (Wymer Systems Estimation and Analysis) Software. Auckland, New Zealand

# Part C
# Empirical Aspects of Network Complexity in the Space-Economy

# Chapter 13
# Effects of a Simple Mode Choice Model in a Large-Scale Agent-Based Transport Simulation

**Dominik Grether, Yu Chen, Marcel Rieser, and Kai Nagel**

## 13.1 Introduction

The traditional transportation planning forecasting process is the four-step process, consisting of the following four steps (for example, Ortúzar and Willumsen 1995):

1. *Trip generation*, where sources and sinks of travel are computed
2. *Destination choice*, where sources and sinks are connected to trips. This results in the so-called origin–destination (OD) matrix
3. *Mode choice*, where the trips are differentiated by mode
4. *Assignment*, where routes are found for the trips, taking into account that much-used streets become slower ("congested assignment").

It has been clear for quite some time now that this approach is at odds with anything that is time dependent. At best, separate runs of the four step process are made for, say, morning peak, mid-day, evening peak, and night. Within the periods, everything is "static" (or steady-state), in the sense flow rates are constant throughout the periods.

The biggest barrier to time dependence is arguably the assignment step, for which a lot of mathematical theory is known (for example, Sheffi 1985). Much of that mathematical theory, however, is no longer valid when physical queues, i.s. spillback that uses up physical space, are introduced DaganzoAssign-w-queues. Physical queues, however, seem indispensable for a more realistic description of the traffic system. One way to address this problem is *dynamic* traffic assignment (DTA) (Peeta and Ziliaskopoulos 2001; Bliemer 2003; Mahut et al. 2003). Although there are different formulations, a standard formulation is to have time-dependent OD matrices, for example, one matrix for every hour. This sequence of matrices is then loaded onto the network, in such a way that traffic that does not

D. Grether (✉)
Transport Systems Planning and Transport Telematics, Technical University Berlin, Berlin, Germany

A. Reggiani and P. Nijkamp (eds.), *Complexity and Spatial Networks*, Advances in Spatial Science,
DOI: 10.1007/978-3-642-01554-0_13, © Springer-Verlag Berlin Heidlberg 2009 

arrive during one time slice is carried into the next time slice, and routes are assigned such that some normative behavioural model (for example, a Nash equilibrium) is reached.

In order to generate these time-dependent OD demand matrices, there seem to be two mainstream approaches:

- Lohse (1997) (also see Lohse et al. 2006), now implemented into the software VISEVA (PTV www page, accessed 2004; Beuck et al. 2007), generates separate OD matrices for different "trip purpose pairs", Trip purpose pairs are, for example, home→work, home→shop, work→leisure, work→home, etc., that is, the trip purposes at *both* ends. These OD matrices are then multiplied, for every time period, with a weighting function that describes how much traffic of this specific trip purpose pair happens at that time period. For example, home→work traffic probably mostly happens in the morning, while work→leisure traffic probably mostly happens in the afternoon. The data for this can be derived from time use surveys.
- The second mainstream approach to generate time-dependent OD demand matrices is activity-based demand generation (for example, Bowman et al. 1999; Bhat et al. 2004; Pendyala and Kitamura 2005; Arentze and Timmermans 2000; Timmermans 2005) generates travellers' daily plans, and transport appears as a derived demand to connect activities at different locations. There are many methods to achieve this, ranging from random utility modelling (Bowman et al. 1999; Bhat et al. 2004) to (partly) rule-based approaches (Arentze and Timmermans 2000).

In both cases, any feedback of congestion effects to the demand generation is done using aggregated quantities such as aggregated link travel times, or zone-to-zone impedances (Ettema et al. 2003; Lin et al. 2008). This can fail rather badly, since the aggregation errors can lead to implausible behavioural responses. For example, a router using link travel times that are aggregated into 15 min bins can, at the onset of congestion, predict rather wrong travel times, and in consequence try to avoid congestion that in synthetic reality does not exist when the vehicle is actually there (Raney and Nagel 2004).

An alternative is to use the iterations which are already done on the level of the route assignment routine and to extend them to other choice dimensions. De Palma and Marchal (2002) describe an early step in this direction, where not only routes but also departure times are adjusted individually for each trip, based on performance in previous iterations. MATSim (MATSIM www page, accessed 2008) takes this approach further:

- The simulation system does not only consider trips, but full daily plans and in consequence individual travellers.
- Additional choice dimensions are added one by one. Time choice has been added in earlier work (Balmer et al. 2005); in this chapter, the addition of mode choice will be described.

This addition of mode choice will be achieved in the following way:

1. Each agent obtains multiple initial plans, one for every mode.
2. The agents try those plans in different settings, modify the time and routing structure of those plans, etc.
3. The agents eventually settle down on a set of plans that suits their needs best.

Conceptually, MATSim agents individually follow genetic algorithms (GA) (Goldberg 1989; Holland 1992), where the MATSim plans correspond to the genes in a GA, the execution of the traffic flow simulation together with the scoring that follows corresponds to the computation of the fitness function in a GA, the MATSim selection between different plans corresponds to selection in a GA, and the algorithms that modify existing MATSim plans correspond to mutation operators in a GA. The MATSim system as a whole, consisting of these adaptive agents, is a co-evolutionary adaptive system (Hraber et al. 1994; Palmer et al. 1994; Arthur 1994; Hofbauer and Sigmund 1998; Drossel 2001). This abstract computational system is then filled with meaning from transport engineering and travel behaviour research. For example, the traffic flow simulation is constructed from transport engineering principles (for example, Gerlough and Huber 1975); the scoring function and the related selection operation follows a utility-based approach (for example, Ben-Akiva and Lerman 1985); and the generation and mutation of the plans follows concepts from travel behaviour research, in particular activity-based demand modelling (for example, Timmermans 2005).

The chapter is organized as follows. Section 13.2 describes the overall approach, concentrating on conceptual aspects, the co-evolutionary adaptation, and the scoring. Section 13.3 then describes the mode choice model. Section 13.4 describes a specific scenario, related to an illustrative study using data from the Zurich metropolitan area. The scenario consists of the geographic and socio-demographic input data and the specific simulation runs that were undertaken. Finally Sect. 13.5 summarizes the results and provides an outlook to future work.

## 13.2 Simulation Structure

The following describes the structure of the simulation that is used. It is the standard structure of MATSim, as described at many places (Raney and Nagel 2006b; Balmer et al. 2005). Readers familiar with the MATSim approach can skip this section.

### *13.2.1 Overview*

Our simulation is constructed around the notion of agents that make independent decisions about their actions. Each traveller of the real system is modelled as an individual agent in our simulation. The overall approach consists of three important pieces:

- Each agent independently generates a so-called *plan*, which encodes its intentions during a certain time period, typically a day.

- All agents' plans are simultaneously executed in the simulation of the physical system. This is also called the *traffic flow simulation* or *mobility simulation*.
- There is a mechanism that allows agents to *learn*. In our implementation, the system iterates between plans generation and traffic flow simulation. The system remembers several plans per agent, and scores the performance of each plan. Agents normally choose the plan with the highest score, sometimes re-evaluate plans with bad scores, and sometimes obtain new plans by modifying copies of existing plans.

A *plan* contains the itinerary of activities the agent wants to perform during the day, plus the intervening trip legs the agent must take to travel between activities. An agent's plan details the order, type, location, duration and other time constraints of each activity, and the mode, route and expected departure and travel times of each leg.

The task of generating a plan is divided into sets of decisions, and each set is assigned to a separate *module*. An agent strings together calls to various modules in order to build up a complete plan. To support this "stringing", the input to a given module is a (possibly incomplete) plan, and the output is a plan with some of the decisions updated. This chapter will make use of two modules only: "activity times generator" and "router". Other modules will be the topic of future work. Once the agent's plan has been constructed, it can be fed into the *traffic flow simulation*. This module executes all agents' plans simultaneously on the network, allowing agents to interact with one another, and provides output describing what happened to the agents during the execution of their plans.

The outcome of the traffic flow simulation (for example, congestion) depends on the planning decisions made by the decision-making modules. However, those modules can base their decisions on the output of the traffic flow simulation (for example, knowledge of congestion). This creates an interdependency ("chicken and egg") problem between the decision-making modules and the traffic flow simulation. To solve this, *feedback* is introduced into the multi-agent simulation structure (Kaufman et al. 1991; Bottom 2000). This sets up an iteration cycle which runs the traffic flow simulation with specific plans for the agents, then uses the planning modules to update the plans; these changed plans are again fed into the traffic flow simulation, etc. until consistency between modules is reached.

The feedback cycle is controlled by the *agent database*, which also keeps track of multiple plans generated by each agent, allowing agents to reuse those plans at will. The repetition of the iteration cycle coupled with the agent database enables the agents to learn how to improve their plans over many iterations.

In the following sections we describe the used modules in more detail.

### *13.2.2 Activity Time Allocation Module*

This module is called to change the timing of an agent's plan. At this point, a simple approach is used which applies a random "mutation" to the duration and end time attributes of the agent's activities. For each such attribute of each activity in an agent's plan, this module picks a random time from the uniform distribution (−30 min + 30 min) and adds it to the attribute. Any negative duration is reset to zero; any activity end time after midnight is reset to midnight.

Although this approach is not very sophisticated, it is sufficient in order to obtain useful results. This is consistent with our overall assumption that, to a certain extent, simple modules can be used in conjunction with a large number of learning iterations (for example, Nagel et al. 2004). Since each module is implemented as a "plugin", this module can be replaced by a more enhanced implementation if desired.

### *13.2.3 Router*

The router is implemented as a *time dependent Dijkstra algorithm*. It calculates link travel times from the events output of the previous traffic flow simulation (see Sect. 13.2.4). The link travel times are encoded in 15 min time bins, so they can be used as the weights of the links in the network graph. Apart from relatively small and essential technical details, the implementation of such an algorithm is straightforward (Jacob et al. 1999; Lefebvre and Balmer 2007). With this and the knowledge about activity chains, it computes the fastest path from each activity to the next one in the sequence as a function of departure time.

### *13.2.4 Traffic Flow Simulation*

The traffic flow simulation simulates the physical world. It is implemented as a queue simulation, which means that each street (link) is represented as a FIFO (first-in first-out) queue with two restrictions (Gawron 1998; Cetin et al. 2003). First, each agent has to remain for a certain time on the link, corresponding to the free speed travel time. Second, a link storage capacity is defined which limits the number of agents on the link. If it is filled up, no more agents can enter this link.

Even though this structure is indeed very simple, it produces traffic as expected and it can run directly off the data typically available for transportation planning purposes. On the other hand, there are some limitations compared to reality – for example, the number of lanes, weaving lanes, turn connectivities across intersections or signal schedules cannot be included into this model.

The output that the traffic flow simulation produces is a list of events for each agent, such as entering/leaving link, left/arrived at activity, and so on. Data for an event includes which agent experienced it, what happened, at what time it happened, and where (link/node) the event occurred. With this data it is easy to produce different kinds of information and indicators like link travel time (which, for example, will be used by the router), trip travel time, trip length, percentage of congestion, and so on.

### 13.2.5 Agent Database: Feedback

As mentioned above, the feedback mechanism is important for making the modules consistent with one another, and for enabling agents to learn how to improve their plans. In order to achieve this improvement, agents need to be able to try out different plans and to tell when one plan is "better" than another. The iteration cycle of the feedback mechanism allows agents to try out multiple plans. To compare plans, the agents assign each plan a "score" based on how it performed in the traffic flow simulation.

Our framework always uses *actual plans performance* for the score. This is in contrast to all other similar approaches that we are aware of. These other approaches always feed back some aggregated quantity such as link travel times and reconstruct performance based on those (for example, URBANSIM www page, accessed 2007; Ettema et al. 2003).

The procedure of the feedback and learning mechanism is described in detail by Balmer et al. (2005). For better understanding, the key points are restated here.

1. The agent database starts with at least one complete plan per agent, with one plan marked as "selected".
2. The simulation executes these marked plans simultaneously and outputs events.
3. Each agent uses the events to calculate the score of its "selected" plan and decides, which plan to select for execution by the next traffic flow simulation. When choosing a plan, the agent database can either:

   – Create a new plan by sending an existing plan to the router, adding the modified plan as a new plan and selecting it,
   – Create a new plan by sending an existing plan to the time allocation module, adding the modified plan and selecting it,
   – Pick an existing plan from memory, choosing according to probabilities based on the scores of the plans. The probabilities are of the form

   $$p_j = \mathrm{e}^{\beta U_j} / \sum_i \mathrm{e}^{\beta U_i},$$

   where $U_j$ is the score (utility) of plan $j$, and $\beta$ is an empirical constant. This is the familiar logit model (for example, Ben-Akiva and Lerman 1985).

4. Next, the simulation executes the newly selected plans, that is, it goes back to 2.

This cycle continues until the system has reached a relaxed state. At this point, there is no quantitative measure of when the system is "relaxed"; we just allow the cycle to continue until the outcome seems stable.

### 13.2.6 Scores (= Utilities) for Plans

In order for adaptation to work in a meaningful way, it is necessary to be able to compare the performance of different plans. This is easiest achieved by assigning scores to plans. This is the same as the fitness function in genetic algorithms, or the objective function in optimization problems. Note once more that every agent has its own scoring function, and attempts to optimize for her-/himself.

In principle, arbitrary scoring schemes can be used (for example, prospect theory; Avineri and Prashker 2003). In this work, a utility-based approach is used. The approach is related to the Vickrey bottleneck model (Arnott et al. 1990), but is modified in order to be consistent with our approach based on complete daily plans (Charypar and Nagel 2005; Raney and Nagel 2006a). The elements of our approach are as follows:

- The total utility of a plan is computed as the sum of individual contributions:

$$U_{\text{total}} = \sum_{i=1}^{n} U_{\text{perf},i} + \sum_{i=1}^{n} U_{\text{late},i} + \sum_{i=1}^{n} U_{\text{travel},i},$$

  where $U_{\text{total}}$ is the total utility for a given plan; $n$ is the number of activities, which equals the number of trips (the first and the last activity on a day are "stitched together"); $U_{\text{perf},i}$ is the (positive) utility earned for performing activity i; $U_{\text{late},i}$ is the (negative) utility earned for arriving late to activity $i$; and $U_{\text{travel},i}$ is the (negative) utility earned for travelling during trip $i$. In order to work in plausible real-world units, utilities are measured in Euro.

- A logarithmic form is used for the positive utility earned by performing an activity:

$$U_{\text{perf},i}(t_{\text{perf},i}) = \beta_{\text{perf}} \cdot t_{*,i} \cdot \ln\left(\frac{t_{\text{perf},i}}{t_{0,i}}\right),$$

  where $t_{\text{perf}}$ is the actual performed duration of the activity, $t_*$ is the "typical" duration of an activity, and $\beta_{\text{perf}}$ is the marginal utility of an activity at its typical duration. $\beta_{\text{perf}}$ is the same for all activities, since in equilibrium all activities at their typical duration need to have the same marginal utility.

- $t_{0,\text{i}}$ is a scaling parameter that is related both to the minimum duration and to the importance of an activity. If the actual duration falls below $t_{0,\text{i}}$, then the utility contribution of the activity becomes negative, implying that the agent should rather completely drop that activity. A $t_{0,\text{i}}$ only slightly less than $t_{*,\text{j}}$ means that the marginal utility of activity $i$ rapidly increases with decreasing $t_{\text{perf,i}}$, implying that the agent should rather cut short other activities. This chapter uses

$$t_{0,i} = t_{*,i} \cdot \exp(-\zeta/t_{*,i})$$

  where $\zeta$ is a scaling constant set to 10 h. With this specific form, the utility at the typical duration,

$$U_{\text{perf},i}(t_{*,i}) = \beta_{\text{perf}} \cdot \zeta$$

is independent of the activity type.[1]

- The (dis)utility of being late is uniformly assumed as:

$$U_{\text{late},i} = \beta_{\text{late}} \cdot t_{\text{late},i},$$

where $\beta_{\text{late}}$ is the marginal utility (in Euro h$^{-1}$) for being late, and $t_{\text{late},i}$ is the number of hours late to activity $i$.

- The (dis)utility of travelling is uniformly assumed as:

$$U_{\text{travel},i} = \beta_{\text{travel}} \cdot t_{\text{travel},i},$$

where $\beta_{\text{travel}}$ is the marginal utility (in Euro h$^{-1}$) for travel, and $t_{\text{travel},i}$ is the number of hours spent travelling during trip $i$.

In principle, arriving early or leaving early could also be punished. There is, however, no immediate need to punish early arrival, since waiting times are already indirectly punished by foregoing the reward that could be accumulated by doing an activity instead (opportunity cost). In consequence, the effective (dis)utility of waiting is already $-\beta_{\text{perf}}$. Similarly, that opportunity cost has to be added to the time spent travelling, arriving at an effective (dis)utility of travelling of $-|\beta_{\text{travel}}| - \beta_{\text{perf}}$.

No opportunity cost needs to be added to late arrivals, because the late arrival time is spent somewhere else. In consequence, the effective (dis)utility of arriving late remains at $\beta_{\text{late}}$. These values ($\beta_{\text{perf}}, \beta_{\text{perf}} + |\beta_{\text{travel}}|$, and $|\beta_{\text{late}}|$) are the values that need to be compared to the values of the parameters of the Vickrey model (Arnott et al. 1990).

### *13.2.7 Discussion of the Scoring Function*

In our investigations, it turns out that the following aspects of the scoring function are of prime importance:

- The typical duration, $t_{*,i}$ of each activity type.
- The height of the utility function at its typical duration, that is, $U(t_{*,i})$, for each activity type.
- The slope of the utility function at its typical duration, for each activity type.

[1]This "consequence" is actually the motivation for the specific mathematical form of the activity performance utility contribution. The reason for this motivation is not relevant to this chapter, but is described in Charypar and Nagel (2005).

- The curvature of the utility function at its typical duration, for each activity type.

In consequence, at first glance it seems that there are four free parameters per activity type. Fortunately, this number can be reduced by the following arguments:

- In order to be optimal, the activity durations need to be selected such that all slopes (= marginal utilities) are the same, at least in the absence of constraints such as opening times or other influences such as strongly variable travel times. This implies that one can, as a first approximation, set all slopes at the typical duration to the same value. This ends up being the marginal utility of leisure time (Jara-Díaz et al. 2004), which can be estimated.
- By the same argument, it should be possible to estimate "typical durations" of activity times from time use surveys: If marginal utilities are the same, then the typical durations need to be set such that the typical durations from time use surveys are recovered – In our current work, the typical durations are directly taken from actual durations from time use surveys in Switzerland (see below).
- As long as activity dropping is not possible, the absolute height of the utility does not matter. This justifies the arbitrary setting of

$$U_{\text{perf},i}(t_{*,i}) = \beta_{\text{perf}} \cdot \zeta.$$

  It also means that the absolute level of our agent score is meaningless, and *only differences between scores can be interpreted as utility differences.*

The curvature at the typical durations remains as the most problematic parameter. This parameter determines the flexibility of an activity: a large curvature means that the marginal utility increases strongly when the activity duration is reduced, implying that time should rather be saved somewhere else. Conversely, the marginal utility *de*creases strongly when the activity duration is increased, implying that additional time should rather be spent somewhere else. The above utility function has a second derivative of $-\beta_{\text{perf}}/t_{*,i}$. This means that, with the above utility function, no free parameter is left to separately adjust the curvature at the typical duration. The second derivative is inversely proportional to the typical duration, meaning that longer activities always have more flexibility than shorter activities.

## 13.3 Mode Choice Model

This section will present and characterize the mode choice model. This will be achieved by two additional elements in MATSim:

- An extension of the scoring function, now taking into account the (dis)utility of travel by non-car modes
- A mechanism to generate non-car plans

All agents will carry on to maximize personal utility but the calculation of this utility depends on chosen mode.

### 13.3.1 Extension of Scoring Function

The disutility of travelling from Sect. 2.6 is

$$U_{\text{travel},i} = \beta_{\text{travel}} \cdot t_{\text{travel},i},$$

where $t_{\text{travel},i}$ is the travel time in hours spent for trip $i$ and $\beta_{\text{travel}}$ is the marginal utility of travel. To include alternative modes, it is sufficient to make the (dis)utility of travel dependent on the mode. A simple approach to do this is to use different valuations of the time for the two modes:

$$U_{\text{travel, mode, } i} = \begin{cases} \beta_{\text{car}} \cdot t_{\text{travel,i}}, & \text{if trip } i \text{ is by car} \\ \beta_{\text{non-car}} \cdot t_{\text{travel},i} & \text{if trip } i \text{ is not by car,} \end{cases}$$

where $\beta_{\text{car}}$ and $\beta_{\text{non-car}}$ are the marginal utilities of travelling by car or not by car (in Euro $\text{h}^{-1}$), respectively, and $t_{\text{travel},i}$ is the number of hours spent travelling during trip $i$. For the time being this leaves out all more complicated aspects of non-car travel valuations, such as changing vehicles, schedule restrictions, waiting times, etc.

The task is now to select values for those marginal utilities. For this, it is important to note once more that $\beta_{\text{car}}$ and $\beta_{\text{non-car}}$ are *not* values of time by themselves, but they are *additional* marginal disutilities caused by travelling, in addition to the marginal opportunity cost of time. This is consistent with econometric approaches (Jara-Díaz and Guerra 2003).

### 13.3.2 Generating non-Car Plans

Besides the separate scoring of the non-car travel, it is necessary to generate plans that use the non-car mode. In all investigations described in this chapter, this is done by giving all travellers an additional initial plan that uses the non-car mode on all trips. The duration of every non-car trip is assumed to take approximately twice as long as the car mode at free speed.[2]

---

[2]The algorithm to construct the trip durations of the non-car mode was later modified to take *exactly* twice as long as the car mode at free speed. This explains differences between this chapter and other publications on the same subject. Eventually, these estimates need to be replaced by real-world data.

This is based on the (informally stated) goal of the Berlin public transit company to generally achieve door-to-door travel times that are no longer than twice as long as car travel times. This, in turn, is based on the observation that non-captive travellers can be recruited into public transit when it is faster than this benchmark (Reinhold 2006). For the purposes of the present chapter, it is assumed that all non-car modes very roughly have the shared characteristics that they are slower than the (non-congested) car mode – this will be further disaggregated in future work. In the same vein, both for car and for non-car trips there are no separate considerations of access and egress.

The non-car plan can undergo time adaptations as all other plans can. In consequence, it is quite possible that an agent will end up having multiple car plans *and* multiple non-car plans. If one mode scores consistently worse than the other mode, most plans of that mode will eventually be deleted. However, the plans deletion mechanism is programmed in a way that the last plan of every mode needs to be kept. In this way, it is ensured that travellers maintain the option to switch modes at all times.

## 13.4 Zurich Scenario

### *13.4.1 Network*

The scenario covers the area of Zurich, Switzerland, which has about 1m inhabitants. It is shown in Fig. 13.1. The network is a Swiss regional planning network, which includes the major European transit corridors. It consists of 24,180 nodes and 60,492 links.

The links have attributes (flow capacity, free speed, number of lanes,. . .) suitable for static traffic assignment. These turned out to be generated with a view towards *national* forecasts, and were thus not sufficiently detailed within the city of Zurich with its dense road network. Thus, all links within a circle with radius 4 km around the centre of Zurich have their attributes modified as follows:

- Links corresponding to primary roads in OpenStreetMap[3] get a capacity of at least 2,000 vehicles per hour. If the original capacity is higher than that, the capacity is not changed.
- Links corresponding to secondary roads in OpenStreetMap keep their original capacity (usually between 1,000 and 2,000 veh $h^{-1}$).
- All other links get a capacity of at most 600 veh $h^{-1}$. If the original capacity is lower, it is not changed.
- A few single links are manually adjusted based on local knowledge.

[3] See http://www.openstreetmap.org.

**Fig. 13.1** Switzerland network, area of Zurich enlarged

## 13.4.2 Population, Initial Demand

The simulated demand consists of all travellers within Switzerland that cross at least once during their day an imaginary boundary around Zurich. This boundary is defined as a circle with a radius of 30 km and with its centre at "Bellevue", a central place in the city of Zurich. To speed up computations, a random 10% sample was chosen for simulation, consisting of 181,725 agents.

The travellers have complete daily activity patterns based on microcensus information (Balmer et al. 2006; Meister et al. 2008). Such activity patterns can include activities of type *home*, *work*, *education*, *shopping*, and *leisure*. Each agent gets two plans based on the same activity pattern. The first plan uses only "car" as transportation mode, while the second plan uses only "non-car".

This demand was then extended with people crossing the borders of Switzerland and travelling within the region of Zurich, either because they live in neighbouring countries but work in Switzerland, because they live in Switzerland but work outside, or because they travel through Switzerland on transit. Again, a 10% sample was taken, adding 5,759 agents to the demand. This part of the demand is important to get more realistic traffic volumes especially on highways. These agents of our population have no option to switch from mode car to non-car. We will refer to them as "transit" traffic in the following paragraphs.

To specify opening and closing times for the facilities where activities are performed, activities are classified by type, that is, it is distinguished between home, work, education, shop and leisure activity types. Opening and closing times for the facilities where those types are performed are shown in Table 13.1.

### 13.4.3 Simulation Runs and Base Case

The simulation is run for 250 iterations, to retrieve a relaxed state in which the initial plans are adapted to the traffic conditions. In each iteration, 10% of the agents adapt routes and 10% adapt activity times. With the remaining probability of 80%, agents select one of the existing plans as described in Sect. 2.5. In doing so they can choose between modes car or non-car. This is done until iteration 200 is reached. In the last 50 iterations route and time adaption is switched off and 100% of the agents select plans based on the experienced performance. Thus all agents stop to try new options and return to the plans that have been experienced as best ones. The used parameter values are summarized in Table 13.2.

For the Zurich region, data from 159 traffic counting stations is available. The hourly measured traffic volumes can be compared with the amount of traffic of the

$$\beta_{\text{non-car}} = -3\,\text{Euro}\,\text{h}^{-1}$$

scenario simulation runs. This comparison is shown in Fig. 13.2. Most important is the curve using squares for data point representation which is calculated for each hour by following formula:

**Table 13.1** Activity opening and closing times used in the scenario

| Activity type | Opening time | Closing time |
|---|---|---|
| Home | 00:00 | 24:00 |
| Work | 06:00 | 20:00 |
| Education | 06:00 | 20:00 |
| Shop | 08:00 | 20:00 |
| Leisure | 00:00 | 24:00 |

**Table 13.2** Behavioural parameters used in the scenario

| Parameter | Value |
|---|---|
| $\beta_{\text{perf}}$ | 6 Euro $\text{h}^{-1}$ |
| $\beta_{\text{car}}$ | −6 Euro $\text{h}^{-1}$ |
| $\beta_{\text{non-car}}$ | −3 and −6 Euro $\text{h}^{-1}$ |
| $\beta$ (existing plans) | 4 |

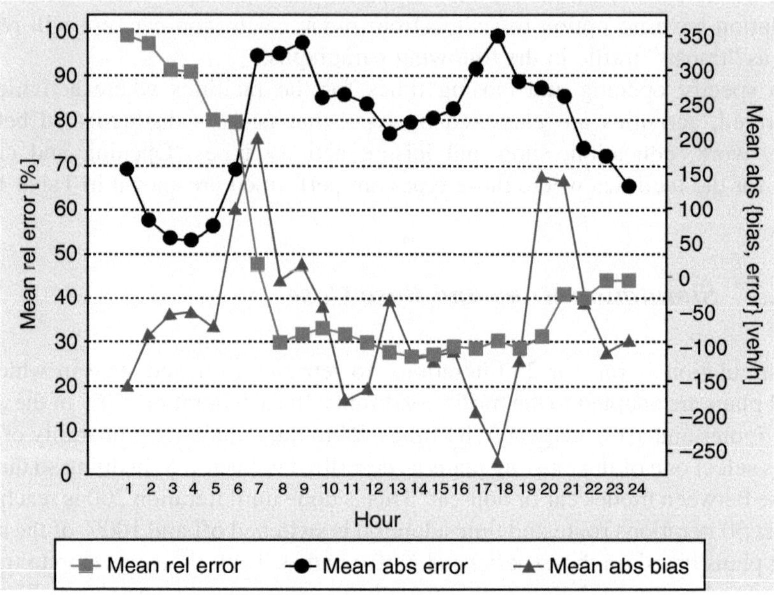


**Fig. 13.2** Realism of the $\beta_{\text{non-car}} = -3\,\text{Euro}\,\text{h}^{-1}$ simulation run. One hundred and fifty nine traffic counting stations provide real traffic counts for the Zurich area. The three curves show mean relative error (*squares*), mean absolute error (*dots*) and the mean absolute bias (*triangles*) when comparing the traffic volumes of the base case with the real values

$$\frac{\text{Simulated traffic volume} - \text{Real traffic volume}}{\text{Real traffic volume}} * 100.$$

During the night, that is from 00:00 a.m. till 07:00 a.m., the simulation deviates from reality with 50 to 100%. However the simulation results for the daytime, that is from 07:00 am till 09:00 pm, have a relative deviation of about 30%. After 09:00 pm the deviation is 40% or slightly higher.

## *13.4.4 Sensitivity*

In order to test the model's sensitivity, exactly the same set-up was run again, from the same initial conditions, but this time with a $\beta_{\text{non-car}}$ of $-6$ Euro $\text{h}^{-1}$. Since at this point the model does not differentiate between public transit and other non-car modes, such a change is a bit difficult to interpret in practical terms, but it might be loosely taken as a price increase of all non-car modes.

**Table 13.3** Results for the Zurich scenarios with a $\beta_{\text{non-car}}$ of −3 and −6

| | $\beta_{\text{non-car}} = -3$ | $\beta_{\text{non-car}} = -6$ | Difference |
|---|---|---|---|
| Size of population | 181,725 | 181,725 | 0 |
| Avg. trip duration (s) | 714 | 716 | +2 |
| Avg. score (EUR) | 177.36 | 176.53 | −0.84 |
| Car rate (%) | 42.31 | 60.25 | +17.94 |
| Non-car rate (%) | 54.62 | 36.68 | −17.94 |
| Transit rate (%) | 3.07 | 3.07 | 0 |

The third column displays the difference, that is, values of $\beta_{\text{non-car}} = -3$ scenario are subtracted from the scenario for $\beta_{\text{non-car}} = -6$

The results of the two simulation runs of the Zurich scenario are summarized in Table 13.3. The first and second column contain the data for the $\beta_{\text{non-car}}$ of −3 and −6, respectively. The third column contains the difference between the $\beta_{\text{non-car}} = -3$ and the $\beta_{\text{non-car}} = -6$ values.

The first line contains the number of agents used for simulation. The next two lines contain indicators for the performance of the system as a whole. While the average trip duration increases a non-significant 2 s, the average score decreases by 0.84 Euro. This is plausible, since an effective price increase in the system without compensation elsewhere needs to lead to a decrease in utility.

### *13.4.5 Winner–Loser Analysis*

It is immediately possible to identify the winners and losers of a policy. An example of such an analysis, which shows the spatial distribution of losers, is Fig. 13.3. The map pictures the greater Zurich area, whereby each dot or cross symbolizes an agent's home location. Colorization is based on the relative change using the $\beta_{\text{non-car}} = -3$ scenario as base and the $\beta_{\text{non-car}} = -6$ scenario as compare case. Crosses stand for agents that lose more than 1% utility.

A look at the spatial distribution of the losers in Fig. 13.3 shows that losers are more likely to reside at the border of the city area. In the base scenario, where $\beta_{\text{non-car}}$ is set to −3, non-car travelling is a profitable option for them. Changing $\beta_{\text{non-car}}$ to −6 forces them to either "pay" more, or to switch to car, which results in more congestion on streets and thus longer travel times for everybody.

This is reflected by Fig. 13.4. Gray dots symbolize home locations of agents that stay at the chosen mode despite the change of $\beta_{\text{non-car}}$. Crosses depict the agents changing from car to non-car while car pictograms stand for the contrary mode swap. One can see that most of the agents living in the city area stay with their original non-car mode. Residents living at the borders of the metropolitan area tend to switch more often from non-car to car. The reason is that most of their trips

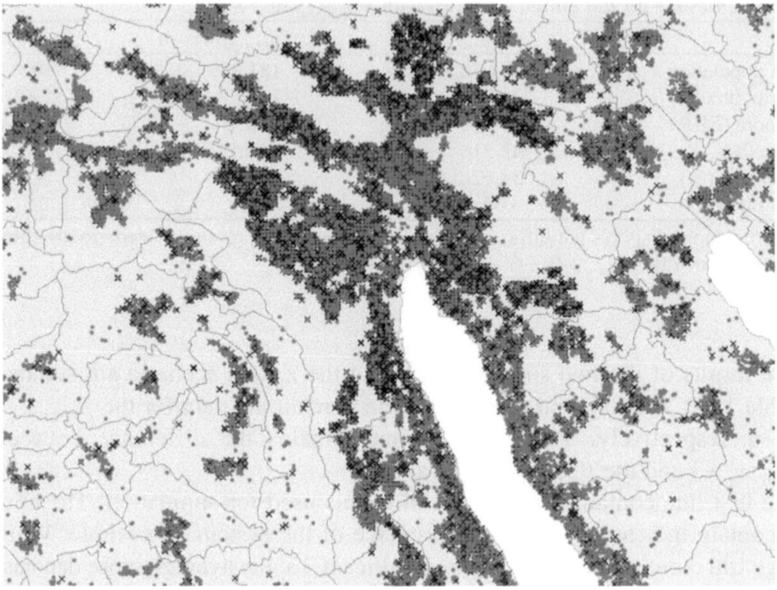

**Fig. 13.3** Map of the greater Zurich area. Each dot or cross locates a home location of an agent. Crosses stand for agents losing more than 1% of utility due to the raise of $\beta_{\text{non-car}}$

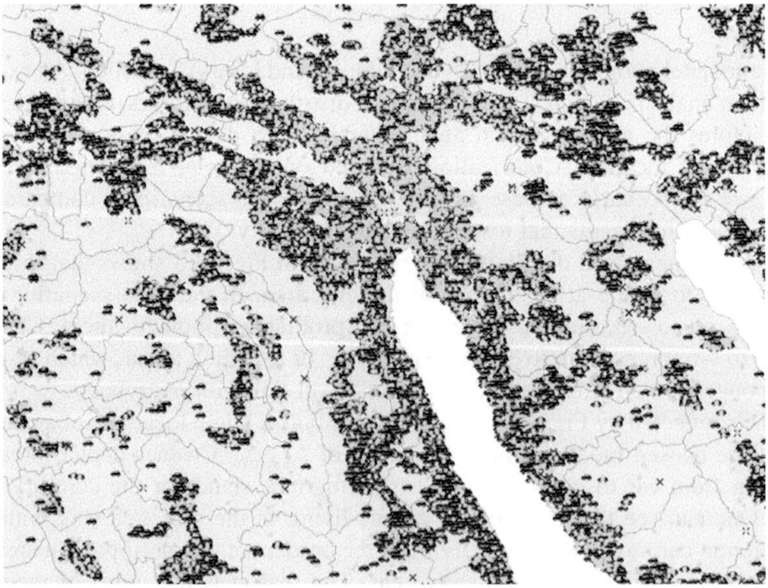

**Fig. 13.4** Map of the greater Zurich area. Dots, car pictograms and crosses symbolize home locations of agents in respect to the mode change due to the raise of $\beta_{\text{non-car}}$. Note that agents living at the border of the metropolitan area are most likely forced to switch from non-car to car (car pictograms)

are longer than the trips of their inner city counterparts. And with longer trips, the change of $\beta_{non-car}$ has a stronger effect, and the time advantage of the car (in non-congested conditions) plays a larger role.

These images are meant as a first illustration of what will be possible with microscopic methods. Future analysis should probe more deeply into the details of the behavioural mechanics. For example, one could imagine the following approach:

1. Introduce the policy measure but force every traveller to behave as before. This should identify those people who presumably feel most threatened by a policy measure; let us call them "directly affected".
2. Allow all agents to re-compute their route choice *once*. In the example discussed in this chapter, this would presumably lead to a relatively large initial mode change reaction.
3. Let the system relax along all relevant choice dimensions by doing a large number of iterations. In the example discussed in this chapter, this would presumably lead to some of the initial mode changes being reversed, because of increased congestion in the car mode. The final result would identify the distribution of winners and losers after the system has adapted to the policy measure. It may be important to find out in how far the gains and losses have shifted from those that were "directly" affected to those that are now indirectly affected (for example, via increased car mode congestion).

## 13.5 Conclusion

In this chapter, a mode choice model for the MATSim framework was presented. The model provides a possibility to analyse car vs. non-car travel decisions. To achieve this, our scoring function was extended by one parameter, $\beta_{\text{non-car}}$. In addition, initial plans using the non-car mode were generated by assuming that they take approximately twice as long as the car in an empty network. Nothing more is needed to simulate mode choice. The parameter $\beta_{\text{non-car}}$ can be interpreted as the agents' disutility of using the non-car mode; it needs to be compared with a similar parameter for the car, $\beta_{\text{car}}$.

The model was applied to the city of Zurich. Starting from a plausible base case, the parameter $\beta_{\text{non-car}}$ was doubled. As expected, car usership went up. Because of the agent-based approach, it was easy to allocate gains and losses to the agents' home locations; the result was shown in a graphical way. Similarly, the geographic distribution of mode switchers was shown.

In the longer run, the simple model for the non-car mode will be replaced by a detailed model that includes the effects of the actual public transit schedule. That model will then be able to compute the populations' reaction to changes in the schedule, the routing, the fare system, etc.

**Acknowledgments** This work was funded in part by the Volvo Research and Educational Foundations within the research project "Environmentally oriented Road Pricing for Livable Cities", and funded in part by the German ministry for research and education (BMBF) within the research project "adaptive traffic control" (ADVEST).

## References

Arentze TA, Timmermans HJP (2000) ALBATROSS: a learning-based transportation oriented simulation system. EIRASS (European Institute of Retailing and Services Studies), TU Eindhoven, NL

Arnott R, de Palma A, Lindsey R (1990) Economics of a bottleneck. J Urban Econ 27(1):111–130

Arthur B (1994) Inductive reasoning, bounded rationality, and the bar problem. Am Econ Rev (Papers and Proceedings) 84:406–411

Avineri E, Prashker JN (2003) Sensitivity to uncertainty: need for paradigm shift. Transport Res Rec 1854:90–98

Balmer M, Raney B, Nagel K (2005) Adjustment of activity timing and duration in an agent-based traffic flow simulation. In: Timmermans HJP (ed) Progress in activity-based analysis. Elsevier, Oxford, UK, pp 91–114

Balmer M, Axhausen KW, Nagel K (2006) A demand generation framework for large scale micro simulations. Transport Res Rec 1985:125–134, doi: 10.3141/1985–14

Ben-Akiva M, Lerman SR (1985) Discrete choice analysis. MIT Press, Cambridge, MA

Beuck U, Nagel K, Justen A (2007) Application of the VISEVA demand generation software to Berlin using publicly available behavioral data. Paper 07–2807. Transport Res Board Annual Meeting, Washington, DC

Bhat CR, Guo JY, Srinivasan S, et al. (2004) A comprehensive econometric microsimulator for daily activity-travel patterns (cemdap). Transport Res Rec 1894:57–66

Bliemer M (2003) Analytical dynamic traffic assignment with interacting user-classes: theoretical advances and applications Using a variational inequality approach. PhD thesis, Technical University Delft, NL

Bottom JA (2000) Consistent anticipatory route guidance. PhD thesis. Massachusetts Institute of Technology, Cambridge, MA

Bowman JL, Bradley M, Shiftan Y et al. (1999) Demonstration of an activity-based model for Portland. In: World transport research: selected proceedings of the 8th world conference on transport research 1998, vol 3. Elsevier, Oxford, pp 171–184

Cetin N, Burri A, Nagel K (2003) A large-scale agent-based traffic microsimulation based on queue model. In: Proceedings of Swiss transport research conference (STRC), Monte Verita, CH. See http://www.strc.ch. Earlier version, with inferior performance values: Transportation Research Board Annual Meeting 2003 paper number 03–4272

Charypar D, Nagel (2005) Generating complete all-day activity plans with genetic algorithms. Transportation 32(4):369–397

Daganzo CF (1998) Queue spillovers in transportation networks with a route choice. Transport Sci 32(1):3–11

de Palma A, Marchal F (2002) Real case applications of the fully dynamic METROPOLIS tool-box: An advocacy for large-scale mesoscopic transportation systems. Netw Spatial Econ 2 (4):347–369

Drossel B (2001) Biological evolution and statistical physics. Preprint arXiv:condmat/0101409v1, arXiv.org

Ettema D, Tamminga G, Timmermans H et al. (2003) A micro-simulation model system of departure time and route choice under travel time uncertainty. In: Proceedings of the meeting of the international association for travel behavior research (IATBR), Lucerne, Switzerland. See http://www.ivt.baug.ethz.ch

Gawron C (1998) Simulation-based traffic assignment. PhD thesis, University of Cologne, Cologne, Germany. URL http://www.zaik.uni-koeln.de/AFS/publications/theses.html
Gerlough DL, Huber MJ (1975) Traffic flow theory. Special report No. 165. Transportation Research Board, National Research Council, Washington, DC
Goldberg DE (1989) Genetic algorithms in search, optimization and machine learning. Addison-Wesley, Reading, MA
Hofbauer J, Sigmund K (1998) Evolutionary games and replicator dynamics. Cambridge University Press, Cambridge
Holland JD (1992) Adaptation in natural and artificial systems. Bradford Books. Reprint edition
Hraber P, Jones T, Forrest S (1994) The ecology of echo. Artif Life 3(3):165–190
Jacob RR, Marathe MV, Nagel K (1999) A computational study of routing algorithms for realistic transportation networks. ACM J Exp Algorithm 4 (1999es, Article No. 6). URL http://jea.acm.org/ARTICLES/Vol4Nbr6/
Jara-D′az S, Munizaga MA, Greeven P et al. (2004) The activities time assignment model system: The value of work and leisure for Germans and Chileans. Presented at the European Transport Conference (ETC), Strasbourg
Jara-Díaz SR, Guerra R (2003) Modeling activity duration and travel choice from a common microeconomic framework. In: Proceedings of the meeting of the International Association for Travel Behavior Research (IATBR), Lucerne, Switzerland. See http://www.ivt.baug.ethz.ch
Kaufman DE, Wunderlich KE, Smith RL (1991) An iterative routing/assignment method for anticipatory real-time route guidance. Technical Report IVHS Technical Report 91–02, University of Michigan Department of Industrial and Operations Engineering, Ann Arbor, MI 48109, May 1991
Lefebvre N, Balmer M (2007) Fast shortest path computation in time-dependent traffic networks. In: Proceedings of Swiss Transport Research Conference (STRC), Monte Verita, CH, September 2007. See http://www.strc.ch
Lin D-Y, Eluru N, Waller ST et al. (2008) Integration of activity-based modelling and dynamic traffic assignment. Transport Res Rec 2076:52–61
Lohse D (1997) Verkehrsplanung, volume 2 of Grundlagen der Straßenverkehrstechnik und der Verkehrsplanung. Verlag für Bauwesen, Berlin
Lohse D, Schiller C, Teichert H et al. (2006) Ein zweiseitig gekoppeltes Modell zur simultanen Berechnung der Verkehrserzeugung, Verkehrsverteilung und Verkehrsaufteilung: theoretischer Hintergrund und praktische Anwendung für ein nationales Modell der Schweiz. Verkehrsforschung Online 3(1):1
Mahut M, Florian M, Tremblay N (2003) Space–time queues and dynamic traffic assignment: a model, algorithm and applications. Paper, Transportation Research Board Annual Meeting, Washington, DC
MATSIM www page. MultiAgent Transport SIMulation. http://matsim.org/, accessed 2008. URL http://www.matsim.org
Meister K, Rieser M, Ciari F, et al. (2008) Anwendung eines agentenbasierten Modells der Verkehrsnachfrage auf die Schweiz. In: Proceedings of Heureka 08, Stuttgart, Germany, March 2008
Nagel K, Strauss M, Shubik M (2004) The importance of timescales: Simple models for economic markets. Physica A 340(4):668–677
Ortúzar J de D, Willumsen LG (1995) Modelling transport. Wiley, Chichester
Palmer RG, Brian Arthur W, Holland JH et al. (1994) Artificial economic life: a simple model of a stock market. Physica D 75:264–274
Peeta S, Ziliaskopoulos AK (2001) Foundations of dynamic traffic assignment: the past, the present and the future. Netw Spatial Econ 1(3):233–265. doi:10.1023/A:1012827724856
Pendyala RM, Kitamura R (2004) FAMOS: The Florida activity mobility simulator. In: Timmermans (2005). PTV www page. Planung Transport Verkehr. See http://www.ptv.de, accessed 2004

Raney B, Nagel K (2004) Iterative route planning for large-scale modular transportation simulations. Future Gener Comput Syst 20(7):1101–1118

Raney B, Nagel K (2006a) An improved framework for large-scale multi-agent simulations of travel behavior. In: Rietveld P, Jourquin B, Westin K (eds) Towards better performing European Transportation Systems. Routledge, London. Similar version TRB preprint number 05–1846

Raney B, Nagel K (2006b) An improved framework for large-scale multi-agent simulations of travel behavior. In: Rietveld P, Jourquin B, Westin K (eds) Towards better performing European Transportation Systems. Routledge, London, pp 305–347

Reinhold T (2006) Konzept zur integrierten Optimierung des Berliner Nahverkehrs. In: Öffentlicher Personennahverkehr. Springer, Berlin, 2006. doi: 10.1007/3–540–34209–5 8

Sheffi Y (1985) Urban transportation networks: Equilibrium analysis with mathematical programming methods. Prentice-Hall, Englewood Cliffs, NJ

Timmermans HJP (ed) (2005) Progress in activity-based analysis. Elsevier, Oxford, UK

URBANSIM www page, accessed 2007. URL http://www.urbansim.org

# Chapter 14
# Complex, Adaptive Systems, Through Time and Across Space

## Alberta Power Generation

**Kevin Seel and Nigel Waters**

> Given how rumours drive markets, and the way investors flock like sheep and follow the words of various gurus [traditional, market equilibrium theory] ... is clearly unrealistic (New Scientist, Editorial, 2008, 199, #2665, p. 5).
>
> Dynamics and surprise are everything. Waldrop (1992, p. 271).
>
> "Imagine what happened to my Tatiana? She up and rejected Onegin ... I never expected it of her!" Pushkin.

## 14.1 Introduction

Complexity does not mean complicated (Nijkamp 2007; Waldrop 1992, pp. 11–12). Confusion arises since complexity is a word in common use but in science it has a special meaning (O'Sullivan 2004). The first part of this chapter will consider the various definitions of complexity that have appeared in the literature. The second part will discuss a case study of the deregulation of the Alberta, Canada, electrical power generation industry, illustrating the dynamics of a complex system.

During the past few decades the geographical, social (especially economics; see the pioneering work of Arthur 1989) and natural sciences (biology, chemistry and physics) have all been subject to calls for a greater use of the methods and approaches of complexity science for, according to Richards (2002, p. 99), "complexity is one of the fastest growing and [most] pervasive branches of science". Despite flickering fires of enthusiasm for the use of these techniques over the years (Mayer 1990), these calls have not produced a consistent and sustained body of research. There is no overwhelming paradigm shift. Two recent articles, one in physical geography and one in human geography have urged a renewed interest

N. Waters (✉)
Department of Geography, George Mason University, Fairfax, VA, USA

A. Reggiani and P. Nijkamp (eds.), *Complexity and Spatial Networks*, Advances in Spatial Science,
DOI: 10.1007/978-3-642-01554-0_14, 

(O'Sullivan 2004; Richards 2002) and now formal statements recommending the development of complexity in the various sub-disciplines of geography are common (for example, in land development, Doak and Karadimitriou 2007, in economic geography, Boschma and Martin 2007, and in the study of space and place, Manson and O'Sullivan 2006, among others).

This chapter will review these two commentaries in particular and provide a critique of complexity as it is being used in geography and geographic information science and allied disciplines. In the past, researchers have adopted a diversity of attitudes toward complexity science. Few were initially supportive, many were openly hostile (Brian Arthur's seminal 1989 paper was rejected by three leading journals: Waldrop 1992), and some were prescient (Hicks arguing in 1939 that it would lead to the "wreckage… of the greater part of economic theory"), while most, it will be argued here, have been far too cautious in assessing the potential of complexity analysis. Even now, it is not the majority of researchers that appreciates the rich cornucopia of rewards that awaits those scholars willing to combine the tools of complexity science with the traditional methods of their disciplinary sciences.

## 14.2 Part 1: Complexity Defined

Manson (2001), states that complexity has been defined in three ways: algorithmic complexity; deterministic complexity and aggregate complexity. Others, including Richards (2002, p. 99) have suggested that complexity "may be viewed simply as incorporating the continuum between 'order' and 'chaos'". Here we will follow Manson's framework in analysing how complexity science has in the past, and how it might in the future, contribute to geographic information science. In particular, we will pay attention to the problem of spatializing system dynamics.

### *14.2.1 Algorithmic Complexity*

Although O'Sullivan (2004, p. 283) subsequently dismisses algorithmic complexity as having "no obvious application to geography", and presumably to social science in general, Manson (2001) is less hasty in his judgment noting that algorithmic complexity has two components. The first component concerns the difficulty of solving a problem couched in mathematical terms. This aspect of algorithmic complexity, Manson suggests, is less useful to geographers and yet solution complexity, frequently expressed in terms of the so-called "big O notation" (Black 2007), has been, for example, a *sine qua non* for those social scientists presenting at the triennial ISOLDE (International Symposium on Locational Decisions, http://isolde.geog.ucsb.edu/isoldeXI_about.php) Conferences ever since their inception in 1978.

ISOLDE researchers (an unusual mix of primarily geographers and operations research specialists) frequently begin their talks by stating how their new algorithm solves a problem in a fast, polynomial time when the problem was previously thought to be intractable other than by complete enumeration. Since there are vast classes of problems in transportation geography (Waters 2005) where this is important it is difficult to support O'Sullivan's dismissive view.

Manson is more positive concerning the second aspect of algorithmic theory, which relates to the simplest computational representation of a system that will reproduce its behaviour. Determination of this manifestation of the system structure is usually aided by information theory which has been more widely deployed in social science research. Manson promotes the importance of the classification of remotely sensed imagery and the impact of ecological structure on biodiversity, a topic that has seen a resurgence of interest with Costanza and Voinov's (2004) work on landscape simulation modeling (see the discussion below). Despite his support for this research, Manson ignores the work of Wilson's entropy maximizing models (1974) and its associated and extensive literature, including the ubiquitous, four-step transportation planning model that became associated with this work.

Manson (2001, p. 406) argues that a major limitation of the application of algorithmic complexity to social and environmental problems is that it "may incorrectly equate data with knowledge". This is similar to Clifford Geertz's plea for "thick description", a concept he popularized but which he acknowledged was originally developed by Ryle (Geertz 1973, p. 6). Interestingly, Ryle in explaining the language of philosophy uses cartography as a simile (see Tanney 2007). It will be shown in the case study in the second part of this chapter that providing electricity users with continuous data on the cost of the electricity consumption can help to optimize the functioning of the system, dampening the effect of the positive feedback loops within the system.

Likewise in Plato's *Phaedrus*, Socrates tells the story of how the god, Theuth, placed his invention of writing before the Egyptian king Thamus. The king argued that the new technology did not guarantee wisdom but merely "a conceit of wisdom" (Rockwell 1999; Postman 1993). From the vantage point of more than two millennia we can see now that writing did indeed provide the opportunity for endless wisdom, although few would deny that the loss of an oral tradition had its own problems and did much to disparage traditional, indigenous knowledge; Clayton and Waters 1999).

Debate over the value of data per se has reached the popular press with discussions over the value of Google's search engine and its ability to provide instantaneous access to vast repositories of data and articles. Carr (2008) asks rhetorically and provocatively whether Google is making us stupid and he too alludes to Plato's *Phaedrus*. That technology changes us is generally accepted in such modern folklore as Fubini's "Law" (Herremans et al. 2007) but in an era where data mining tools and software are widely used throughout industry and indeed academia (Miller and Han 2008; Miller 2007) it is surprising to see Manson ignoring this treasure trove of applications of the tools of complexity and related forms of analysis.

Again the popular press appears to be rushing in enthusiastically where academic angels fear to tread. Wired Magazine (2008, p. 7) introduced a series of short articles on data mining of massive petabyte data sets with this comment: "Solving scientific problems used to require grand theories. Now it just requires number crunching. Welcome to the Petabyte Age". Examples that are provided include the Europe Media Monitor (EMM 2008). This website tracks news in 35 languages worldwide. A graph is provided that shows the top ten stories over the previous four hours. The EMM website may itself become part of the news system since news organizations may visit the site to know the stories they should reference, thus setting up a positive feedback loop. Indeed this is exactly what occurred when, in September, 2008, a Google news bot crawled on to the Sun Sentinel's website and picked up a 6-year-old story from December 10, 2002, stating that United Airlines was seeking bankruptcy protection. The article was accessed by Income Security Advisors and distributed to a Bloomberg stock market information site. Since the story was assumed to be current United's stock fell 75% from $12 to $3 before rebounding when the truth became known.[1]

Consequently, we disagree with Manson (2001) and argue that data is indeed knowledge, it is just a different kind of knowledge. This aspect of complexity is already extensively exploited and will have increasing importance in the coming years. Excellent examples are provided in the Handbook of Geographic Information Science (Fotheringham and Wilson 2007) by Skupin and Fabrikant (2007) and by Gahegan (2007). Their chapters on Spatialization and multivariate geovisualization show how data mining techniques can reveal patterns in the data with no understanding of the intrinsic characteristics of the variables (Waters 2009). Fubini's so called "Law", mentioned above, argues that people initially use new technologies to do the same things as before only faster. Eventually they come to use the technology to do new things and these new things change the way society functions and indeed change society itself: hence first we change technology and then technology changes us (Herremans et al. 2007).

In the case study discussed in Part 2 of the chapter we show how data relating to the real time cost of electricity can be used by consumers to curb peak demand and dampen the boom-and-bust cycles in the construction of generating capacity.

### *14.2.2 Deterministic Complexity*

For Manson (2001) deterministic complexity has four primary characteristics: (1) the use of deterministic mathematics and mathematical attractors; (2) feedback processes, both positive and negative; (3) sensitivity to initial conditions; and (4) chaos. It is deterministic complexity that would appear to hold most promise

---

[1] See http://www.alootechie.com/content/google-chicago-tribune-blame-each-other-collapse-united-airlines-stock-price.

for the social sciences and is the basis of our case study of the Alberta electrical power generation industry.

With respect to the first of these characteristics, May (1976, p. 460) provided a seminal treatment describing simple mathematical models of, for example, population growth (14.1):

$$X_{t+1} = \alpha X_t (1 - X_t), \tag{14.1}$$

where $X_t$ is the current population and $X_{t+1}$ is the future population that is dependent on $X_t$ $(0 < X < 1)$ and on $\alpha$, a growth rate parameter in the range: $0 < \alpha < 4$. When $\alpha$ lies in the range 1–3, the population of such a system exhibits equifinality in that it will settle on a given value equivalent to $(1–1/\alpha$ regardless of the initial population value at time t (Manson 2001). This final steady state value is known as an attractor. The population will die out if $\alpha$ is less than 1 and grow without check if $\alpha$ is larger than 4.

Despite the importance of these findings they are not included in the earliest texts on modelling in geography and related sciences (for example, Thomas and Huggett 1980; Burghes and Borrie 1981). Indeed their importance was not realized by the broader academic community until the 1980s. Bennett and Chorley (1978, pp. 395–397) are among the pioneers who did include an extensive discussion of the application of these models in physical–ecological systems.

The second aspect of deterministic flexibility includes feedback processes. Where there are negative feedback processes the system will exhibit stable behaviour such as that shown by an attractor. Positive feedback produces a situation where a population may grow until the system collapses or where the population declines at an exponential rate until it ceases to exist. Equation (14.1) describes a simple system and additional variables can both make the system more complex and introduce additional loops that exhibit complexity in their behaviour making prediction of the system's state increasingly difficult. Ford (1999a) describes a variety of real world systems that are suitable for modelling using a system dynamics approach and this is the approach used in our case study below.

When $\alpha$ is a little over 3.8 the system becomes completely random and chaotic and has no discernible attractor (Manson 2001; May 1976, p. 462). Since a deterministic equation describes the system behaviour it is not truly chaotic but is described as deterministically chaotic. Other values of $\alpha$ greater than 3 and less than 4 allow the system to oscillate between various attractors and become highly sensitive to small changes in the initial value of $\alpha$.

This third characteristic of deterministic complexity has been described in Wilson's work (Wilson 1981) on catastrophe theory and the study of rapid jumps in the system behaviour known as bifurcations. Despite the promise of this research in the early 1980s it did not spawn the expected paradigm shift and few researchers exploited the field, though Wolfram (2002) is a notable exception. Even the study of dynamic systems along with their structure and feedback processes while hugely popular at the time of the publication of the Club of Rome Report by Meadows et al. (1972) resulted in waves of criticism (Cole et al. 1973; Vargish 1980).

Once you have "cried wolf" it is difficult to regain trust and once the direst predictions of the original limits to growth study had proven unlikely to be realized it appeared hard for Meadows et al. (1992) to gain support for the more modest prognostications that were produced by their World3 models. These outcomes included the suggestion that human use of resources and production of pollutants had by the early 1990s reached unsustainable levels; that in a world where it was "business as usual" there would be an uncontrollable per capita decline in energy use, and food and industrial production; and that to prevent this there would have to be a reduction in population growth and material consumption together with an increased efficiency in the factors of production, both energy and materials. Meadows et al.'s 1992 book, *Beyond the Limits*, sounded a note of optimism, declaring that a sustainable future was possible but that it would take a revolution in our social systems and systems of productions, a revolution that would need to be as profound as the agricultural and industrial revolutions that went before. Moreover, it would require a revolution that was as rapid as prior revolutions were gradual.

What *Limits to Growth* and *Beyond Limits to Growth* provide is, in the words of Geoffrion (1976), insight and not numbers. This is a view echoed by Nijkamp and Reggiani (1995, p. 185) when they note the importance of the logical structure of the system dynamics models used by Meadows and her colleagues. In the years since the second of these volumes was published, there has been little cause for optimism and even the simplest and most straightforward of Meadows' findings have gone unheeded as population growth continues its unchecked rise.

In the same paper, Nijkamp and Reggiani describe the mechanics of the bifurcation process and review the literature on catastrophe theory but despite noting the initial popularity of this theory and despite citing an extensive literature they too indicate that applications of the methodology were difficult (Nijkamp and Reggiani 1995, pp. 185–186):

> A weak element in catastrophe models is that the identification, explanation and estimation of critical turning points is extremely difficult since their occurrence is too irregular to be captured with sufficient statistical evidence by normal time series. As a consequence catastrophe theory has often been used for illustrative expositions rather than for predictive purposes...

To some extent recent work with system dynamics models has made the study of rapid change in system outputs and behaviour easier to study (Ford 1999a; and see the discussion of the Alberta electrical power generation industry below) but realistic models that capture the structural properties of real world systems are time consuming to build.

Chaos and fractals, the fourth characteristic of deterministic complexity, have received widespread attention in the geographical literature. Perhaps the best explanation of the usefulness of these tools, at least in physical and environmental geography, is given by Phillips (1999, p. 19) who notes that "deterministically chaotic systems are sensitive to minor variation in initial conditions and to small perturbations, such that miniscule changes or variations grow over time". To say the least, this is disconcerting since minor errors and lack of precision in input parameters can produce drastically different output as recounted by Edward Lorenz (1963, 2002) and his description of the so-called "butterfly effect" and as explained in most descriptions of chaos theory (Richards 2002).

In urban geography the most exhaustive treatments have come from Batty and Longley (1994) who have shown that the boundaries of cities exhibit self-similarity across all scales. A recent discussion by Batty incorporates a more explicit treatment of the concepts of complexity as they relate to cities (Batty 2005). The property of scale invariance has been exploited in the recent analysis of networks of all types and especially for analysing the structure of the internet (Schintler et al. 2005; Waters 2006).

### *14.2.3 Aggregate Complexity*

Aggregate complexity relates to the interactions among components of a system. For O'Sullivan this is the most beguiling of Manson's definitions of complexity. It might be argued that the greater the interaction between entities in a system the greater the likelihood of system homogeneity. New modelling methodologies such as geographically weighted regressions are only necessary when the relationships between dependent and independent variables vary across space (Waters et al. 2007) and this is less likely to occur in a well integrated system. Across small well, integrated neighbourhoods variables that explain voting behaviour in the Canadian 2006 Federal Elections have been shown to be similar (Mawa 2009) but over larger regions and across the country powerful explanatory models may show considerable differences (Li 2009) making simple, global solutions unacceptable.

## 14.3 Part 2: Complexity Applied

### *14.3.1 Modelling Complexity through Time and Across Space: A Case Study of the Deregulated Alberta Electrical Power Generation Industry*

#### 14.3.1.1 Modelling Approaches: Privileging Space or Privileging Time

Costanza and Sklar (1985, p. 47) observe that complexity can be modelled and articulated in three ways: "Dynamic models are those articulated in time: spatial models are articulated in space; and compartment models are articulated in the system components (state variables)". Even today most models are articulated (realized and constructed) primarily in one of these three ways but most commonly either across space or through time. Susan Crow in a paper presented to the Fourth GIS and Environmental Modeling Conference held in Banff in 2000 (Crow 2000) describes the mechanics of these two primary methods of modelling complexity.

The first of these approaches, that which privileges space, can be implemented using ESRI's Model Builder (http://www.esri.com). The spatial modelling is explicit within the GIS framework but time is handled through the "stacks of maps" approach. The second method where time is paramount can be modeled using Costanza and Voinov's (2004) spatial modeling environment (SME). Here a system dynamics model is implemented for a given spatial unit and the model is then duplicated across a set of spatial cells. This is the approach used in the present study where three regional cells are modelled separately and then linked together.

We will now describe a similar approach to the modelling of the Alberta, Canada, electrical power generation industry. The model uses system dynamics together with an explicit spatial component to provide insights into how the deregulated electrical power generation market will evolve over the long term. Hirsh (1999) provided an early discussion that deregulation in the US electrical utility system might not be the panacea that eagonomics would have postulated. Indeed it might be argued that in moving to deregulated power markets Alberta paid little heed to the difficulties encountered in the UK and California (Watts 2001) and other jurisdictions and also ignored the lessons learned in the real estate, construction, mining, oil and gas and airline industries among others.

Building on the work of Baumol and Benhabib (1989), Berry (1991) has suggested that economies may be chaotic systems that can be described by deterministic, first order, nonlinear difference equations that produce extremely complex behaviours over time. Baumol and Benhabib (1989) state that there are various reasons for studying chaos including the importance of studying uncertainty in that small changes in system variables can have dramatic impacts on system behaviour (the butterfly effect) and the characterization of the full range of possible system behaviours including dramatic oscillations, the so-called boom-and-bust cycles. In addition, models that exhibit chaotic behaviour, as with all dynamic models, can be used to disprove universally held propositions such as "the allegation that profitable speculation is always and necessarily stabilizing.... Similarly, it was shown that slight lags in response can undermine apparently rational countercyclical policy" (Baumol and Benhabib 1989, p. 80). Our case study below supports both of these conclusions. Berry uses moving average techniques and Baumol and Benhabib's methodology for identifying chaotic behaviour to support his analysis of long-wave rhythms in the US economy. Writing in 1991 his best prediction was that the labour market would "bottom out" in 1995 with the next growth cycle beginning "in the first decade of the twenty-first century" (Berry 1991, p. 191). With hindsight we know that the global melt down was delayed by a little more than a decade, perhaps partially justifying King's criticism of Berry's book (King 1993). Nevertheless Berry's thesis that a dramatic collapse would occur has proven correct. That it has been exacerbated by investor "herd behaviour" or flocking (Waldrop 1992) and a lack of regulation of, for example, the derivatives market[2] provides support for the modelling approach used below and some of our conclusions.

It might have been suspected that decisions that were made by private market agents would be taken with their own selfish interests and profit maximization in mind rather than deferring to the interests of the public. What was perhaps less obvious is that these decisions were often neither rational nor well informed and reflected short term behaviour. Market participants have been influenced by perception, "herd" mentality and other psychological factors. The decisions of those involved in the power supply industry and the construction of electrical power generation plants often interact through complex, highly dynamic and counter-intuitive patterns of feedback that involve all of the other players in the system. One difficulty that the industry faced was the long lead times that power construction requires. Once a need is determined, once the demand is apparent for increased power generation and a decision is made to build a new power plant, that plant must be designed, the site must be selected, environmental regulations must be satisfied and the construction and commissioning of the plant have to occur, all before the power becomes available to the consumer. Consequently the time between the signal to act, the decision to act and the action being completed may be in the order of several years.

Other factors have confounded the supply of deregulated electrical power. These include the unusual properties of the "commodity" of electricity which is created and destroyed on demand, for the most part cannot be stored and is subject to complex processes of economics, engineering and physics, all of which interact in unpredictable ways. In modelling the behaviour of the deregulated system in Alberta a number of research questions were addressed. First, if the market functioned as intended why did investors not bring capacity into service to benefit from the extremely high power prices that existed in the year 2000 and later? Second, could investors be expected to bring online new power plants in a steady and timely fashion to keep pace with growth in Alberta's demand? Third, would power plant construction, by contrast, appear in a series of waves of boom and bust similar to other deregulated industries? The null hypothesis would be that the market would function as desired with stable prices (Fig. 14.1) while the alternative hypothesis would see "boom and bust" cycles appear in the amount paid for electricity (Fig. 14.2).

#### 14.3.1.2 A System Dynamics Simulation Model of the Deregulated Alberta Electrical Market

To build a realistic model of the deregulated Alberta electrical power generation market a system dynamics model was developed. This is the approach that was used

[2] See http://www.washingtonpost.com/wp-dyn/content/story/2008/10/14/ST2008101403344.html, the Washington Post article.

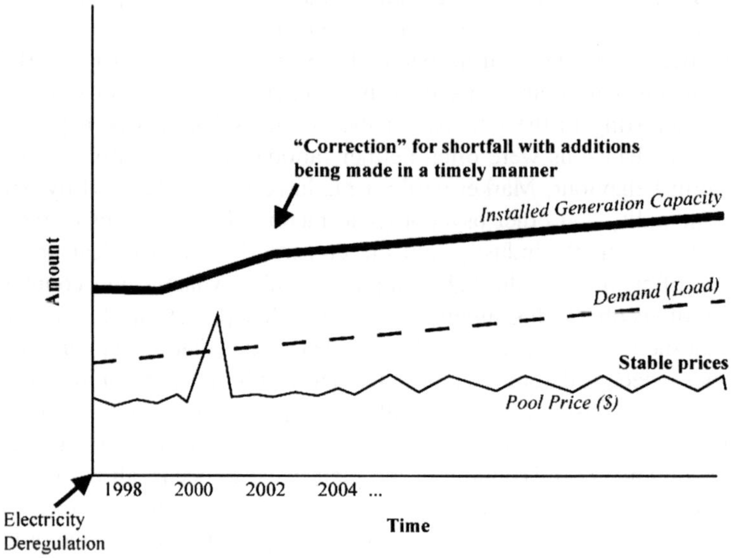


**Fig. 14.1** Alberta market functions as desired

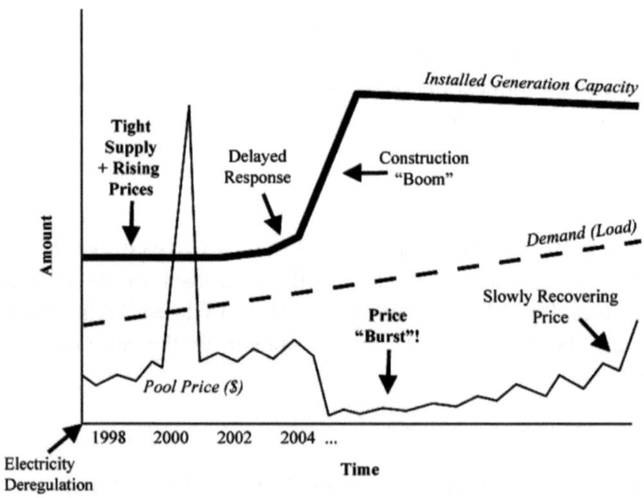


**Fig. 14.2** The Alberta market functioning with "boom and bust" cycles

in *The Limits to Growth* studies and derives from the work of Forrester (1971). The assumptions are that the system behaviour is determined by system structure and that this structure is comprised of interconnected two-way feedback loops. Creating a realistic model required: a detailed representation of the actual physical system that incorporated the key feedback dynamics between modelled elements; depiction of the various forms of investor behaviour and a method for testing strategies to minimize the effect of the anticipated boom and bust cycles. Following Geoffrion (1976) we were seeking insight not numbers. The model captures the supply side, wholesale markets and transmission components of the Alberta electrical power generation system as described in the Alberta Advisory Council on Electricity (AACE) Report (2002). The model was given the acronym APPCON (Alberta Power Plant CONstruction model) contains 4,000 lines of code and is programmed using the iThink software from High Performance Systems (1990). For complete details see Seel (2004).

The APPCON model incorporates the primary components of the spatial structure of the Alberta electrical power generation system including three regional models (for north eastern, central and southern Alberta with two transmission corridors between the first and second and the second and third of these regions) and interprovincial transfers, both east and west. Generation types modelled include coal, gas turbines, cogeneration, wind and hydro although there is no representation of any large hydro plant. Figure 14.3 shows the primary elements of the modular design of the model while Fig. 14.4 shows how the model is spatialized between the three regions. Figure 14.5 provides an overview of the core model dynamics showing the key feedback processes as causal loop diagrams.

Figure 14.6 shows all the primary feedback loops within the structure of the APPCON model. Two of these loops are "reinforcing", positive feedback loops while the remaining three are "balancing", negative feedback loops. The positive feedback loops are those that model Arthur's (1989, 1996) phenomenon of increasing returns. The behaviour of the model is both moderated and "complexified" by the influence of the two negative feedback loops. As Waldrop (1992, p. 36) has noted, when discussing Arthur's work: "that's why you get patterns in any system: a rich mixture of positive and negative feedbacks...it's the mix of these two forces that produces the complex pattern...". In the figures below the "+" signs indicate positive feedback and the "−" sign a negative feedback relationship between the variables. The double hatch marks on the connector arrows indicate a time delay, an important factor in reproducing suboptimal responses of the investors in the model.

Figure 14.6, illustrating the R1 positive feedback loop, shows how new generating capacity is added to the power generation system in response to increases in demand. As demand for electricity in one of the three regions increases, the Alberta Power Pool dispatches additional capacity thus lowering the capacity reserve margin. When the annual capacity margin drops to 15% of total installed capacity this is considered to be the signal for new investment in capacity to be introduced (CERI 2002). This does not immediately translate into new construction as new plants must be designed, financed and approved by the investors. The delay may be from one to several years as the new plant must be proposed, sited, permitted,

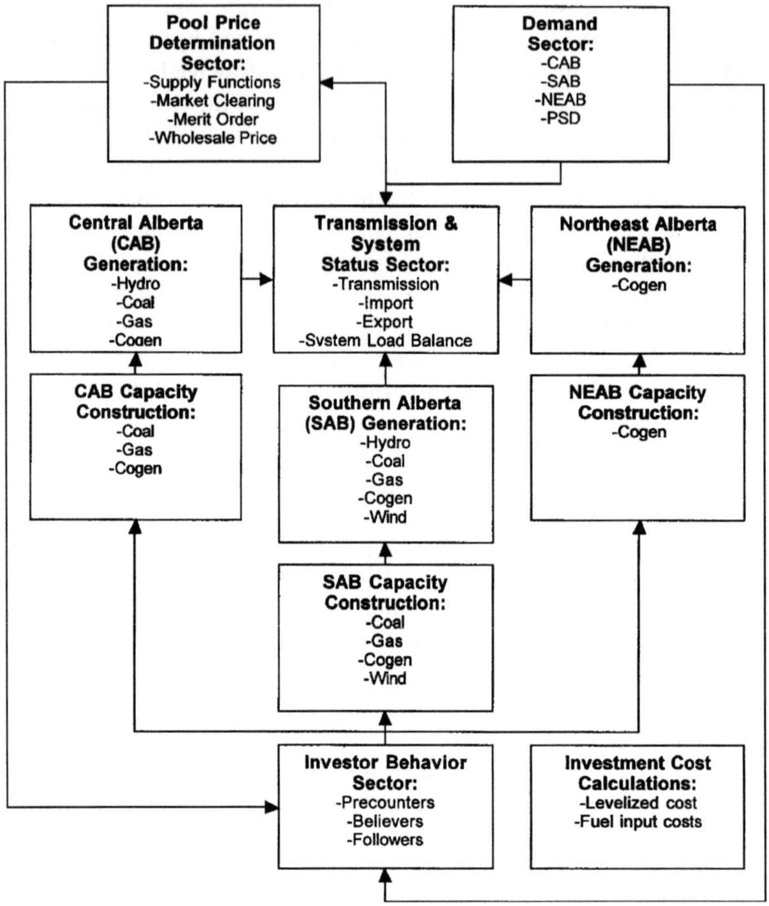


**Fig. 14.3** High level structure showing modular design of the APPCON model

constructed and commissioned before it is made operational and the new capacity becomes available to the consumer.

Linked to the R1 loop and shown in Fig. 14.6 is the reinforcing, positive feedback that describes how new generation additions are a function of potential profitability. Dispatched generation is organized in a "merit order" from least to most expensive. As demand continues to rise the dispatched power becomes increasingly expensive making investment in new generation capacity increasingly attractive. Since there is a lag of one to several years between the "signal" for increased capacity and the arrival of that capacity to the consumer the use of increasingly expensive generating capacity will continue making investment appear more and more desirable. The negative feedback loops are also shown in Fig. 14.6. The first, the B1 loop, models the introduction of new generating capacity

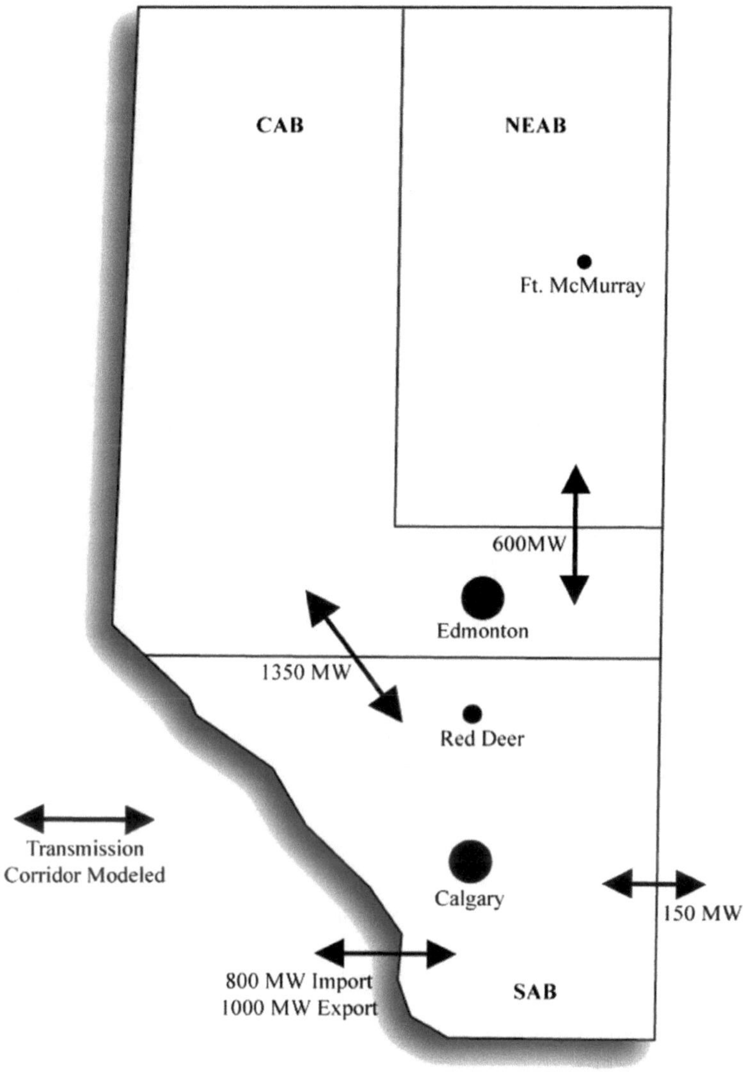


**Fig. 14.4** Alberta regions and transmission corridors modeled in APPCON

and will add to the reserve margin reducing the attraction of investing it yet more capacity.

The second balancing loop, B2, illustrates investor behaviour. As new generation is proposed and enters construction, it will have an influence on the propensity of investors to enter the market. The amount off influence depends on the level of "conservatism" of the investor. Three levels of conservatism are modelled, precounters, followers and believers, listed in order from most to least conservative.

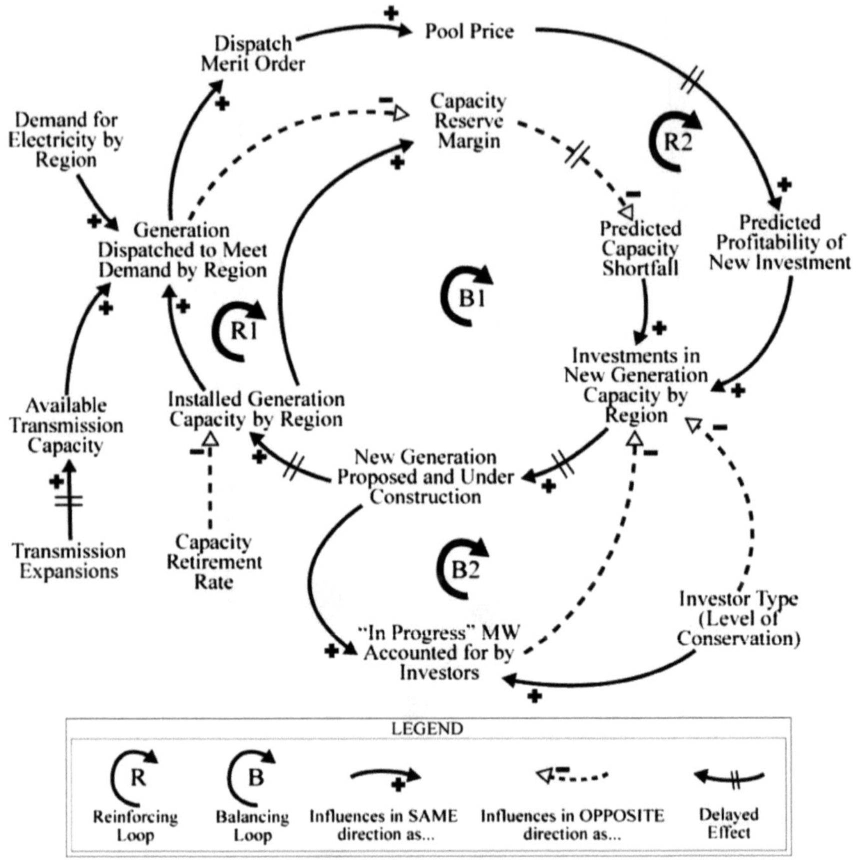


**Fig. 14.5** Causal loop diagram: primary feedback loops in the APPCON model

More is said about investor behaviour below. The model, as constructed to this point, has the potential to generate undesirable "boom-and-bust" price cycles. As demand for power rises with industry expansion and population growth, price will also rise, creating the price boom. This triggers new power plant construction that, due to construction lags, may lead to subsequent over expansion of the power supply as investors react to the opportunities provided by price spikes and hence the potential for the boom-and-bust cycles to appear. Two policy mitigation strategies were embedded in the model. The first part of this loop tests the effects of a real-time, conservation response by consumers to price spikes that occur during times of peak load. In this loop consumers respond to price spikes by curtailing demand, essentially by delaying appliance use and thus reducing peak demand and preventing the capacity reserve margin dipping below the critical 15% level. Although CERI (2002) has suggested that consumers already exhibit some of this behaviour,

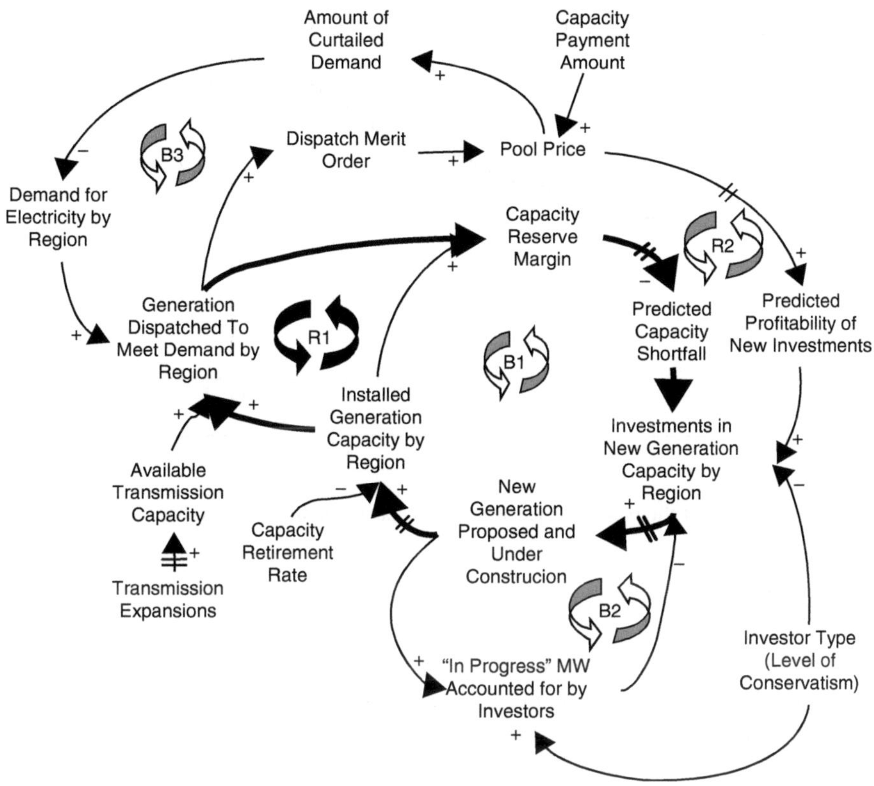


**Fig. 14.6** Causal loop diagram: showing all five feedback loops

this aspect of the model had the goal of testing a more formal and widespread consumer response to price spikes.

A second intervention strategy incorporated into the B3 loop was a continuous capacity payment (CCP) that would provide additional incentives for investors to bring capacity online earlier by providing additional revenue to cover the capital and other fixed costs not covered by the energy price. The CCP helped to spread the costs of price spikes over all hours of the day and over each day of the year.

### 14.3.2 *Modelling Investor Behaviour*

Following Ford (2001) three types of investor behaviours were included in the APPCON Model. The first group were the *believers* who are the most aggressive in that their investment behaviour ignored all power generating capacity that had been either approved or was under construction. *Precounters* were the most rational or

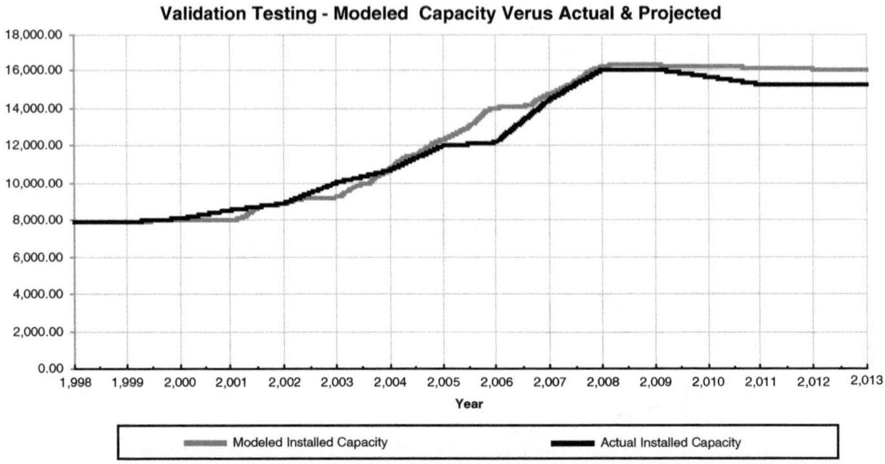


**Fig. 14.7** Validation of the base case scenario with capacity additions (90% fit)

conservative of the investor types modelled in that they took into account all capacity that was on hold, under approval or under construction. Finally, the third group of investors were the *followers* who exhibited a "herd mentality" or the "flocking" behaviour discussed by Waldrop (1992, p. 42). These investors would only commit after others had invested in plants that were already under construction.

Figure 14.7 shows one of a series of validation runs (see Seel 2004, for others) in which a base case scenario is projected into the future and additional capacity is added to the system. Over the historical record the model provides a 90% fit. Three experimental scenarios were initially explored: 100% precounters, 100% believers and a 50:50 mix of followers and precounters. The simulations were carried out over a 15 year time period using four time steps per day. The approval process took 12 months for all gas fired generation, 24 months for coal fired and 6 months for wind turbines. With gas and wind turbines 5% of the applications were refused while for coal this figure was 10%. Delays prior to construction were built into the process, 3 months for wind, 6 months for gas and coal fired generation. Construction itself took 6, 18, 36 and 48 months for wind, gas, cogeneration and goal power plants, respectively. Energy imports were set at 800 MW from British Columbia and 150 MW from Saskatchewan. All figures were based on historical analysis and experimentation and sensitivity analysis of earlier runs of the simulation model for the recent historical record.

#### 14.3.2.1 Results of the Simulations

For the 100% precounter scenario there is a correction in the southern Alberta (SAB) region in 2003 and then capacity comes online in a steady progression similar to what was expected under the null hypothesis shown in Fig. 14.1. The 100% believers' scenario exhibits two construction booms, the first in 2002–2005 and the second in 2011–2013. The mixed scenario with an even split of followers and precounters produces a huge boom in the years 2002–2003 and then an "echo" in 2006–2007.

Investigations of the product mix using the base case scenario exhibited a long "golden age" for gas up until 2007 and then a re-emergence of coal as a primary method of generation for the baseload.

The 2000–2001 price spikes occurred in all scenarios and are perhaps unavoidable but the presence of investors that assumed the follower behaviour caused the greatest dampening of the spikes while precounters resulted in the highest overall prices. Using a $10 per megawatt capacity payment produced the smoothest introduction of new capacity ahead of demand. The complexity of the behaviour of the system is revealed by the fact that values either above $10 or below this figure had little effect in smoothing out the price spikes.

Most interesting of all was the introduction of price sensitive demand into the model. This resulted in up to 750 MW of load being curtailed in the peak demand periods in 2001 and 2003 using the base case scenario. Consequently there was a much smaller, delayed construction boom that required almost 1,600 MW less capacity by 2013.

### *14.3.3 Discussion*

It appears likely that something similar to the rational, precounter mindset was initially anticipated when conceiving Alberta's deregulated power generation market. However, the APPCON model has shown that the presence of other investor behaviours can have a dramatic effect on market evolution and result in varying degrees of boom and bust cycles. The boom-and-bust outcome, and thus the alternative hypothesis (Fig. 14.2), was shown to exist for the base case scenario and also when it was modified with the believer and precounter: follower scenarios. The desired, null hypothesis (Fig. 14.1) that avoided the boom and bust oscillations was produced in the 100% precounter scenario and in the base case simulation that introduced capacity payments.

Rhetorically, our first research question may be posited again: why did investors fail to bring into production capacity that would take advantage of the extremely high prices that emerged in the year 2000 and later. The answer appears to be that the price spikes that occurred in 2000 and 2001 were unavoidable due to the long lead times that were required to construct new capacity in the post-deregulated market. It may be concluded, in concert with Ford (1999b), that the momentum of previous energy policies and the long delays before new remedies become fully effective implies that short-term supply problems and price volatility during the transition process cannot be solved but may only be weathered.

The second research question asked whether investors would bring online new power plant generation in a smooth and steady flow. The APPCON model showed that the market exhibited counter intuitive behaviour, or a "surprise" element, and that the boom-and-bust cycles could be mitigated through the use one of the following two strategies: first, a capacity payment of $10 per megawatt or second the introduction of conservation strategies by consumers in periods of peak demand (such a strategy assumes that the consumers are informed of the cost of electricity at all times).The ability of system dynamics models to yield results that have this surprise component is one of their most useful features (Forrester 1991; Mass 1991; and see Thompson et al. 1990, for a typology of surprise).

## 14.4 Conclusion

This chapter has reviewed and examined past studies of complexity in the social sciences with particular emphasis on those applications in geography where the ramifications of differences in behaviour across space may be significant. In the second part of the chapter one of the most promising methodologies discussed in the first part, system dynamics, has been extended to incorporate spatial variations in the deregulated market for electrical power in Alberta, Canada. This market exhibits many of the characteristics of those systems described in the discussion of complexity, including positive and negative feedback loops and counter-intuitive behaviour. A spatially explicit, system dynamics approach allows for an understanding and remediation of undesirable aspects of the price responses of this system. It is hoped that such approaches are a way forward for research in other systems that resist traditional, reductionist forms of analysis.

**Acknowledgments** The authors would like to acknowledge helpful comments from an anonymous referee and assistance in the preparation of a number of the figures by Krystal Workman.

## References

Alberta Advisory Council on Electricity (2002) Report to the Alberta Minister of Energy. Edmonton, Alberta

Arthur WB (1989) Competing technologies, increasing returns, and lock-in by historical events. Econ J 99:116–131

Arthur WB (1996) Increasing returns and the New World of Business. Harv Bus Rev 74(4):100–109

Batty M (2005) Cities and complexity: understanding cities with cellular automata, agent-based models, and fractals. MIT Press, Cambridge, MA

Batty M, Longley P (1994) Fractal cities: a geometry of form and function. Academic Press, New York

Baumol WJ, Benhabib J (1989) Chaos: significance, mechanism and economic applications. J Econ Perspect 3:77–105

Bennett RJ, Chorley RJ (1978) Environmental systems: philosophy, analysis and control. Methuen, London
Berry BJL (1991) Long-wave rhythms in economic development and political behaviour. The Johns Hopkins University Press, Baltimore, MD
Black P (2007) Big O notation. Dictionary of algorithms and data structures. Online at: http://www.nist.gov/dads/HTML/bigOnotation.html Accessed: July 14, 2007
Boschma R, Martin R (2007) Editorial: constructing an evolutionary economic geography. J Econ Geogr 7:537–548
Burghes DN, Borrie MS (1981) Modelling with differential equations. Ellis Horwood, Chichester, UK
Carr N (2008) Is Google making us stupid? The Atlantic 302(1):56–63
CERI (2002) Electricity price forecast for the Alberta integrated electrical system. Canadian Energy Research Institute, Calgary, Alberta
Clayton D, Waters N (1999) Distributed knowledge, distributed processing, distributed users: integrating case based reasoning and GIS for multicriteria decision making In: Thill J-C (ed) Multicriteria decision-making and analysis: a geographic information sciences approach. Ashgate, Brookfield, USA, pp 275–308
Cole H, Freeman C, Jahoda M et al. (1973) Thinking about the future: a critique of the limits to growth. Clark, Irwin and Co., Toronto
Costanza R, Sklar F (1985) Articulation, accuracy, and effectiveness of mathematical models: a review of freshwater wetlands applications. Ecol Model 27(1):45–68
Costanza R, Voinov A (eds) (2004) Landscape simulation modelling: a spatially explicit dynamic approach. Springer, New York
Crow S (2000) Spatial modeling environments: integration of GIS and conceptual modeling frameworks. http://www.colorado.edu/research/cires/banff/pubpapers/2/
Doak J, Karadimitriou N (2007) (Re)development, complexity and networks: A framework for research. Urban Studies 44:#2, 209–229
EMM (2008) Europe Media Monitor. http://emm.jrc.it/overview.html#labs;accessed July 14, 2008
Ford A (1999a) Modeling the Environment: an introduction to system dynamics modeling of environmental systems. Island Press, Washington, DC
Ford A (1999b) Cycles in competitive electricity markets: a simulation study of the Western United States. Energ Policy 27:637–658
Ford A (2001) Waiting for the boom: a simulation study of power plant construction in California. Energ Policy 29:847–869
Forrester JW (1971) World dynamics. MIT Press, Cambridge, MA
Forrester JW (1991) System dynamics and the lesson of 35 years. Available at: http://sysdyn.clexchange.org/sdep/papers/D-4224–4.pdf
Fotheringham AS, Wilson JP (eds) (2007) The handbook of geographic information science. Blackwell, Oxford, UK
Gahegan M (2007) Mulitivariate geovisualization, Chap. 16 In: Fotheringham, Wilson (eds) The handbook of geographic information science. Blackwell, Oxford, UK, pp 292–316
Geertz C (1973) The interpretation of cultures. Basic Books, New York
Geoffrion AM (1976) The purpose of mathematical programming is insight not numbers. Interfaces 7(1):81–92
Herremans L, Murch R, McKenzie J (2007) The realities and challenges of bringing global perspectives into the classroom through technology, Chap. 1. In: Cooke L (ed) Frontiers in higher education. Nova Publishers, Hauppauge, New York
Hicks JR (1939) Value and capital: an inquiry into some fundamental principles of economic theory. Clarendon Press, Oxford
High Performance Systems (1990) STELLA II documentation. High Performance Systems, Hanover, NH
Hirsh R (1999) Power loss: the origins of deregulation and restructuring in the American electric utility system. MIT Press, Cambridge, MA

King LJ (1993) Book review of Berry, 1991, op. cit. Ann Assoc Am Geogr 83:175–177

Li MX (2009) Predicting the pattern of votes in the 2006 Canadian Federal Election using socioeconomic variables in a multiple regression and geographically weighted regression. Unpublished MGIS Report, Department of Geography, University of Calgary

Lorenz EN (1963) Deterministic nonperiodic flow. J Atmos Sci 20(2):130–141

Lorenz EN (2002) Predictability – a problem partly solved, Chap 3 In: Palmer T, Hagedorn R (eds) Predictability of weather and climate. Cambridge University Press, Cambridge, UK

Manson SM (2001) Simplifying complexity: a review of complexity theory. Geoforum 32: 405–414

Manson SM, O'Sullivan D (2006) Complexity theory in the study of space and place. Environ Plann A 38:677–692

Mass NJ (1991) Diagnosing surprise model behavior: a tool for evolving behavioral and policy insights (1981). Syst Dynam Rev 7:68–86

Mawa JU (2009) Modeling the 2006 Federal Election outcomes with census 2006 variables. Unpublished MGIS Report, Department of Geography, University of Calgary

May RM (1976) Simple mathematical models with very complicated dynamics. Nature 261 (5560):459–467

Mayer L (1990) An introduction to quantitative geomorphology. Prentice-Hall, Englewood Cliffs, NJ

Meadows DH, Meadows DL, Randers J et al. (1972) The limits to growth. Universe Books, New York

Meadows DH, Meadows DL,Randers J (1992) Beyond the limits: confronting global collapse, envisioning a sustainable future. McClelland and Stewart, Toronto, Canada

Miller HJ (2007) Geographic data mining and knowledge discovery. Chap 19 In: Wilson JP, Fotheringham AS (eds) Handbook of geographic information science. Blackwell, Oxford, UK, pp 352–366

Miller HJ, Han J (2008) Geographic data mining and knowledge discovery, 2nd edn. Taylor and Francis, New York

Nijkamp P (2007) Comment, post workshop discussion. Institute Para Limes, Workshop, September 20–22, 2007, Lochem/Barchem, Holland

Nijkamp P, Reggiani A (1995) Non-linear evolution of dynamic spatial systems: the relevance of chaos and ecologically based models. Reg Sci Urban Econ 25:183–210

O'Sullivan D (2004) Complexity science and human geography. Trans Inst Br Geogr NS 29: 282–295

Phillips JD (1999) Earth surface systems: complexity, order and scale. Blackwell, Malden, MA

Postman N (1993) Technopoly. Knopf, New York

Richards A (2002) Complexity in physical geography. Geography 87(2):99–107

Rockwell G (1999) Is humanities computing an academic discipline? Seminar presentation to a workshop of that name held at the University of Virginia in Charlottesville, November, 1999. http://jefferson.village.virginia.edu/hcs/rockwell.html; last accessed July, 14, 2008

Schintler LA, Gorman SP, Reggiani A et al. (2005) Scale-free phenomena in communications networks: a cross-Atlantic comparison. In: Reggiani A, Schintler LA (eds) Methods and models in transport and telecommunications. Springer, Berlin, pp 201–220

Seel KC (2004) 'Boom and bust' cycles in power plant construction: a simulation study of the temporal and geographical aspects of the Alberta competitive electrical industry. Unpublished PhD Dissertation. Department of Geography, University of Calgary

Skupin A, Fabrikant SI (2007) Spatialization, Chap. 5. In: Fotheringham S and Wilson JP (eds) The handbook of geographic information science. Blackwell, Oxford, UK, pp 61–79

Tanney J (2007) Gilbert Ryle. Stanford Encyclopedia of Philosophy. http://plato.stanford.edu/entries/ryle/;last accessed July 14, 2008

Thomas RW, Huggett RJ (1980) Modelling in geography: a mathematical approach. Harper & Row, London

Thompson M, Ellis R, Wildavsky A (1990) Cultural theory. Westview Press, Boulder, CO

Vargish T (1980) Why the person sitting next to you hates *limits to growth*. Technol Forecast Soc Change 16:179–189

Waldrop MM (1992) Complexity: the emerging science at the edge of order and chaos. Simon and Schuster, New York

Waters NM (2005) Transportation GIS: GIS-T, Chap 59. In: Longley P, Goodchild M, Maguire D, and Rhind D (eds). Geographical information systems and science, Second Edition, Abridged, Wiley, New York

Waters NM (2006) Network and nodal indices: measures of complexity and redundancy: a review, chap 2. In: Reggiani A, Nijkamp P (eds) Spatial dynamics, network and modelling. Edward Elgar, Cheltenham, UK and Northampton, MA, USA, pp 13–33

Waters N (2009) The handbook of geographic information science, edited by John P. Wilson and A. Stewart Fotheringham. J Reg Sci 49 (2), 382–5

Waters NM, Hansen CV, Sun H (2007) Web-based GIS to explore media influence and election processes. ESRI Users Conference, San Diego, August 2007; available online at: http://gis.esri.com/library/userconf/proc07/papers/papers/pap_1845.pdf

Watts PC (2001) Heresy? The case against deregulation of electricity generation. Electricity J 14 (4):19–24

Wilson AG (1974) Urban and regional models in geography and planning. Wiley, New York

Wilson AG (1981) Catastrophe theory and bifurcation: applications to urban and regional systems. Croom Helm, London

Wired Magazine (2008) The petabyte age. Wired, 16, #7, 106–121. http://www.wired.com/science/discoveries/magazine/16–07/pb_intro;last accessed July 14, 2008

Wolfram S (2002) A new kind of science. Wolfram, Media Inc., Champaign, IL

# Chapter 15
# Measuring and Visualizing Urban Network Dynamics

## A GIS and Graph-Theoretic Approach

**Laurie A. Schintler and Giacomo Galiazzo**

## 15.1 Introduction

An urban area is a complex, dynamic system of networks through which information, capital and power propagate across and within nodes of activities. While innovations in information technology are making it easier for transactions in these networks to occur over greater distances, the importance of spatial proximity in such networks is still very much relevant. Economic, social and other types of benefits drive activities to co-locate, where one may view the process as one of preferential attachment. The physical agglomeration of activities that arises out this process, at any point in time, is what we characterize in this chapter as the "backbone" of region. We hypothesize that such a feature is not static, but rather, it shifts in space over time in response to changing constraints and circumstances.

In this chapter, we develop a technique to identify and visualize the backbone of an urban area. The approach creates a giant component based on the gridded distribution of activities in a region, from which a backbone is extracted based on the structurally most important grid cells, or nodes, that comprise the component. The conceptual underpinning that is used to establish the designation is based on social network theory. According to our model, a backbone is ultimately defined by the grid cells in the giant component that have the highest level of centrality. To demonstrate the viability of the approach and the insight that can be gained through its application, we apply the technique to the Metropolitan Statistical Areas (MSAs): Cedar Rapids (Iowa), Chicago and Phoenix. We look specifically at population and employment distributions for 1990 and 2000 using U.S. Census data summarized at the census tract level.

L.A. Schintler (✉)
School of Public Policy, George Mason University, Fairfax, VA, USA

A. Reggiani and P. Nijkamp (eds.), *Complexity and Spatial Networks*, Advances in Spatial Science,
DOI: 10.1007/978-3-642-01554-0_15, © Springer-Verlag Berlin Heidelberg 2009

## 15.2 Overview of Methodology

Again, the backbone of a region is identified through two processes: the designation of a giant component and the ranking of nodes in the component according to some measure of centrality. A giant component is, by definition, a set of regularly-shaped cells that are minimally connected, where the network topology of the component is defined by the adjacency of the cells in space. In our case, a component is defined more specifically by the set of cells that are minimally linked at some critical threshold for the density of the activity of interest. In particular, the densities of the cells that comprise the component are either equivalent or higher than the critical cut-off point.

One of the requisite inputs to the process is a grid that contains densities of the activity of interest – for example, population or employment. Two sources of information for producing such a grid include remote sensing or satellite imagery and polygon-level data.

Remote sensing and satellite imagery provide visual representations of structures, vegetation and other physical features in a region. There are different approaches for using the contents of the images to draw information on a distribution of activities. Li and Weng (2005) summarize some of these techniques and how they are used to derive population density distributions. One method mentioned in the article involves counting dwelling units on the imagery and then applying a statistic on the average number of persons per dwelling unit to arrive at a spatial distribution for the population. Pixel counts or other remote sensing variables can also be used to characterize a population distribution in a region.

The use of remote sensing and satellite imagery, though, has some drawbacks. First, the accuracy of the activity distribution that is extracted depends heavily on the spatial resolution of the image and on the assumptions that are used to translate structures into people. Second, the process of extraction can be computationally intense and time consuming. Of course, this issue is becoming increasingly irrelevant with advancements in computing power and innovations in image processing techniques and algorithms. A final concern is that the temporal and geographic breadth of high resolution imagery is still very limited.

An alternative to image processing techniques is to use data that is summarized based on polygons. At least for locations in the United States, such data is widely available for a variety of geographies. A continuous density surface can be created from such data by doing spatial interpolation using the polygons (Tobler 1979). In this chapter, do interpolation on the weighted centroids of polygons. More specifically, kriging is employed as a method of interpolation. Kriging offers some advantages over other simple polynomial interpolation or methods based on inverse distance weights. It also provides minimum variance in the prediction errors and it can be used to interpolate between the surfaces of two distinct variables (Goodvearts 1997; Wackernagel 2003). The latter is referred to as cokriging.

As a means to establish the critical threshold that goes into defining a component, we utilize three-dimensional imagery. Fig. 15.1 illustrates the approach for

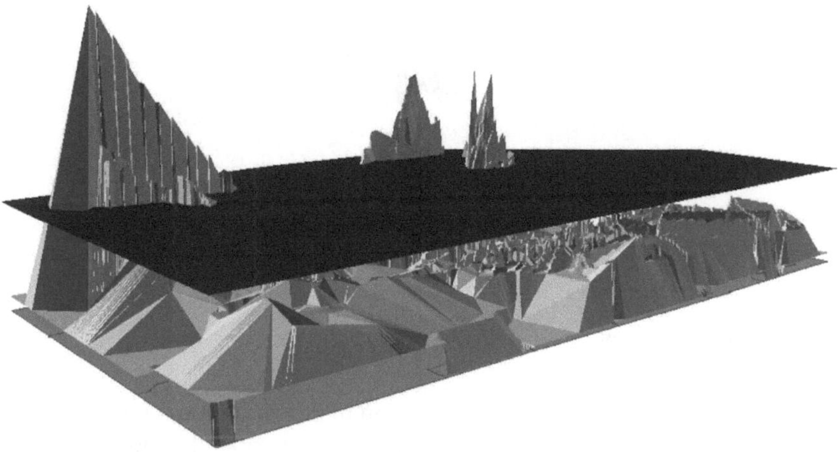

**Fig. 15.1** The cut-off threshold and activity surface for a hypothetical region

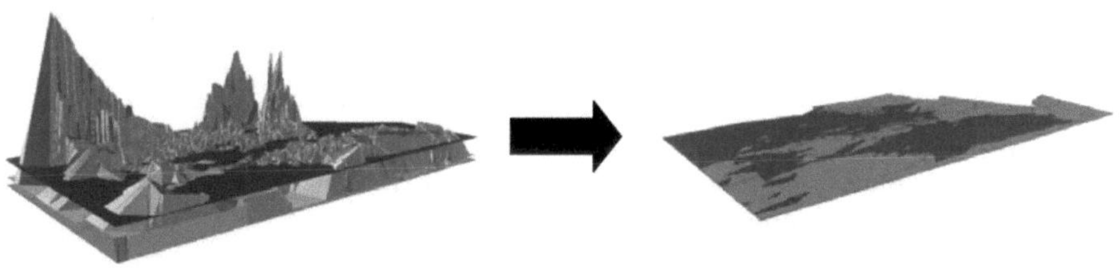

**Fig. 15.2** The emergence of the fully-connected lattice from a cut-off threshold

a hypothetical region and density distribution. The heights of the peaks in the diagram reflect densities at different locations, with steeper spikes indicating higher values, and the dark plane that intersects the surface represents the density threshold. In Fig. 15.1, there are three parts of the surface that are higher than threshold and those parts are not physically connected with one other. In this case, a lower threshold must be considered in order to arrive at a connected surface, where the lattice structure underlying the resultant surface represents a network with the least level of connection among its parts. The left-hand graphic in Fig. 15.2 shows the point at which the plane intersects the surface in such a way that a giant component is defined and the conversion of the component to a regular quadrant grid is shown to the right.

The structural properties and relative importance of the nodes (or quadrants) that define the network can be established through graph-theoretic measures of centrality. With these measures, a subset of the cells with the highest values, are extracted to define a backbone. Figure 15.3a, b illustrates the process, with the resultant backbone indicated in black on the map depicted in Fig. 15.3a, b.

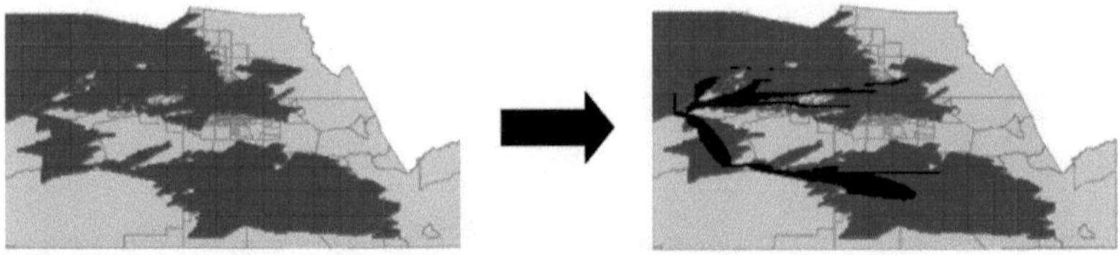

a b

Fig. 15.3 Giant component and backbone for hypothetical region

## 15.3 Case Studies

In our experiments, betweenness is used as an indicator of nodal importance. The betweenness of a node measures how many shortest paths that node is involved respect to the total number of shortest paths through the network (Freeman 1979; Wasserman and Faust 1994). Rook's rule is used to establish the adjacency of cells in the quadrant grid for the graph analysis. Two sets of analyses are performed using the Cedar Rapids, Chicago and Phoenix metropolitan areas. The first examines how population backbones in those cities have shifted over the time period from 1990 to 2000 and the second does a cross-sectional comparison of population and employment backbones for the year 2000. In all cases, a backbone is ultimately defined by the top 10% cells ranked according to betweenness in the respective component.

### *15.3.1 Comparison of Population Backbones Over Time*

The population density surfaces used in this analysis are based on U.S. Census population data summarized at the census tract level. To generate the surfaces, kriging was applied to the centroids of the tracts weighted by population.

#### 15.3.1.1 Cedar Rapids

The interpolated population surfaces for Cedar Rapids show that the city experienced some growth between 1990 and 2000, with more areas of the region filling in over time. Figures 15.4a, b show the surfaces for the two time periods. In Fig. 15.4c, the components of the two interpolated surfaces are compared. "Light grey" represents the population surface for 1990, "the darkest shade of grey" is the surface for 2000 and "the medium shade of grey" is the overlap between the two surfaces. The minimal amount of "medium grey" in Fig. 15.4c implies that there were substantial changes in the spatial distribution of population in Cedar Rapids over time.

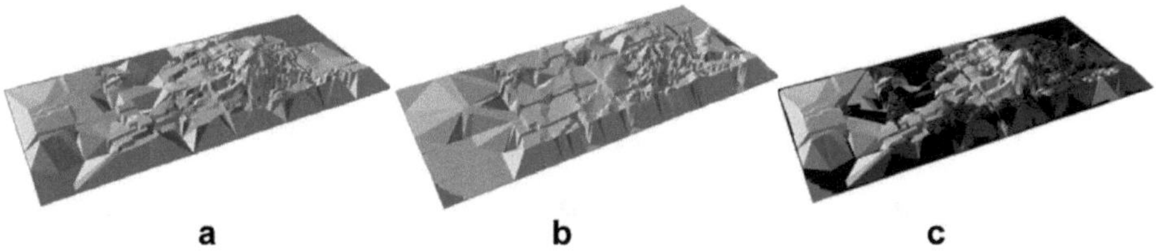


**Fig. 15.4** Population surface maps and components for Cedar Rapids (1990 and 2000)

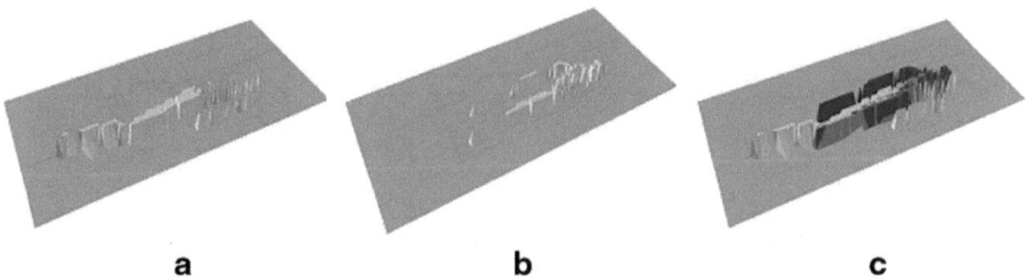


**Fig. 15.5** Population backbones for Cedar Rapids (1990 and 2000)

A comparison of the backbones for the two time periods show shifts in the location and pattern of the structurally most important parts of the city. The backbone for 1990 is characterized as a single row that extends from the western parts of the city to points east (see Fig. 15.5a). In contrast, the pattern for 2000 is more truncated and seems to follow double branch structure (Fig. 15.5b). The shifts in the patterns are reinforced in Fig. 15.5c, which overlays the two backbones and shows in green the lack of overlap between the two structures.

#### 15.3.1.2 Chicago

The population density surfaces for Chicago, like Cedar Rapids, show population growth over the decade being studied (see Fig. 15.6a, b). The surfaces also indicate a shift in population away from the downtown and suburbs located inward to locations on the western periphery of the city. This conclusion is reinforced in Fig. 15.6c, which compares the two components for population.

The analysis shows substantial shifts in the population backbones. In 1990, the areas identified as being structurally important are fragmented, with some located in the northern parts of the city and others to the east and southwest (Fig. 15.7a). The pattern changes quite considerably in 2000, where the backbone is exhibited as a continuous band stretching from the eastern portions of Chicago to points west and north (Fig. 15.7b). The shift is visible in Fig. 15.7c, which depicts very little overlap in the two backbones.

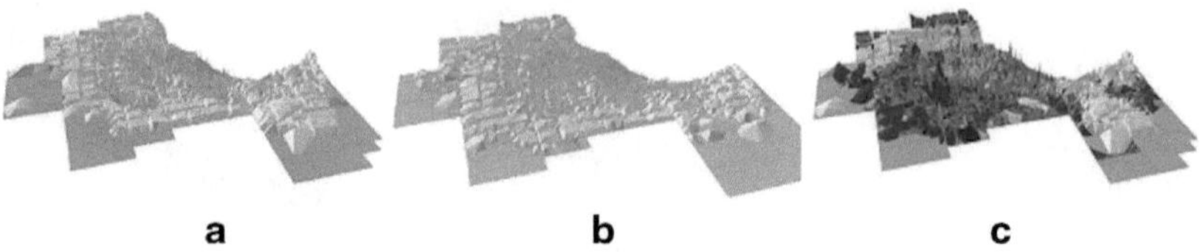


**Fig. 15.6** Population surfaces and components for Chicago (1990 and 2000)

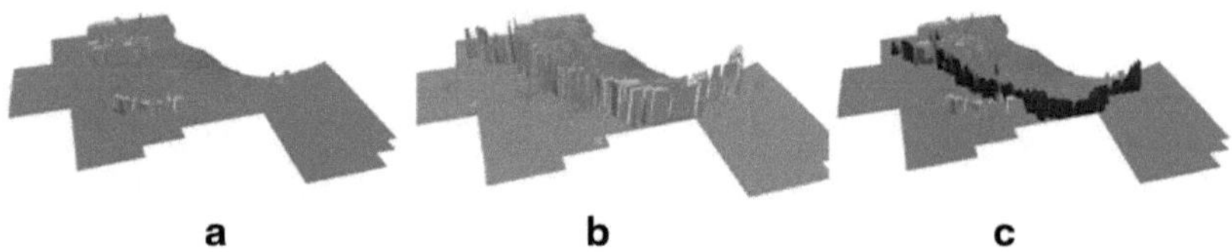


**Fig. 15.7** Population backbones for Chicago (1990 and 2000)

#### 15.3.1.3 Phoenix

Population in Phoenix grew quite considerably during 1990 and it expanded outward in nearly all directions from the city, as depicted in Figs. 15.8a, b. Figure. 15.8c shows very little overlap in the population components and further support for a shift in population from the inner core to the periphery.

The backbones for Phoenix for 1990 and 2000 (see Figs. 15.9a, b) appear to have similar structures, with both taking the shape of a ring. Over that time period, however, the backbone appears shifts slightly outward. Further evidence of this shift is shown in Fig. 15.9c.

### *15.3.2 Comparison of Population and Employment Backbones*

This section summarizes the results of the cross-sectional study, which compares population and employment patterns within each of the three cities. The analysis uses 2000 census tract level employment figures from the Census Transportation Planning Package, as well as the population data that was used in the preceding analysis. Every census tract is, thus, assigned two values: one for population and the other for employment.

For each city, we used kriging to generate interpolated surfaces for both population and employment and cokriging to generate a third surface using the isotropic data. The cokriging uses population as the primary variable. The cokriging surface represents a correction brought to the population interpolation by the information

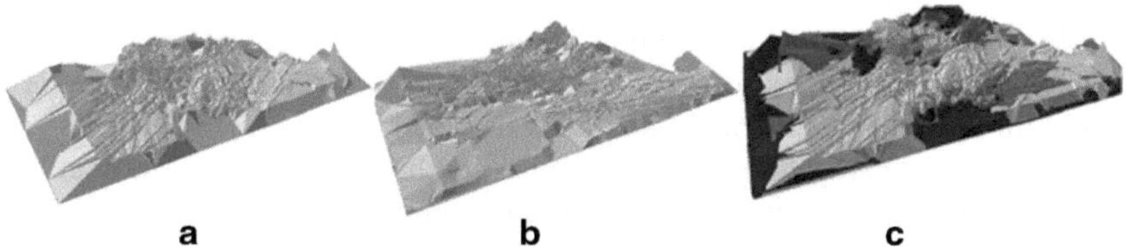

Fig. 15.8 Population surfaces and components for Phoenix (1990 and 2000)

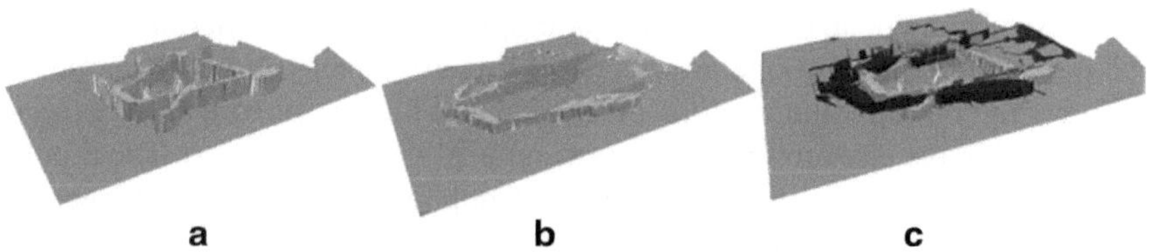

Fig. 15.9 Population backbones for Phoenix (1990 and 2000)

Fig. 15.10 Activity surfaces and backbones for Cedar Rapids (2000)

included in the employment dataset. The cokriging results show that the auto- and the cross-variograms are not proportional to the same basic model at any of the data locations (Isaaks and Srivastava 1989). Therefore, this results in a surface that is different from the kriging surfaces that are generated separately for population and employment.

#### 15.3.2.1 Cedar Rapids

The kriging results for Cedar Rapids show that the patterns of population and employment in that city were highly spatially coincident in 2000. One slight exception is in the downtown area, where employment is more prominent than population. A generally strong spatial correlation between population and employment is supported by the cokriging surface, which is shown in Fig. 15.10a. In that diagram, the "lightest grey" represents the surface for population, "the darkest shade" is that for employment and the "medium shade" is the cokriging surface.

There is also strong a strong spatial association between the backbones associated with population and employment, as depicted in Fig. 15.10b. In particular, there are common parts along the center of the southern band running east and west and along portions of the northern band running parallel to that.

#### 15.3.2.2 Chicago

In the analysis of Chicago, the patterns of employment and population were found to be quite different from one another. The kriged surfaces show that population forms a ring through the middle of the city, while employment is primarily concentrated in the downtown and western suburbs (see Fig. 15.11a). The topological analysis shows that the structurally important population nodes in Chicago form a ring around the central city and employment forms a buffer around the population. Additionally, some isolated structurally important employment nodes at locations in the far western suburbs (Fig. 15.11b).

#### 15.3.2.3 Phoenix

In Phoenix, the population and employment surfaces created with the kriging technique appear to be somewhat different from one another (see Fig. 15.12a). While there are some areas of overlap between the two activities in the middle portion of the city, population is nearly absent from the central city and employment is lacking from the far periphery of the city. The patterns for the employment and population backbones are generally consistent with what was found for the two activity's respective components. Structurally important areas of employment form a ring inside a ring for population (see Fig. 15.12b).

Figure 15.13 shows the three auto- and cross-covariance surfaces that were used as weights for the development of the kriging surfaces for Phoenix. From these diagrams, it is possible to verify that the weight that employment brings to the

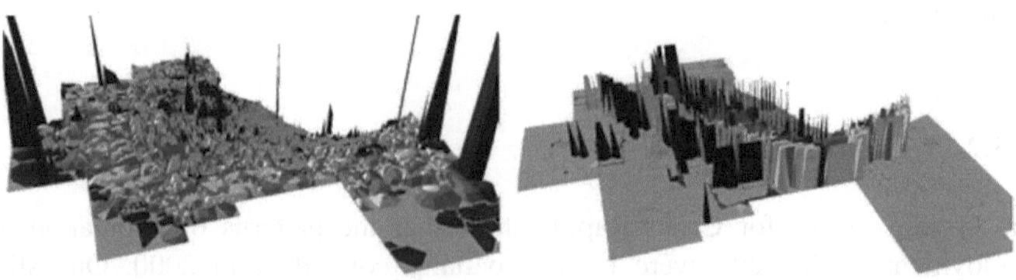

a b

**Fig. 15.11** Activity surfaces and backbones for Chicago (2000)

**Fig. 15.12** Activity surfaces and backbones for Phoenix (2000)

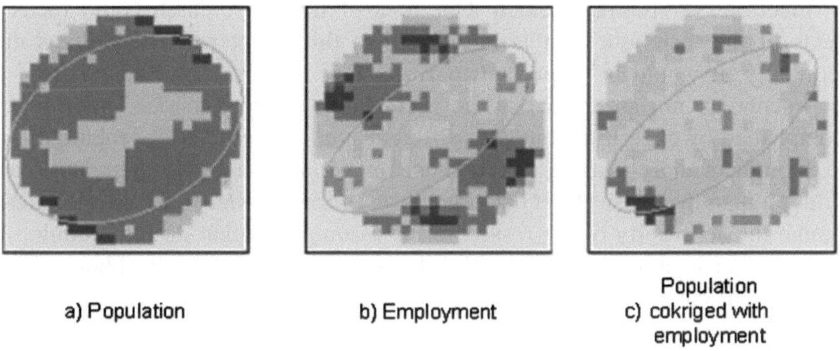


**Fig. 15.13** Auto- and cross-covariance surface for Phoenix

cokriging surface is considerable. This is especially evident from the anisotropy component.

#### 15.3.2.4 Measures of Correlation for the Components and Backbones

Measures of spatial correlation were also generated using the quadrant values from the kriging and the betweenness scores associated with the graph-theoretic analysis. We generated two sets of coefficients: one for the kriging results and the other for the betweenness scores (see Table 15.1).

For Cedar Rapids, the coefficient for the population and employment components show moderate correlation (0.28), whereas, for the structurally important locations defined by the backbones, there is a stronger correlation. The correlation coefficients for Chicago show similar results to those found for Cedar Rapids in the case of the population and employment components. An interesting contrast with Cedar Rapids, though, is that for Chicago the correlation between structurally important employment and population clusters is extremely small (0.04). Lastly, for Phoenix, there is a moderate correlation (0.53) between the population and employment components. The association between the respective backbones is very minimal (0.06).

**Table 15.1** Correlation coefficients for the components and backbones

| Surfaces | Component | | | Backbone | | |
|---|---|---|---|---|---|---|
| | Cedar rapids | Chicago | Phoenix | Cedar rapids | Chicago | Phoenix |
| Employment and population | 0.56 | 0.53 | 0.53 | 0.28 | 0.05 | 0.06 |
| Population and cokriging | 0.97 | 0.79 | 0.64 | 0.84 | 0.13 | 0.16 |
| Employment and cokriging | 0.55 | 0.56 | 0.60 | 0.27 | 0.06 | 0.12 |

## 15.4 Directions for Future Research

The exploratory analysis presented in this chapter raises some interesting questions and directions for future research. One area of further research could focus on aspects of the technique that is introduced in this chapter. First, perhaps most importantly, it would be important to verify that the results we found from the experiments are not an artifact of the vector-to-raster conversion. This might be accomplished by experimenting with different spatial interpolation techniques, but also through an analysis using employment and population data with greater spatial resolution than the census tract data that we used for the analysis. It is important to examine the sensitivity of the results to the assumptions that go into defining the backbone. For example, how does the shape and location of a backbone change with different cut-offs for the betweenness scores – for example, top 20% versus 10%? Also, how might the results change when other measures of nodal importance are used?

It is also important to understand how the results of the analysis should be interpreted and how they are consistent with other studies of activity distribution in metropolitan areas (for example, Griffith and Wong 2007). Potential universalities in the results when more cities are considered could also be examined. For instance, would we find other cities to show a value of 0.5 for the correlation between their employment and population components, as we found for the cities of Cedar Rapids, Chicago and Phoenix? Further, what is the role that constraints play in defining what is seen in a city in terms of the location and shape of an activity backbone? To what extent does land use regulations and zoning, physical features such as lakes or mountains or road infrastructure play a role in characterizing where structurally important concentrations are located? Lastly, further thought should be given to the practical interpretation of a backbone and what it can tell us about the spatial network dynamics of a city.

## References

Freeman LC (1979) Centrality in social networks: conceptual clarification. Soc Networks 1:215–239

Goodvaerts P (1997) Geostatistics for natural resources evaluation. Oxford University Press, Oxford

Griffith D, Wong D (2007) Modeling population density across major U.S. cities: a polycentric spatial regression approach. J Geogr Syst 9:53–75
Isaaks EH, Srivastava RM (1989) An introduction to applied geostatistics. Oxford University Press, Oxford
Li G, Weng Q (2005) Using landsat ETM imagery to measure population density in Indianapolis, Indiana, USA. Photo Grammetric Imagery Remote Sens 71:8 947–958
Tobler W (1979) Smooth pycnophylactic interpolation for geographical regions. JASA 74:519–530
Wackernagel H (2003) Multivariate geostatistics: an introduction with application. Springer, Berlin
Wasserman Se, Faust K (1994) Social network analysis: methods and applications. Cambridge University Press

# Chapter 16
# Spatial Autocorrelation in Spatial Interaction

## Complexity-to-Simplicity in Journey-to-Work Flows

**Daniel A. Griffith**

## 16.1 Introduction

Carey (1858) and Ravenstein (1885) first proposed, through analogy, the gravity model of Newtonian physics as a description for economic and social spatial interaction, with Sen and Smith (1995) furnishing a comprehensive treatment of this model more than a century later. In the late 1960s, Wilson spelled out an entropy maximizing derivation of the gravity model, including the use of row and column totals as additional information for modeling purposes (that is, the doubly-constrained version), followed by a utility maximization derivation of it by Niedercorn and Bechdolt (1969). Flowerdew and Atkin (1982) and Flowerdew and Lovett (1988) articulated linkages between the Poisson probability model and spatial interaction. Within this same time interval, Anas (1983) established a linkage between the doubly-constrained gravity model and a logit model of joint origin-destination choice, which indirectly relates to a Poisson specification that includes a separate indicator variable for each origin and each destination (that is, 2n 0–1 binary variables, each having a single 1 and n-1 0s). Curry (1972; also see Curry et al. 1975, 1976) followed by Griffith and Jones (1980), first raised the issue of spatial autocorrelation effects embedded in spatial interaction. These investigations were followed by a formulation of the network autocorrelation concept (see Black 1992; Black and Thomas 1998; Tiefelsdorf and Braun 1999). More recently, LeSage and Pace (2008), Griffith (2008), and Fischer and Griffith (2008) have returned to the issue of spatial autocorrelation effects embedded in spatial interaction, specifying spatial autoregressive and spatial filter versions of the unconstrained gravity model, but in terms of attribute geographic distributions. Chun (2007) moves beyond this conceptualization to that of more explicitly spatially autocorrelated flows.

---

D.A. Griffith
School of Economic, Political and Policy Sciences, University of Texas at Dallas, Dallas, TX 86080, USA

A. Reggiani and P. Nijkamp (eds.), *Complexity and Spatial Networks*,
Advances in Spatial Science,
DOI: 10.1007/978-3-642-01554-0_16, © Springer-Verlag Berlin Heidelberg 2009

This chapter addresses an extension of the Griffith and Fischer-Griffith specification to the doubly-constrained spatial interaction model in a manner that is informed by the work of Chun. Questions of interest include:

1. How can network and geographic distribution spatial autocorrelation effects be distinguished?
2. Can eigenvectors portraying geographic distribution spatial autocorrelation effects also capture network autocorrelation effects?
3. What role do the origin and destination weights play in accounting for spatial autocorrelation in a doubly-constrained spatial interaction model?
4. Can eigenvectors portraying spatial autocorrelation be efficiently and effectively approximated for sizeable sparse geographic weights matrices?

Explorations of these questions are presented in terms of a selected journey-to-work dataset (Texas, USA), in an attempt to glean new insights furnished by spatial filtering methodology. An ultimate goal is to better understand complexity features of spatial autocorrelation latent in spatial interaction whose recognition is inspired by linear combinations of eigenfunctions derived from surface partitionings portraying this spatial autocorrelation. In doing so, specification simplicity is sought.

## 16.2 Conceptual Foundation

Spatial filtering of georeferenced counts data involves specifying a geographically heterogeneous mean and variance in order to capture spatial autocorrelation (Griffith 2002). Spatial filtering seeks to transform a variable containing spatial dependence into one free of spatial dependence by partitioning the original georeferenced attribute variable into two synthetic variates: a spatial filter variate capturing latent spatial dependency that otherwise would remain in the response residuals, and a nonspatial variate that is free of spatial dependence. Griffith (2000) proposes a transformation procedure that depends on the eigenfunctions of matrix $(\mathbf{I}-\mathbf{11}^T/n)\mathbf{C}(\mathbf{I}-\mathbf{11}^T/n)$ – where **I** denotes the identity matrix, **1** is an n-by-1 vector of ones, T denotes matrix transpose, and **C** is a binary 0–1 geographic weights matrix (that is, $c_{ij} = 0$ if areal units i and j are neighbors, and $c_{ij} = 0$ otherwise; $c_{ii} = 0$) – a term appearing in the numerator of the Moran Coefficient (MC) spatial autocorrelation index, and is based on the following theorem (Griffith 2003):

> The first eigenvector, say $\mathbf{E}_1$, is the set of numerical values that has the largest MC achievable by any set for the spatial arrangement defined by the geographic connectivity matrix **C**. The second eigenvector is the set of values that has the largest achievable MC by any set that is uncorrelated with $\mathbf{E}_1$. The third eigenvector is the third such set of values. And so on. This sequential construction of eigenvectors continues through $\mathbf{E}_n$, the set of values that has the largest negative MC achievable by any set that is uncorrelated with the preceding (n−1) eigenvectors.

As such, Griffith (2000) argues that these eigenvectors furnish distinct map pattern descriptions of latent spatial autocorrelation in georeferenced variables.

The spatial filter is constructed by using judiciously selected eigenvectors as regressors (for example, selected with a stepwise Poisson regression routine from a candidate set representing at least a minimum level of spatial autocorrelation, perhaps of a particular nature), which results in spatial autocorrelation being filtered out of the residuals of georeferenced variables, with the regression residuals representing spatially independent variable components. Of note is that these eigenvectors representing distinct map patterns are both mutually orthogonal and mutually uncorrelated in their numerical form, a property that is corrupted by the weighting involved in computing Poisson regression parameter estimates.

Analyzing the spatial autocorrelation contained in the geographic distributions of origin and destination attributes overlooks their links through the network of flows that constitutes spatial interaction. These origin/destination variables can be linked to the individual flows data dyads through Kronecker products: $\mathbf{E}_K \otimes \mathbf{1}$ and $\mathbf{1} \otimes \mathbf{E}_K$, which result in the addition of an origin and a destination term (that is, $e_{ik} + e_{jk}$). The net result essentially is two sets of indicator variables, the first set containing one variable for each origin and the second set containing one for each destination. Consequently, the same spatial autocorrelation effects are captured by these eigenvectors as are by the 2n indicator variables associated with a doubly-constrained model specification.

Chun (2007) argues that a more conceptually appealing specification posits, for example, correlation between the flow between the pair of locations i and j in a tessellation and each flow that originates at each location that is a neighbor of origin i, and terminates at each location that is a neighbor of destination j (that is, the spatial filter term takes on the form $e_{ik}e_{jk}$). Asymptotically, this specification results in the $n^2$-by-$n^2$ connectivity matrix for flows being defined by the Kronecker product $\mathbf{C}_{n^2} = \mathbf{C}_n \otimes \mathbf{C}_n$, where $\mathbf{C}_{n^2}$ is the $n^2$-by-$n^2$ binary 0–1 connectivity matrix for the $n^2$ spatial interaction flows, and $\mathbf{C}_n$ is the n-by-n connectivity matrix for the tessellation of n origins/destinations. As with most geographic weights matrices, $\mathbf{C}_{n^2}$ is very sparse. It also is quite sizeable. For example, the 254-by-254 $\mathbf{C}_n$ matrix for Texas becomes a 64,516-by-64,516 $\mathbf{C}_{n^2}$ matrix. Current desktop computer technology suggests that this is sizeable, often being unable to handle $\mathbf{C}_{n^2}$ matrices larger than 10,000-by-10,000 (that is, cases involving 100-by-100 $\mathbf{C}_n$ matrices). Although this upper limit will continue to increase as computer technology continues to advance, the Kronecker product $\mathbf{C}_{n^2} = \mathbf{C}_n \otimes \mathbf{C}_n$ allows this restriction to be circumvented. This Kronecker product specification yields the origin and destination variates $\mathbf{E}_i \otimes \mathbf{E}_j$ and $\mathbf{E}_j \otimes \mathbf{E}_i$ for constructing spatial filters. Because a minimum level of spatial autocorrelation should be represented by a given eigenvector, say 0.25 (that is, roughly 5% of the variance of an eigenvector's values is attributable to redundant information resulting from the presence of spatial autocorrelation), and the eigenvalues of $\mathbf{C}_n \otimes \mathbf{C}_n$ are the pairwise products of the individual eigenvalues of $\mathbf{C}_n$, this threshold value should be increased to 0.5 (that is, $0.5^2 = 0.25$) for each of the eigenvectors in a Kronecker product. The net result of this specification is a dramatic reduction in computational intensity.

## 16.3 An Empirical Grounding

Texas is partitioned into 254 counties, resulting in 64,516 flow dyads (of which 52,243 had a 0 flow in 2000). The total journey-to-work flows for 2000 by county within the state total 9,229,209. Inter-county distance is calculated with the geometric centroids of these counties. The following doubly-constrained entropy maximizing model specification has been calibrated [using a modified version of the algorithm outlined by Foot (1981, p. 90)]:

$$F_{ij} = \kappa A_i O_i B_j D_j e^{-\gamma d_{ij}},$$

where $F_{ij}$ is the journey-to-work flow between counties i and j, $O_i$ is the total flows originating in county (that is, workers), $D_j$ is the total flows terminating in county j (that is, jobs), $d_{ij}$ is the inter-county centroid distance separating counties i and j, $A_i$ are the n origin-balancing factors, $B_j$ are the n destination-balancing factors, $\gamma$ is the distance-decay parameter, and $\kappa$ is a constant of proportionality. Iterative calibration resulted in the average distance traveled to work for the observed and predicted flows matching to ten decimal places, and estimates of the $A_i$ and $B_j$ terms, whose largest values were restricted to being exactly 1 (which results in $\kappa$ being other than 1) for nonlinear estimation stability reasons.

Next, a Poisson regression was executed. The sum variable $LN(O_i) + LN(D_j)$ was included as an offset variable, and the three covariates included were $d_{ij}$, $LN(A_i)$, and $LN(B_j)$ – terms are expressed in their logarithmic forms because Poisson regression estimates the expected value of $LN(F_{ij})$. Estimation results, which are identical to those obtained with the preceding iterative algorithm (except for rounding error), include

| | |
|---|---|
| $\hat{\kappa}$ | 1.0273 |
| $A_i$ exponent | 1.0003 |
| $B_j$ exponent | 0.9997 |
| $\hat{\gamma}$ | 3.6735 |

The associated pseudo-$R^2$ for the predicted and observed flows is 0.995, which decreases to 0.942 when the six extreme flows (that is, >200,000) are set aside, but considerable overdispersion is present (deviance statistic = 106.5). Figure 16.1 displays a plot of the predicted and observed flows. Of note is that $n^2$ observations (that is, flow pairs) are used to estimate 2n + 4 parameters.

The $A_i$ and $B_j$ values relate to the origin- and destination-specific indicator variables for a generalized linear model (GLM) specification (Fig. 16.2) as follows:

$$LN(A_i) \equiv 8.30152 + 0.99981 I_{origin}, \quad \text{and}$$

$$LN(B_j) \equiv -8.33247 + 1.00001 I_{destination},$$

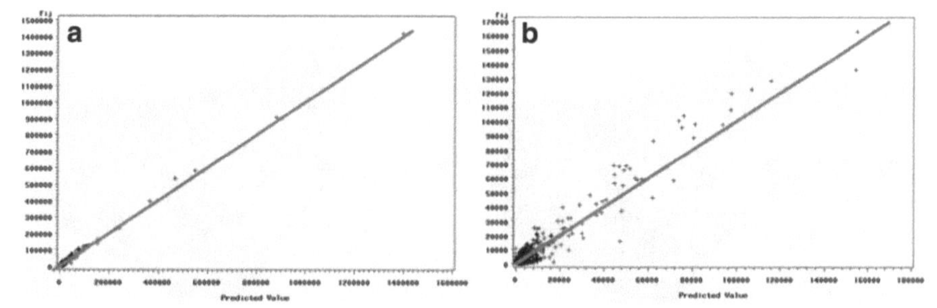


**Fig. 16.1** Scatterplots of the predicted (*vertical axis*) and observed (*horizontal axis*) 2000 inter-county journey-to-work flows within Texas generated with a doubly-constrained gravity model specification. The gray lines denote the lines of perfect correspondence. *Left* (**a**): for all 64,516 flows. *Right* (**b**): for 64,510 flows, after the six outliers have been removed

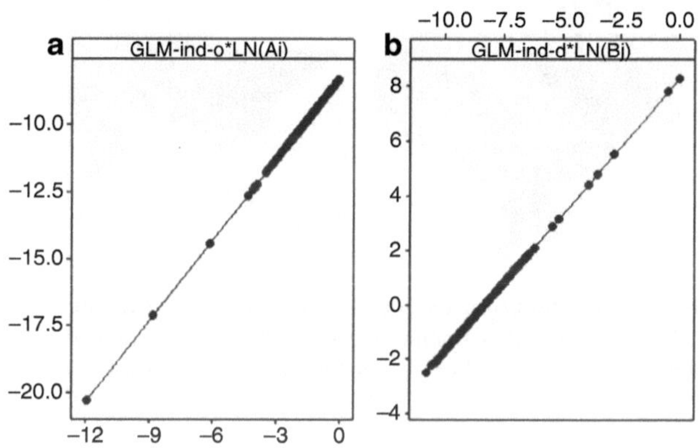


**Fig. 16.2** Scatterplots of the balancing factors versus the GLM indicator variable coefficient estimates. The gray lines denote perfect correspondence. *Left* (**a**): GLM coefficients versus LN($A_i$). *Right* (**b**): GLM coefficients versus LN($B_j$)

where the Poisson regression model was specified with no intercept, 254 origin indicator variables, and 253 destination indicator variables (that is, the 254th had its coefficient set to 0). An equivalent specification is to include an intercept term and set the 254th origin indicator variable coefficient to 0.

The logarithms of the variables $O_i$, $D_j$, $A_i$ and $B_j$ were found to mimic a bell-shaped curve reasonably well, but not perfectly. Maps of these four variables appear in Fig. 16.3. Visual inspection of the maps portraying the total number of workers and of jobs reveals that not only are they similar, but they also highlight the three major urban areas of Austin, Dallas, and Houston. Visual inspection of the maps portraying the $A_i$ and $B_j$ balancing factors reveals a swath of high emissivity through the centre of the state, with low emissivity in the major metropolitan areas and the more remote western counties, and an east-west differentiation of attractivity, with the western counties being more attractive while the eastern counties are more self-serving.

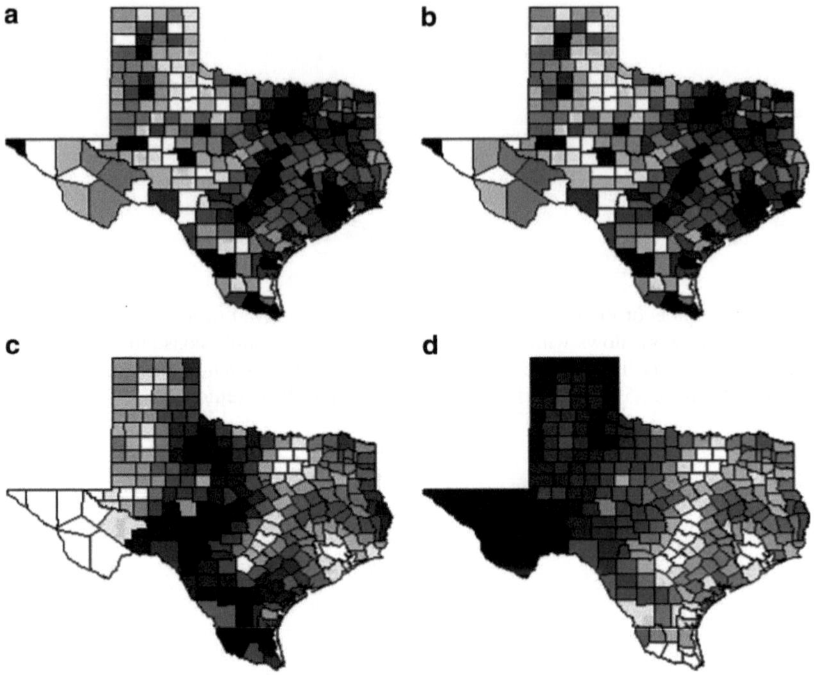


**Fig. 16.3** Geographic distributions of the journey-to-work flows covariates; light gray to black denotes low to high. *Top left* (**a**): total number of workers by county. *Top right* (**b**): total number of jobs by county. *Bottom left* (**c**): $A_i$ values by county. *Bottom right* (**d**): $B_j$ values by county

## 16.4 Spatial Autocorrelation in the Texas Journey-to-Work Flows Data

Levels of spatial autocorrelation, as well as spatial filter descriptions of the $A_i$, $O_i$, $B_j$ and $D_j$ terms, are reported in Table 16.1. The log-transform of each variable displays weak-to-moderate positive spatial autocorrelation. The products $A_iO_i$ and $B_jD_j$ represent the doubly-constrained balancing of flows to ensure that the number of workers predicted as leaving a county exactly equals the number of workers observed in that county, and the number of jobs predicted for a county exactly equals the number of jobs observed in that county. The $A_i$ and $B_j$ values are computed as distance-discounted functions of neighboring values, resulting in their containing spatial autocorrelation by construction (that is, they are autoregressive in mathematical form). Especially the $B_jD_j$ term appears to involve a canceling of at least some spatial autocorrelation effects. Because the Kronecker products $\mathbf{E}_K \otimes \mathbf{1}$ and $\mathbf{1} \otimes \mathbf{E}_K$ construct eigenvectors that behave like origin and destination indicator variables, no eigenvectors can be selected by a stepwise Poisson regression analysis as additional covariates for the doubly-constrained gravity model specification.

**Table 16.1** Summary of spatial filter construction for the origin and destination geographic distributions

| Features | $LN(O_i)$ | $LN(D_j)$ | $LN(A_i)$ | $LN(B_j)$ | $LN(A_iO_i)$ | $LN(B_jD_j)$ |
|---|---|---|---|---|---|---|
| Initial MC | 0.433 | 0.366 | 0.476 | 0.658 | 0.481 | 0.273 |
| Pseudo-$R^2$ | 0.529 | 0.466 | 0.654 | 0.807 | 0.615 | 0.427 |
| Residual $z_{MC}$ | 0.53 | 0.06 | 4.66 | 5.47 | 2.68 | 1.44 |
| S-W | 0.958 | 0.959 | 0.634 | 0.748 | 0.923 | 0.841 |
| Spatial filter MC | 0.876 | 0.872 | 0.692 | 0.799 | 0.784 | 0.668 |
| Common eigenvectors | | | $E_1, E_9, E_{18}, E_{23}, E_{57}$ | | | |
| | | | | $E_8$ | | |
| | | | $E_{29}$ | | | |
| | | $E_{16}, E_{20}$ | | | | $E_{16}$ |
| | $E_3, E_4, E_5, E_{14}, E_{25}, E_{34}$ | | | | $E_3, E_4, E_5, E_{14}, E_{25}, E_{34}$ | |
| Shared eigenvectors | | | | $E_6, E_{10}, E_{15}, E_{21}, E_{26}, E_{32}, E_{44}, E_{53}, E_{54}, E_{58}$ | | |
| | $E_7, E_{12}, E_{13}, E_{22}, E_{30}, E_{47}, E_{56}$ | | $E_{11}, E_{35}, E_{41}, E_{60}, E_{62}$ | | $E_{47}$ | |
| | | | $E_{27}, E_{40}, E_{48}, E_{63}, E_{65}$ | | | $E_{12}, E_{20}, E_{30}, E_{56}, E_{59}$ |
| Miscellaneous eigenvectors | $E_{38}, E_{44}$ | | $E_{33}, E_{37}, E_{38}, E_{46}$ | $E_{12}, E_{13}, E_{30}, E_{56}, E_{59}$ | $E_{22}, E_{33}, E_{37}, E_{38}$ | $E_{35}, E_{46}, E_{60}$ |
| Unique eigenvectors | | $E_{17}, E_{28}, E_{45}, E_{51}$ | $E_{19}, E_{24}, E_{42}$ | | | |

This finding is in stark contrast to the construction of origin and destination spatial filters for an unconstrained gravity model specification (that is, the balancing factors are not included). Eigenvectors selected, from a candidate set of 65 portraying positive spatial autocorrelation, for origin- (45) and destination-based (48) spatial filter descriptions using flows as the response variable are reported in Table 16.2, and represent moderate-to-strong positive spatial autocorrelation (respectively, MC = 0.773 and 0.776, and Geary Ratio (GR) = 0.232 and 0.179). Three of the common eigenvectors (that is, $E_1$, $E_9$, and $E_{57}$) also are common eigenvectors in Table 16.1. Figure 16.4 displays the geographic distributions of

**Table 16.2** Summary of spatial filter construction for the unconstrained gravity model flows

| Eigenvectors | Origin spatial filter | Destination spatial filter |
|---|---|---|
| Common | $E_1$–$E_5$, $E_7$, $E_9$, $E_{12}$, $E_{15}$, $E_{17}$, $E_{20}$, $E_{26}$, $E_{27}$, $E_{29}$, $E_{30}$, $E_{32}$, $E_{33}$, $E_{36}$–$E_{38}$, $E_{44}$, $E_{46}$, $E_{48}$, $E_{50}$, $E_{52}$–$E_{54}$, $E_{57}$, $E_{60}$, $E_{61}$, $E_{63}$ | |
| Unique | $E_6$, $E_{13}$, $E_{18}$, $E_{19}$, $E_{24}$, $E_{31}$, $E_{39}$–$E_{41}$, $E_{55}$, $E_{56}$, $E_{58}$, $E_{62}$, $E_{65}$ | $E_8$, $E_{10}$, $E_{11}$, $E_{14}$, $E_{16}$, $E_{21}$–$E_{23}$, $E_{25}$, $E_{28}$, $E_{34}$, $E_{35}$, $E_{43}$, $E_{45}$, $E_{49}$, $E_{51}$, $E_{64}$ |

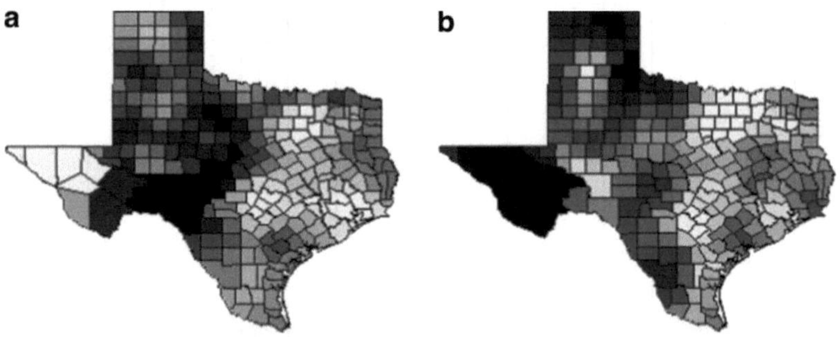

**Fig. 16.4** Geographic distributions of the flows-based spatial filters; light gray to black denotes low to high. *Left* (**a**): origin-based spatial filter. *Right* (**b**): destination-based spatial filter

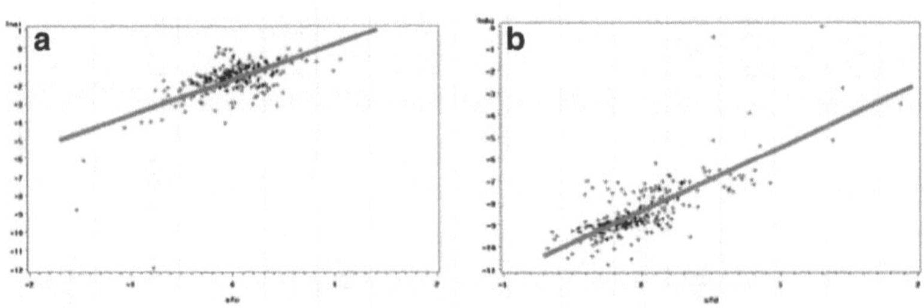


**Fig. 16.5** Scatterplots of the log-balancing factors (*vertical axis*) versus the corresponding flows-based spatial filters (*horizontal axis*). The gray lines denote the lines of perfect correspondence. *Left* (**a**): origin pairs. *Right* (**b**): destination pairs

these two spatial filters, and Fig. 16.5 displays the relationships between these spatial filters and their corresponding log-balancing factors; the origin pair has an $R^2$ of 0.355, whereas the destination pair has an $R^2$ of 0.601. Unfortunately, these spatial filters fail to preserve origin and destination totals; to do so, they would have to correlate perfectly with their corresponding log-balancing factors. These attribute-based spatial filters also fail to fully account for spatial autocorrelation in the flows (see Fischer and Griffith 2008).

## 16.5 Some Features of Complexity in Spatial Flows

An unconstrained gravity model furnishes a respectful benchmark description of the Texas journey-to-work data using Poisson regression, yielding a pseudo-$R^2$ value of 0.984. Removing the six extreme flows reduces this value to 0.878 (see Fig. 16.6). It also has considerable overdispersion (deviance statistic = 119.7), and has a typical Poisson V-shaped prediction spread with increasing flow size. Because this extra-Poisson variation could not be captured with a negative binomial model specification, quasi-likelihood techniques were employed for estimation purposes. The estimated origin and destination totals exponents are, respectively, 0.4901 and 0.6695; both indicate that, on average, the total number of workers and the total number of jobs in a county need to be markedly deflated in order to predict inter-county flows. The distance decay parameter estimate is 3.8519.

The simple unconstrained gravity model overlooks any spatial autocorrelation that is present. Respecifying it to include spatial filters that account for spatial autocorrelation results in a Poisson regression model that includes 45 eigenvectors constituting the origin-base spatial filter, and 48 eigenvectors constituting the destination-based spatial filter; these eigenvectors were selected with a backward stepwise Poisson regression procedure.[1] Each of these candidate eigenvectors has

**Fig. 16.6** Scatterplots of the predicted (*horizontal axis*) and observed (*vertical axis*) 2000 inter-county journey-to-work flows within Texas generated with an unconstrained gravity model specification. The gray lines denote the lines of perfect correspondence. *Left* (**a**): for all 64,516 flows. *Right* (**b**): for 64,510 flows, after the 6 outliers have been removed

[1] None of the 88 candidate eigenvectors portraying negative spatial autocorrelation were selected.

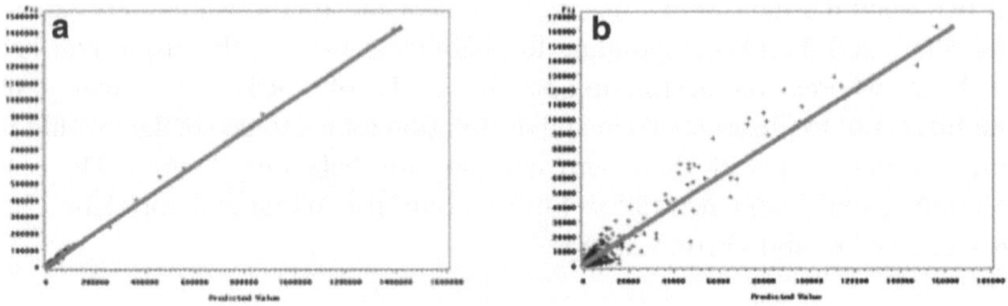

**Fig. 16.7** Scatterplots of the predicted (*horizontal axis*) and observed (*vertical axis*) 2000 inter-county journey-to-work flows within Texas generated with an unconstrained gravity model specification containing origin- and destination-based spatial filters. The gray lines denote the lines of perfect correspondence. *Left* (**a**): for all 64,516 flows. *Right* (**b**): for 64,510 flows, after the 6 outliers have been removed

an adjusted MC ($MC/MC_{max}$, where $MC_{max} = 1.10825$ is the maximum MC for the surface) of at least 0.25. Here the pseudo-$R^2$ is 0.994, which reduces to 0.936 when the six extreme flows are not included (see Fig. 16.7). The estimated origin and destination totals exponents are, respectively, 0.6533 and 0.7671, again indicating a need to deflate these numbers; but they both have moved closer to a value of 1 (the assigned value when they are included as offset values in the doubly-constrained specification). And, the distance decay parameter estimate is 3.6377, a noticeable decrease from the corresponding simple gravity model result. Unfortunately, considerable overdispersion persists here (deviance statistic = 109.5).

## 16.6 Additional Features of Complexity in Spatial Flows

Respecifying the spatial interaction model to include both indicator variables and selected eigenvectors $\mathbf{E}_i \otimes \mathbf{E}_j$ and $\mathbf{E}_j \otimes \mathbf{E}_i$ more directly addresses spatial autocorrelation in flows. Now, using the criterion of $MC/MC_{max} > 0.5$ (that is, $0.5^2 = 0.25$ for the Kronecker product vectors), the number of candidate eigenvectors extracted from $\mathbf{C}_n$ is 35, resulting in $35^2 = 1{,}225$ candidate eigenvectors for matrix $\mathbf{C}_{n^2}$. Parameter estimation still requires the sum $LN(O_i) + LN(D_j)$ to be included as an offset variable; the 2n indicator variable coefficients need to be estimated simultaneously with the spatial filter based upon spatial interactions.

The spatial-interaction-based spatial filter is constructed with 723 of the 1,225 candidate eigenvectors (Fig. 16.8). This spatial filter only marginally increases the pseudo-$R^2$ value, increasing it to 0.998, which decreases to 0.977 when the six extreme flows are set aside (Fig. 16.9). The most noticeable results are: a marked decline in overdispersion (deviance statistic = 35.3, which still is excessive); and, a marked decline in the distance decay parameter, to 1.1146 (the pairwise 90%, 95% and 99% confidence intervals for this estimate and that for the doubly-constrained specification do not overlap). A comparison of Figs. 16.7 and 16.9 reveals that the predicted and observed flows now are more similar, with much less deviation in the smaller values, once spatial autocorrelation is accounted for. Meanwhile, the new

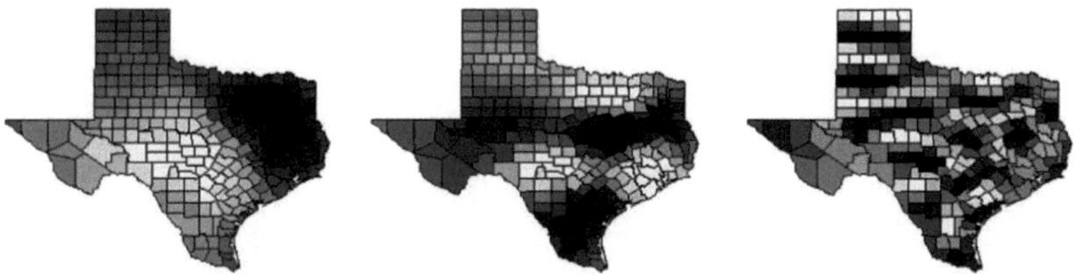

**Fig. 16.8** Geographic distributions of the three common eigenvectors; light gray to black denotes low to high. *Left* (**a**): $E_1$ (global pattern). *Middle* (**b**): $E_9$ (regional pattern). *Right*: $E_{57}$ (local pattern)

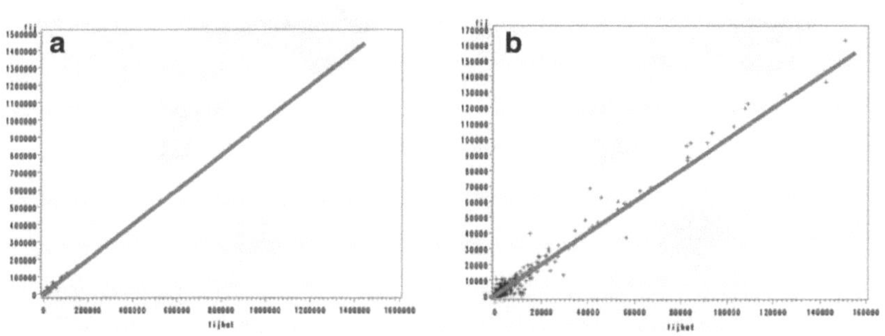


**Fig. 16.9** Scatterplots of the predicted (*horizontal axis*) and observed (*vertical axis*) 2000 inter-county journey-to-work flows within Texas generated with a doubly-constrained gravity model containing a spatial filter spatial autocorrelation adjustment specification. The gray lines denote the lines of perfect correspondence. *Left* (**a**): for all 64,516 flows. *Right* (**b**): for 64,510 flows, after the 6 outliers have been removed

indicator variable coefficient estimates fail to conform to a bell-shaped curve (see Figs. 16.10a, c; the most conspicuous deviations are in their tails), and fail to align with their doubly-constrained gravity model counterparts (Fig. 16.10b); this finding holds for their respective spatial filter descriptions, too (Figs. 16.10d, e). The spatial autocorrelation displayed by the geographic distributions of these coefficients remains positive and moderate, with 36 vectors included in the origin and 33 vectors included in the destination spatial filter (see Table 16.2; respectively, these spatial filters have MC = 0.453 and 0.447); globally, this level is roughly equivalent to that for the doubly-constrained gravity model origin balancing factor result, whereas it is substantially less than that for the doubly-constrained gravity model destination balancing factor result. But the map patterns are quite different (Figs 16.4a, b vs. 16.10f, g). A comparison of results reported in Tables 16.2 and 16.3 reveals that many of the common eigenvectors across the spatial filters span both pairs. In both cases the residual spatial autocorrelation is nonsignificant (respectively, $z_{MC} = 0.6$ and 1.5), and each of the constructed spatial filters has a MC = 0.7 and account for about two-thirds of the geographic variance displayed by the coefficients.

The spatial-interaction-based spatial filter is constructed with a linear combination of products of pairs of origin and destination eigenvectors, and hence has $n^2$ elements. Frequencies of these eigenvectors across the selected Kronecker products

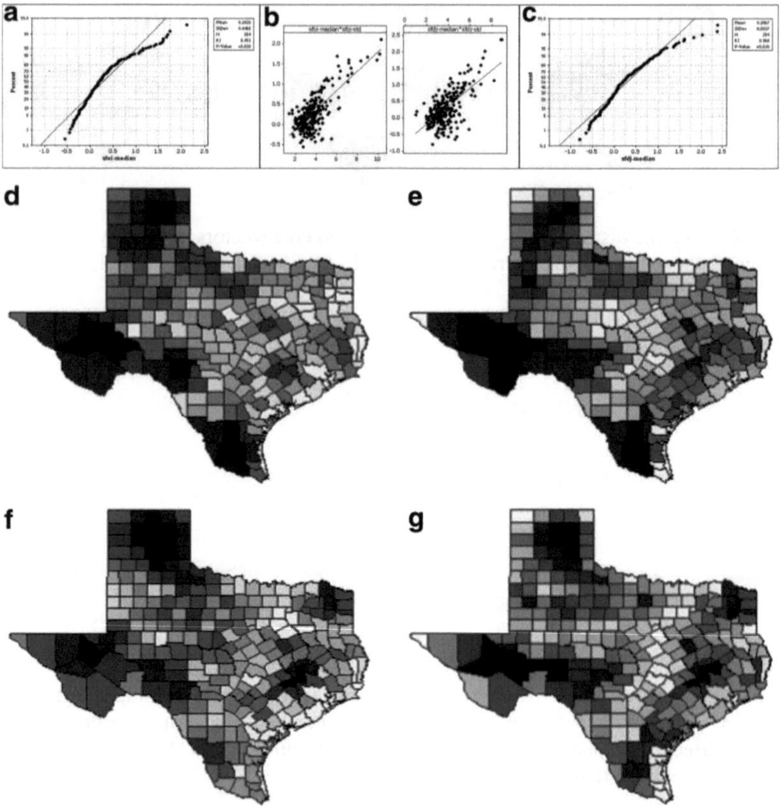


**Fig. 16.10** The spatial-interaction-based spatial filters; light gray to black denotes low to high for the maps. *Top left* (**a**): normal quantile plot for origin spatial filter. *Top middle* (**b**): scatterplots of median spatial filters versus their respective standard deviations (left – origin; right – destination). *Top right* (**c**): normal quantile plot for destination spatial filter. *Middle left* (**d**): geographic distribution of the median origin spatial filter. *Middle right* (**e**): within-county standard deviation for origin-aggregated spatial filter. *Bottom left* (**f**): geographic distribution of the median destination spatial filter. *Bottom right* (**g**): within-county standard deviation for destination-aggregated spatial filter

appear in Table 16.4. Summarizing this synthetic vector by reducing it to an origin as well as a destination version is complicated by the mathematical property of a zero mean for each of the individual eigenvectors: calculating means by origin and by destination yields 0 for all areal units. Therefore, the median together with the standard deviation are used here to construct summary versions; these two measures covary in this case (Figs. 16.11a–c). Because each eigenvector has a variance of 1/n by construction, these variances primarily relate to the sum of squared regression coefficients estimated for each Kronecker product eigenvector. Of note is that fragmented, rather than relatively smooth, geographic patterns are displayed by these spatial filters, as well as their respective variances (Figs 16.11d–f). Because

**Table 16.3** Summary of the spatial-interaction-based spatial filter construction for the doubly-constrained gravity model specification

| Eigenvectors | Origin spatial filter | Destination spatial filter |
|---|---|---|
| Common | $E_2$, $E_4$, $E_5$, $E_8$, $E_{11}$, $E_{12}$, $E_{15}$, $E_{16}$, $E_{19}$, $E_{22}$, $E_{23}$, $E_{29}$, $E_{31}$, $E_{34}$, $E_{38}$, $E_{39}$, $E_{46}$, $E_{47}$, $E_{49}$, $E_{54}$, $E_{56}$, $E_{57}$, $E_{61}$, $E_{63}$ | |
| Unique | $E_1$, $E_6$, $E_7$, $E_{10}$, $E_{14}$, $E_{17}$, $E_{20}$, $E_{25}$, $E_{32}$ | $E_3$, $E_{13}$, $E_{18}$, $E_{21}$, $E_{24}$, $E_{27}$, $E_{30}$, $E_{35}$–$E_{37}$, $E_{44}$, $E_{62}$ |

**Table 16.4** Frequencies of eigenvectors appearing in the Kronecker products

| $C_n$ eigenvector number | Origin frequency | Destination frequency | $C_n$ eigenvector number | Origin frequency | Destination frequency |
|---|---|---|---|---|---|
| 1 | 26 | 19 | 19 | 23 | 19 |
| 2 | 23 | 20 | 20 | 18 | 14 |
| 3 | 23 | 25 | 21 | 21 | 21 |
| 4 | 16 | 19 | 22 | 20 | 24 |
| 5 | 23 | 18 | 23 | 21 | 27 |
| 6 | 18 | 26 | 24 | 15 | 25 |
| 7 | 27 | 29 | 25 | 17 | 19 |
| 8 | 22 | 17 | 26 | 25 | 20 |
| 9 | 21 | 20 | 27 | 17 | 19 |
| 10 | 19 | 14 | 28 | 18 | 24 |
| 11 | 23 | 22 | 29 | 22 | 24 |
| 12 | 18 | 23 | 30 | 17 | 21 |
| 13 | 24 | 13 | 31 | 20 | 18 |
| 14 | 17 | 11 | 32 | 22 | 14 |
| 15 | 21 | 20 | 33 | 24 | 28 |
| 16 | 21 | 24 | 34 | 22 | 23 |
| 17 | 23 | 17 | 35 | 21 | 26 |
| 18 | 15 | 20 | | | |

this fragmentation may be an artifact of how these spatial filters are being summarized here, future research needs to address this particular visualization topic.

Finally, the indicator variable coefficients continue to preserve the row and column totals that serve as constraints in a doubly-constrained gravity model specification. The spatial filter constructed here accounts for spatial autocorrelation in flows between neighboring counties to an origin and neighboring counties to the origin's destination. The indicator variables account for flows from a given origin to all destinations, and all origins into a given destination, and as such has been kept separate from the posited spatial autocorrelation structure employed here. These indicator variables better align observed and estimated flows (Figs. 16.6 and 16.1), with the spatial filter improving upon this alignment as well as minimizing impacts of outlier flows (Figs. 16.7 and 16.9). Although the final Poisson regression model estimation here involves 1,233 parameters, the average number of degrees of freedom per parameter estimate is roughly 52, which should be sufficient for robust results. Of note is that none of the eigenvectors tends to dominate the constructed

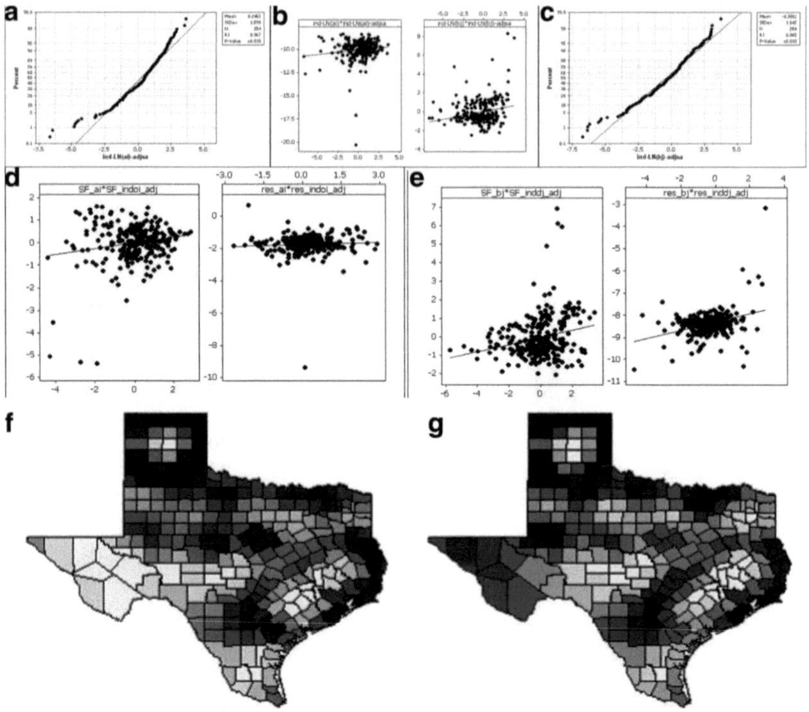


**Fig. 16.11** Spatial filters for the amalgamated indicator variable coefficients adjusted for spatial-interaction-based spatial autocorrelation; light gray to black denotes low to high for the maps. *Top left* (**a**): normal quantile plot for origin spatial filter. *Top middle* (**b**): scatterplots of adjusted and unadjusted amalgamated indicator variable coefficients (left – origin; right – destination). *Top right* (**c**): normal quantile plot for destination spatial filter. *Middle left* (**d**): scatterplots of the adjusted and unadjusted origin spatial filters, and their corresponding adjusted and unadjusted residuals. *Middle right* (**e**): scatterplots of the adjusted and unadjusted destination spatial filters, and their corresponding adjusted and unadjusted residuals. *Bottom left* (**f**): geographic distribution of the origin spatial filter. *Bottom right* (**g**): geographic distribution of the destination spatial filter

spatial filter, which may well be why its origin and destination visualizations (Figs 16.11d and 16.11f) are so fragmented

## 16.7 Implications and Concluding Comments

Results for the spatial-interaction-based spatial filter gravity model are an improvement upon those for the doubly-constrained specification. The decrease in the distance decay parameter estimate for the Texas journey-to-work flows by accounting for spatial autocorrelation in these flows is consistent with Curry's (1972) arguments that the presence of unaccounted for spatial autocorrelation biases the estimated distance decay parameter in a simple gravity model specification. One implication from the analysis reported here is that the constructed spatial filters account for a considerable amount of extra-Poisson variation displayed by flows in geographic space. Estimating a doubly-constrained gravity model yields balancing factors that automatically account for most, if not all, spatial autocorrelation effects attributable to the geographic distributions of relevant origin and destination attributes; but it fails to adequately account for spatial autocorrelation in the flows themselves. And, spatial autocorrelation latent in flows includes global, regional, and local map pattern components (see Table 16.4).

With regard to the four initial posed questions upon which the Texas journey-to-work experiment findings shed light:

- Network spatial autocorrelation can be handled successfully with a Kronecker product of the eigenvectors of matrix $\mathbf{C}_n$, essentially introducing an interaction effect between them, whereas spatial autocorrelation latent in geographic distributions can be handled by including origin- and destination-specific indicator variables (which actually are from the sum of Kronecker products) that preserve row and column total constraints.
- Employing a Kronecker product of the eigenvectors portraying geographic distribution spatial autocorrelation effects also captures network autocorrelation effects.
- The origin and destination weights of a doubly-constrained gravity model account for spatial autocorrelation effects attributable to the geographic distributions of relevant origin and destination attribute variables.
- Kronecker products of the eigenvectors of matrix $\mathbf{C}_n$ can be used to portray spatial autocorrelation in spatial interaction flows, allowing efficient and effective approximation of the eigenvectors needed to be extracted from sizeable sparse geographic weights matrices.

All in all, findings from this case study support the use of spatial filtering methodology to account for spatial autocorrelation effects in geographic flows data.

In conclusion, experimental findings reported here confirm Curry's (1972) conjecture that spatial autocorrelation biases the estimation of distance decay effects uncovered with geographic flows data; for the 2000 Texas journey-to-work data, this bias is by a factor of about 3, and hence is nontrivial.

## 16.8 Future Research

The analysis summarized here suggests two important themes for future research whose findings would contribute to simplifying the complexity of spatial interaction data. First, a more thorough exploration is needed of relationships between origin- and destination-specific indicator variables and Kronecker products of the eigenvectors of matrix $\mathbf{C}_n$ to account for spatial autocorrelation latent in it. This contention builds upon the innovative technical specification outlined by LeSage and Pace (2008), and the novel conceptual specification suggested by Chun (2007). And it results in a model specification that is much simpler than that proposed by Bolduc et al. (1992, 1995, 1997). Another journey-to-work case study for Germany (Griffith 2009) is underway to further address this issue. But more research is needed about this topic.

Second, a mixture of positive and negative spatial autocorrelation contributes substantially to geographic complexity, and separating effects of these two factors would contribute to a better understanding of the complexity of spatial interaction data. Accordingly, the theme of negative spatial autocorrelation should be addressed in future work. No evidence could be found with the Texas journey-to-work flows case study indicating that negative spatial autocorrelation materialized in it. Confirming an absence of negative spatial autocorrelation effects in general, if this is the case, would contribute to an understanding of spatial interaction complexity. Nevertheless, because negative spatial autocorrelation materializes under conditions of geographic competition, the notion of competing destinations associated with spatial interaction suggests that negative spatial autocorrelation could materialize in geographic flows. A spatial filter that is a linear combination of eigenvectors representing negative spatial autocorrelation should be able to capture this effect. More research, both conceptual in nature and with case studies, is needed about this topic.

## References

Black W (1992) Network autocorrelation in transportation network and flow systems, Geogr Anal 24:207–222

Black W, Thomas I (1998) Accidents on Belgium's motorways: a network autocorrelation analysis, J Transp Geogr 6:23–31

Bolduc D, Laferriere R, Santarossa G (1992) Spatial autoregressive error components in travel flow models, Reg Sci Urban Econ 22(3):371–385

Bolduc D, Laferriere R, Santarossa G (1995) Spatial autoregressive error components in travel flow models: an application to aggregate mode choice, In: Anselin L, Florax RJ (eds) New Directions in Spatial Econometrics, Springer, Berlin, pp 96–108

Bolduc D, Fortin B Gordon S (1997) Multinomial probit estimation of spatially interdependent choices: an empirical comparison of two techniques, Int Reg Sci Rev 20(1):77–101

Carey H (1858) Principles of social science. Lippincott, Philadelphia

Chun Y (2007) Behavioral specifications of network autocorrelation in migration modeling: an analysis of migration flows by spatial filtering, unpublished doctoral dissertation, Department of Geography, The Ohio State University

Curry L (1972) Spatial analysis of gravity flows, Reg Stud 6:131–147

Curry L, Griffith D, Sheppard E (1975) Those gravity parameters again. Reg Stud 9:289–296

Fischer M, Griffith D (2008) Modelling spatial autocorrelation in spatial interaction data: an application to patent data in the European Union, J Reg Sci 48:969–989

Flowerdew R, Aitkin M (1982) A method of fitting the gravity model based on the Poisson distribution, J Reg Sci 22:191–202

Flowerdew R, Lovett A (1988) Fitting constrained Poisson regression models to interurban migration flows, Geogr Anal 20:297–307
Foot D (1981) Operational urban models: an introduction. Methuen, NY
Griffith D (2000) A linear regression solution to the spatial autocorrelation problem. J Geogr Syst 2:141–156
Griffith D (2002) A spatial filtering specification for the auto-Poisson model. Stat Probab Lett 58:245–251
Griffith D (2003) Spatial autocorrelation and spatial filtering: gaining understanding through theory and scientific visualization. Springer, Berlin
Griffith D (2007) Spatial structure and spatial interaction: 25 years later, Rev Reg Stud 37 (1):28–38
Griffith D (2009) Modeling spatial autocorrelation in spatial interaction data. J Geogr Syst
Griffith D, Jones K (1980) Explorations into the relationship between spatial structure and spatial interaction, Environ Plan A 12:187–201
LeSage J, Pace R (2008) Spatial econometric modelling of origin-destination flows, J Reg Sci 48:941–967
Niedercorn J, Bechdolt B (1969) An economic derivation of the 'gravity law' of spatial interaction. J Reg Sci 9:273–282
Ravenstein E (1885) The laws of migration. J R Stat Soc 48:241–305
Sen A, Smith T (1995) Gravity models of spatial interaction behavior. Springer, Berlin
Sheppard E, Griffith D, Curry L (1976) A final comment on mis-specification and autocorrelation in those gravity parameters. Reg Stud 10:337–339
Tiefelsdorf M, Braun G (1999) Network Autocorrelation in Poisson Regression Residuals: Inter-district Migration Patterns and Trends within Berlin. Paper presented at the 11th European Colloquium on Quantitative and Theoretical Geography, Durham City, England, September 3–7
Wilson A (1967) A statistical theory of spatial distribution models. Transp Res 1:253–269

Flowerdew R, Lovett A (1988) [illegible] Poisson regression [illegible]
[illegible]
[illegible] spatial [illegible]
[illegible]
[illegible] understanding through [illegible]
[illegible]

# Chapter 17
# Complex Networks Analysis of Commuting

## Recent Advances and a Research Agenda

**Andrea De Montis, Alessandro Chessa, Michele Campagna, Simone Caschili, and Giancarlo Deplano**

## 17.1 Introduction

The emerging new science of networks is providing an elegant paradigm for the characterization of the broad area of complex systems. New research perspectives have been opened in the study of many real phenomena and processes, and recently fields like urban, regional, and environmental sciences have gained new insights from the tools provided by network science. The complex networks analysis (CNA) becomes a useful framework in these fields to disentangle problems of a complex and unpredictable nature.

At the end of the last millennium, the availability of large data sets and the parallel explosion of computer processing power have made a systematic and intensive application of CNA to the study of very large networks(Pastor-Satorras and Vespignani 2004; Albert and Barabási 2002) possible. According to CNA, complex behaviours are signalled by the emergence of some characteristics that can be featured in terms of statistical properties.

CNA provides new insights into the study of new classes of networks that behave differently with respect to the usual random graph networks, where each pair of nodes is connected with a certain probability p. For example, small world networks differ from random networks in that the clustering coefficient – an index of local connectedness of a node – is remarkably higher then the expected random case, while the diameter of the system – equal to the maximum distance between a pair whatsoever of nodes – scales very slowly with the size (measured by the number of nodes of the system); scale free networks differ in that the probability distribution of their degree – a measure of the number of first neighbours of a node – is much broader than the random case, with a divergent variance. Scale free networks growth mechanism can be described by the so called preferential attachment

A. De Montis (✉)
Dipartimento di Ingegneria del Territorio, Sezione Costruzioni e Infrastrutture, Università degli Studi di Sassari, Sassari, Italy

A. Reggiani and P. Nijkamp (eds.), *Complexity and Spatial Networks*,
Advances in Spatial Science,
DOI: 10.1007/978-3-642-01554-0_17, 

model (Barabási and Albert 1999): new nodes tend to grasp the highest advantage from the system they join by linking with nodes that present a very large degree value, the "hubs" of the network.

Beyond the simulations, CNA has been applied to a number of real systems, such as food webs, human interactions, the Internet, the world wide web, the spread of diseases, population genetics, genomics and proteomics. In each of these cases, the analyst inspects active systems characterised by elements (the nodes) connected through different kinds of interactions (the links). For a review of these applications, see Albert and Barabási (2002) and Newman (2003).

Recently also in many fields grouped under the label of regional science, a number of scholars have applied the paradigm of complex network analysis to model urban (Batty 2001; Jiang and Claramunt 2004), regional and socio-economic systems (Barrat et al. 2004a; Latora and Marchiori 2003; Schintler et al. 2005; O'Kelly 1998; Reggiani et al. 2009). Some authors have inspected the influence of geographical space on the network properties (Gorman and Kulkarni 2004; Gastner and Newman 2004; Crucitti et al. 2006; Campagna et al 2007).

Some applications refer to the study of infrastructures and of commuting behaviour (Guimera et al 2003a; Latora and Marchiori 2002; Chowell et al. 2003; Sen et al. 2003; Strano et al. 2007; Porta et al. 2008).

In this chapter, we aim at developing a research agenda on our operative CNA tools, applied to commuting systems, with a glance at policy making and planning. The contents of this chapter are outlined as follows. In the next section, we recall some results recently obtained by applying CNA to study topological, traffic and spatial properties of commuting in insular Italy. In the third section, based on the obtained results, we propose a research agenda on the following main topics: (a) integration between geographic information science (GIS) and CNA to study the spatial properties of a system, (b) the evolution of networks in time, (c) the adoption of CNA as a tool to compare systems, and (d) community detection on real networks. In the fourth section, we present our concluding remarks as well as some outlook reflections.

## 17.2 Complex Network Analysis Applied to Commuting: A Review of Recent Advances

### *17.2.1 The Sardinian Inter-Municipal Commuting Network (SMCN)*

De Montis et al. (2007) has studied the inter-municipal commuting system of the Island of Sardinia, Italy. In that study, the authors inspected workers' and students' daily movements, by adopting a network representation, the Sardinian inter-Municipal Commuting Network (SMCN) characterised by N = 375 vertices,

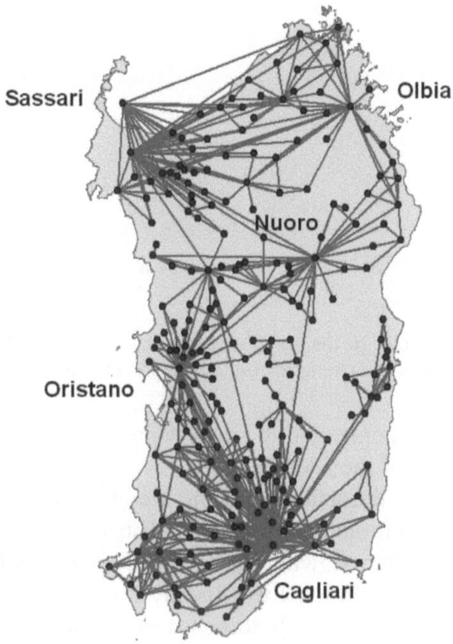


**Fig. 17.1** Geographical and topologic representation of the Sardinian inter-Municipal Commuting Network (SMCN) (De Montis et al. 2008)

each one corresponding to a town, and E = 8124 edges, each one representing how many people commute between two extreme towns.

The authors have analysed the SMCN by representing it as an undirected graph. They have gone beyond the study of the mere topology, by also considering its traffic: through a weighted undirected network (Barrat et al. 2004a) representation they have processed the regional origin destination table, a dataset that reports the daily work and study-led movements among Sardinian municipalities.

Hereafter, we report the most important results obtained in that study.

In Fig. 17.1, the Sardinian inter-Municipal Commuting Network (SMCN) is represented: in this simplified picture, the nodes (black points) correspond to the towns, while the links correspond to a flow value larger than 35 commuters between two towns (De Montis et al. 2008).

The analyses of the topological features show that the SMCN belongs to the class of random networks, since the curve of the probability distribution of the degree $k$ is bell-shaped with a mean value $<k> \sim 40$. The study of the clustering coefficient (Watts and Strogatz 1998) averaged over the degree $k$, $C(k)$, a measure of the level of connectedness among first neighbours of a node, uncovers properties common in other technological networks. In particular, small (with small $k$) municipalities are locally densely interconnected, while large municipalities provide a large set of connections for remote regions otherwise disconnected. This evidence is confirmed also by the analysis of the average degree of the nearest neighbours (Barrat et al. 2004a), which signals a disassortative mixed behaviour: the hub towns

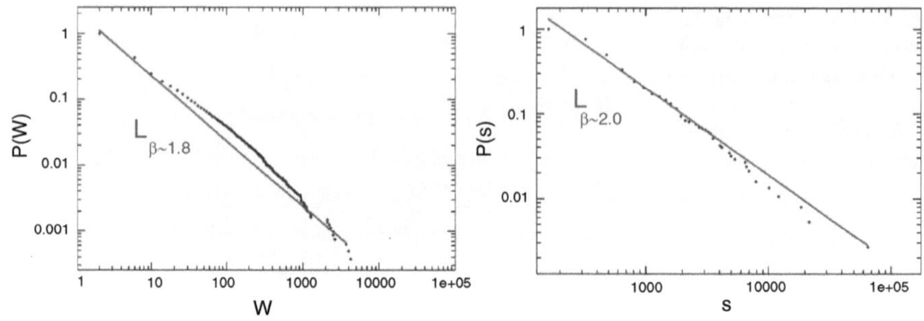


**Fig. 17.2** Log–log plot of the complementary cumulative probability distribution of the weight $w$ (*on the left*) and of the strength $s$ (*on the right*) (De Montis et al. 2007)

are preferentially connected to small degree (less central) municipalities acting as star-like vertices of the SMCN.

On the side of the analysis of the weighted network, the study finds that the complementary cumulative probability distributions of both weight $w$ (equal to the number of commuters flowing between two towns) and strength $s$ (equal to the total traffic handled by a municipality) display a power-law regime over a wide spectrum of degree values (Fig. 17.2). In this case, no characteristic value of the distribution is found and the SMCN can be included in the class of scale-free weighted networks.

The spectrum of the strength $s$ averaged over the values of the degrees reveals a super linear behaviour implying that the higher the number of connections to a town the larger the traffic per connection handled. This means that there are probably hidden properties that control and show the behaviour of the network.

The inspection of the disparity of a node, which measures possible inequalities in the distribution of the traffic flow among the connections of each node, confirms the actual structure of the real network: a fairly high number of commuters are exchanged between hub towns through a very small number of backbone connections constituting the dorsal "highways" of that system as a topological network.

In the next section, a selection of results is reported about the comparison between the Sicilian and the Sardinian inter-Municipal Commuting Networks (SiMCN and SMCN).

### 17.2.2 The Sicilian vs. The Sardinian Inter-Municipal Commuting Network

De Montis et al. (2009) have compared the commuting systems of the two main Italian islands (Sardinia and Sicily) adopting a CNA to discover similarities and dissimilarities in those systems.

In Tables 17.1 and 17.2 we summarise the topological properties of the two networks under study. In Table 17.1, $N$ stands for the number of nodes, $E$ for the number of edges, $k$ for the degree, and $l$ for the average shortest path length between

**Table 17.1** Comparative overview of topological properties of the SMCN and SiMCN (De Montis et al. 2009)

| | $N$ | $E$ | $k_{min}$ | $k_{max}$ | $<k>$ | $<l>$ | $l_{max}$ |
|---|---|---|---|---|---|---|---|
| SMCN | 375 | 8124 | 8 | 279 | 40 | 2.0 | 3 |
| SiMCN | 391 | 9993 | 1 | 280 | 51 | 1.98 | 4 |

**Table 17.2** Comparative overview of topological properties of the SMCN and SiMCN, part 2 (De Montis et al. 2009)

| | $P(k)$ | $C(k)$ | $<C>$ |
|---|---|---|---|
| SMCN | Bell shaped | Downward sloping | 0.26 |
| SiMCN | Bell shaped | Downward sloping | 0.52 |

**Table 17.3** Comparative overview of the traffic properties of the SMCN and the SiMCN (De Montis et al., 2009)

| | $<w>$ | $w_{max}$ | $P(w)$ | $P(s)$ |
|---|---|---|---|---|
| SMCN | 27 | 13953 | Power law with exp $\sim$ 1.8 | Power law with exp $\sim$ 2.0 |
| SiMCN | 37.6 | 10233 | Power law with exp $\sim$ 2.1 | Power law with exp $\sim$ 2.2 |

two nodes whatsoever measured in terms of number of edges. In Table 17.2, $P(k)$ stands for the probability distribution of the degree $k$, and $C(k)$ for the spectrum of the clustering coefficient, measuring the level of local connectedness of a town with a given degree $k$.

The curve of the probability distribution of the degree $P(k)$ is bell shaped for both Sardinia and Sicily with a characteristic and defined mean value: hence those networks belong to the broader class of random graphs. A similar trend of the clustering coefficient spectrum $C(k)$ confirms a common property of the local structure in many infrastructure networks: hub towns tend to connect otherwise disconnected regions, while small degree $k$ towns are locally very densely connected. The average clustering coefficient $<C>$ for the SiMCN is nearly twice as high as it is for the SMCN: this is a sign of a difference in the local structure. Overall in Sardinia a star-like scheme dominates the topology, while in Sicily it is loop-like.

Some traffic properties are outlined in Table 17.3 and pictured in Fig. 17.3. It is possible to note that the weights are very heterogeneous, having a maximum value three orders bigger than the mean value. The complementary cumulative probability distributions of the weights in both cases present a power law behaviour over a broad range of values. The behaviour of the complementary cumulative probability distribution of the strength $s$ fits again a power law line with a slope exponent close to two in both the SMCN and the SiMCN: these systems can be classified as scale free weighted networks.

This implies that it is not possible to find a typical value characterizing the probability distribution of both the weight and the strength. This statistical evidence supports the emergence of two phenomena in both the insular systems: it is not possible to find typical values both of the commuter flow between a pair of towns

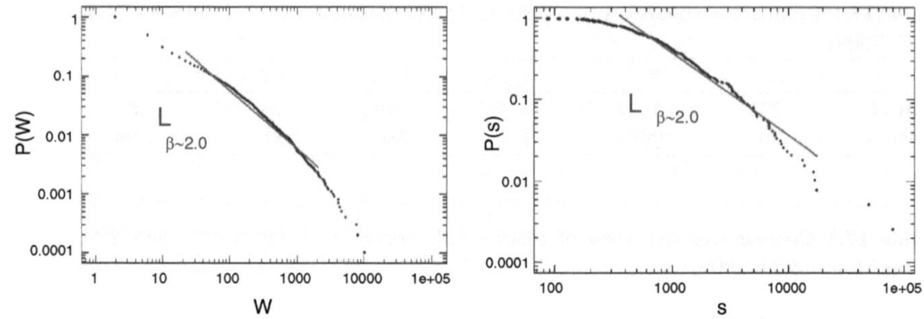


**Fig. 17.3** Log–log plot of the complementary cumulative probability distribution of the weight *w* (*on the left*) and of the strength *s* (*on the right*) for the SiMCN (De Montis et al. 2009)

**Table 17.4** Comparative overview of interplay properties of the SMCN and SiMCN (De Montis et al. 2009)

| | <*s*>(*k*) | <Pop>(*k*) | <Pop>(*S*) |
|---|---|---|---|
| SMCN | Upward sloping with exp ~ 1.9 | Upward sloping with exp ~ 1.7 | Upward sloping with exp ~ 0.9 |
| SiMCN | Upward sloping with exp ~ 1.8 | Upward sloping with exp ~ 1.4 | Upward sloping with exp ~ 0.8 |

whatsoever and of the total amount of commuter traffic handled. The distribution of these two traffic variables can be considered scale invariant.

Regarding the properties described in Table 17.4 on the analysis of the interplay between traffic, topologic and demographic characteristics, the similarities between the SMCN and SiMCN can be detected. In both cases, a super-linear behaviour in the spectrum of the strength $s$ is found with respect to the degree $k$: in these networks the traffic per connection increases when the degree $k$ increases. In other terms, the higher the topologic centrality of a town is, the higher its traffic centrality is (almost twice as much in both systems). The analysis of the interplay between topological, traffic and demographic properties of the networks reveals a common behaviour, since for both networks town population *pop* (ISTAT 2001b) scales faster than the degree $k$: the higher the degree, the higher the population (nearly twice as much). This trend might be a sign of hidden phenomena connected to economies of scale typical of the historical regional development.

From these preliminary results, we would posit that the Sardinian and the Sicilian inter-municipal commuting networks display similar general and local statistical properties in many cases. This implies that, as far as this study is concerned, similar geographical settings lead to common network properties.

### *17.2.3 The Influence of Space on Commuter Networks*

In order to analyse the influence of spatial location on commuting networks, Campagna et al. (2007) have approached the characterization of the spatial features

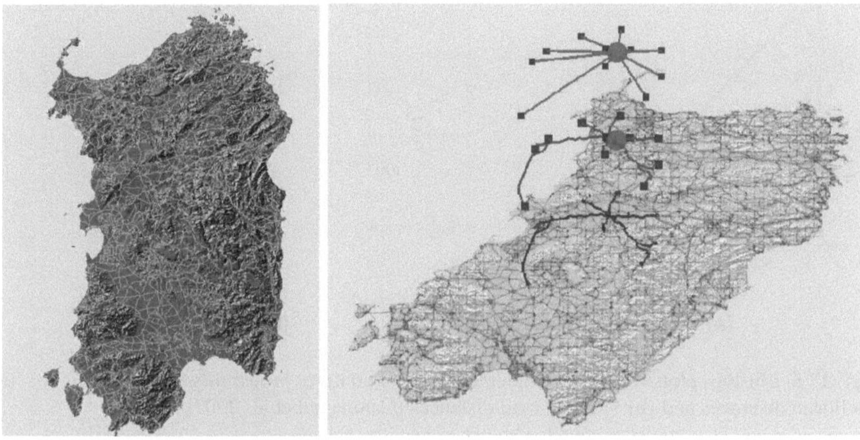

**Fig. 17.4** On the left, a global view of the Sardinian inter-urban road network. On the right, 3D visualizations of the SMECN and SMSCN models (Campagna et al. 2007)

of the SMCN in terms of analysis of the mutual location between municipalities by studying the statistical distribution of commuting distances (Fig. 17.4).

For this purpose, they have considered two networks that are similar to the SMCN, as they display the same set of nodes representing Sardinian towns as well as the same number of links. By contrast, these systems are different from the SMCN, when regarded as weighted networks, since they show two diverse attributes (the weights of the network) for the set of edges. In the case of the Sardinian inter-Municipal Euclidean distance Commuting Network (SMECN), the weight is equal to the length of the segment ideally connecting two towns, while in the Sardinian inter-Municipal Shortest-path distance Commuting Network (SMSCN), to the length of the shortest road path between two towns.

The results of analysis of the complementary cumulative probability distribution of the weights – that is, Euclidean distances ($W_e$) and of the Shortest road distances ($W_r$) – are illustrated in Fig. 17.5. The shape of the curve, for both these networks, indicates the emergence of an exponential decay behaviour.

A different result has been obtained from the inspection of the strength $S_e$ (strength of SMECN) and $S_r$ (strength of SMSCN). The strength of a node combines the information about its connectivity and the intensity of each connection. In the case of spatial distances, this measure represents the total distance covered by town's commuters to reach it. As Fig. 17.6 shows, the complementary cumulative probability distribution of both $S_e$ and $S_r$ display a heterogeneous behaviour in which a not negligible number of hubs are present.

In Fig. 17.7, we report the results obtained from the analysis of the strength $<S_e>(k)$ and $<S_r>(k)$ averaged over all nodes with a given degree $k$. It is possible to observe a super linear behaviour over a wide range of degrees with an exponent $\beta \approx 1.3$ for both networks. In this case, the sum of the distances covered through the

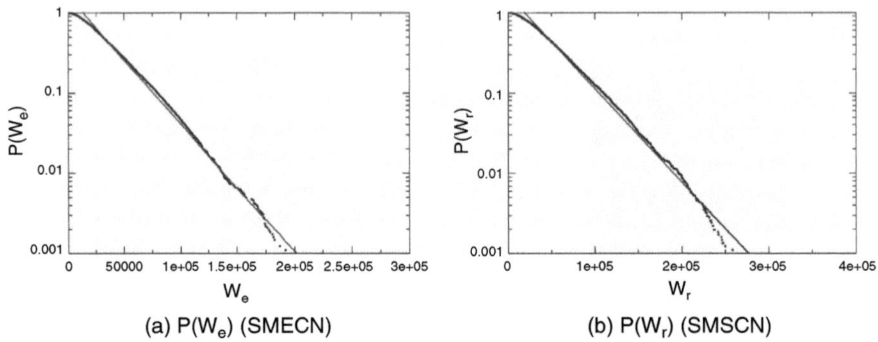


**Fig. 17.5** Lin-log plots of the complementary cumulative probability distributions of (a) Euclidean distances and (b) Shortest road distances (Campagna et al. 2007)

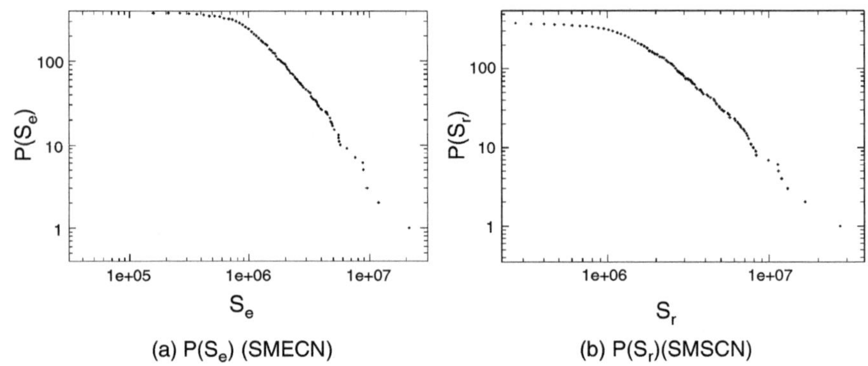


**Fig. 17.6** Log–log plot of the complementary cumulative probability distribution of $S_e$ and of $S_r$ (Campagna et al. 2007)

available connections – that is, a proxy measure of the commuting footprint of each town – grows slightly faster than the number of connections.

Independency of strength and degree would yield a value $\beta = 1$ (Barrat et al. 2004a); by contrast this result reveals that there is a super linear correlation between the space (in terms of distances) and the topology in both networks. A comparison of this result with the corresponding findings obtained by De Montis et al. (2007), who measure an exponent $\beta \approx 1.9$, signals that in the present case the strength grows slower than the degree. In the case of the SMCN, the total amount of traffic handled by a town increases on average more (almost twice as much) than the corresponding number of first neighbours of that town. On the contrary, in the SMECN and SMSCN this correlation is not strong as much as in the previous case: the sum of the distances covered through the available connections grows (only) slightly more than the number of connections.

**Table 17.5** SMCN, SMECN, and SMSCN: a synthesis of the relevant properties (Campagna et al. 2007)

| | Description | Code | *W* | Dataset | Source |
|---|---|---|---|---|---|
| A- spatial network | Sardinian inter-municipal commuting network | SMCN | Commuters' traffic between pairs of municipalities | ODT | ISTAT (1991) |
| Spatial networks | Sardinian inter-Municipal Euclidean distance Commuting Network | SMECN | Euclidean distance between pairs of municipalities | Geographical coordinates of town centres | Centro Interregionale (2007) |
| | Sardinian inter-municipal shortest road distance commuting network | SMSCN | Shortest path road length for each pair of municipalities | Geographical coordinates of town centres and roads geodatabase | Centro Interregionale (2007) |

**Table 17.6** Statistics of the SMCN, SMECN, and SMSCN (Campagna et al. 2007)

| | Code | *P*(*W*) | *P*(*S*) | *S*(*k*) |
|---|---|---|---|---|
| A- spatial network | SMCN | Power-law (scale free network) | Power-law (scale free network) | Super-linear behaviour (line fit exponent $\beta \sim 1.9$) |
| Spatial networks | SMECN | Exponential (random network) | Power-law (scale free network) | Super-linear behaviour (line fit exponent $\beta \sim 1.3$) |
| | SMSCN | Exponential (random network) | Power-law (scale free network) | Super-linear behaviour (line fit exponent $\beta \sim 1.3$) |

In Table 17.6, we summarise the results of the analysis above.

As a general result, the analysis confirms the results found by De Montis et al. (2007) and reveals strong connections between the traffic properties of the system, in terms of commuting flows, and geographical properties.

In particular, it is now worthwhile to develop on a difference emerging between a-spatial and spatial networks. The SMCN, an a-spatial network, displays heterogeneous probability distributions *P*(*W*) and *P*(*S*), where *w* stands for the number of commuters associated to each edge and *s* for the total amount of commuters handled in each node (town). By contrast, the spatial networks SMECN and SMSCN seem to belong to a hybrid class of systems: in fact, they are both homogeneous, with respect to the probability distribution of the distance based weights $P(W_e)$ and $P(W_r)$, and heterogeneous, with respect to the probability distribution of the distance based strengths $P(S_e)$ and $P(S_r)$.

The ambivalence can be explained as follows. Sardinian road system is closed, as bounded by the coastal fringe, so that road distance values range approximately

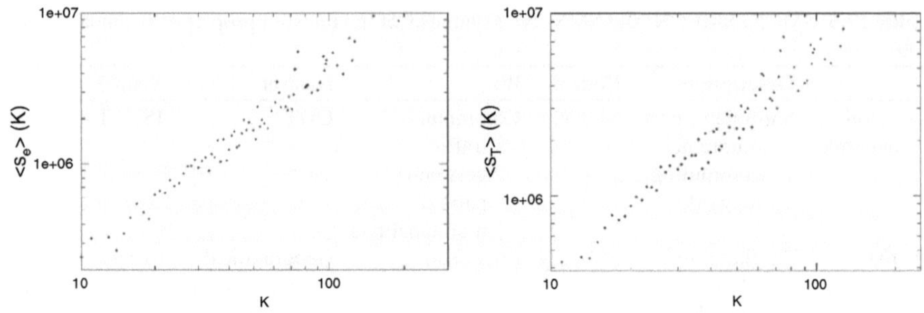


(a) SMECN (b) SMSCN

**Fig. 17.7** Log–log plot of average strength as a function of the degree (Campagna et al. 2007)

between 1 and 400 km. This may be one of the main causes leading to statistical distributions $P(W_e)$ and $P(W_r)$ that are scattered randomly around a typical mean value. On the other side, distance based strengths stand as measures of the total length of the roads covered by commuters around each town and depend on a more volatile combination of traffic and topological local features. This is the main reason why in this case the corresponding probability values $P(S_e)$ and $P(S_r)$ span over quite a broad range and fit power law distributions.

## 17.3 Complex Network Analysis Applied to Commuter Systems: A Research Agenda

Motivated by the results outlined above, in this section we propose a research agenda to fully exploit the potential of CNA to model and analyse territorial phenomena for policy making and planning.

Hereafter we focus on the following research questions (see Fig. 17.8): (a) integration between geographic information science (GIS) and CNA to study the spatial properties of a system, (b) the evolution of networks in time, (c) the adoption of CNA as a tool to compare systems, and (d) community detection on real networks.

### *17.3.1 Integration between GIS and CNA*

In this section, we aim at giving a tentative framework of GIS–CNA integration to show possible ways to take into account the spatial dimension and to help to enrich the understanding of those phenomena which are clearly influenced by space, yet are too often described only in terms of a-spatial variables.

Network modelling in a GIS environment has been widely used to represent real world physical infrastructures such as roads, cables, or pipelines. In these cases, the

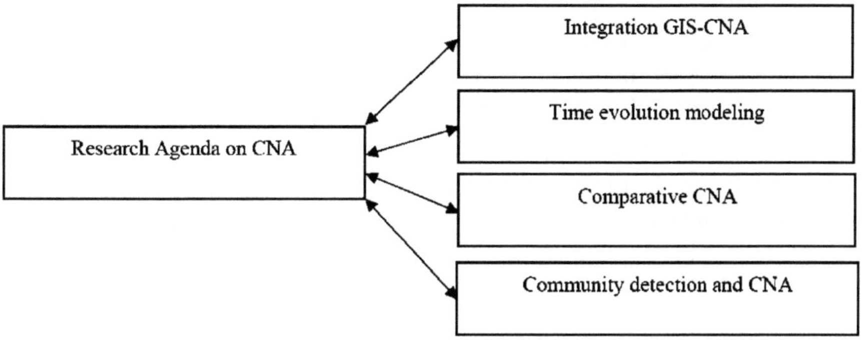


**Fig. 17.8** Research agenda on CNA: a flow chart

components are features characterised as having a spatial dimension prevailing over the others.

CNA has indeed been used to model different systems according to different relationships with space: "pure" spatial physical structures (for example, transport network systems, pipelines), phenomena for which space matters but is not explicitly considered in the model (for example, World Wide Web network), and phenomena for which the spatial dimension is of minor relevance and can reasonably be omitted (for example, movie co-acting network).

While on the one hand GIS have been proven to supply reliable support in physical network modelling, such as in transport planning and utility management, and are widely adopted, the diffusion of the GIS application to CNA is less common in those disciplines or problems where the spatial dimension of the network is considered to be of lesser importance.

The integration of models typical of CNA in a GIS environment can thus be developed according to different approaches each of which underlines a different kind of support GIS offer to CNA. It spans from the representation of the CNA results, to feeding CNA input dataset, or to modelling the complex network itself. In more detail, possible integrations can be outlined as follows:

- Spatial Representation of CNA analysis output: in the simplest case, the results of a CNA can be represented in cartographic form. An example of this kind of CNA-GIS integration can be found in the work of Colizza et al. (2007, Fig. 17.6), where results of CNA are plotted with GIS software.
- GIS feeding CNA models, type I: in this case, spatial properties of the nodes/links are stored in a geodatabase and exported as input data for CNA algorithms (such as link weights in weighted complex networks analysis): when weights in a complex weighted network model show spatial dependence, they can be calculated through overlay or other spatial analysis functions to take into account the influence of surrounding objects/fields. An example of this approach is offered by Campagna et al. (2007) for the commuter network at the regional level described in Sect. 17.2.3 and by Chiricota et al. (2008) for the national commuter network of France.

- GIS feeding CNA models, type II: when the system under analysis is inherently spatial in essence, the resulting network model can be fully represented and analysed in a GIS environment. In this case, some programming may be developed within the GIS to calculate CNA measures.

Although further work is under development by the authors to better formalise the GIS–CNA integration, the overview given is intended as an early call to stimulate both scientists and GIS developer communities to be aware of and possibly join our early efforts.

### *17.3.2 Time Evolution of Complex Networks*

A second topic of interest concerns the inspection of a complex network evolving in time. Until the end of the 1990s, the Erdös and Rényi random model (E-R model) was undoubtedly the best model to represent complex networks. Yet, the E-R model is based on two assumptions: the number of elements (nodes) is fixed *a priori* and it does not change during its life-time; there is also no distinction among nodes.

By contrast, Barabási and Albert (1999) brought up a new network modelling discussion about the rudiments of E-R model by clarifying two properties: they evolve in time and they have preferential attachments. Looking at real world phenomena, it is often possible to observe much the same behaviour. Starting from this similarity between complex network and real world phenomena, how is it possible to model the time evolution of these systems? Is it reasonable to give a deterministic or just qualitative description of real world phenomena?

Using the aB-A model we are able to describe systems characterised by preferential attachment and high connectivity. Some of the most enthralling questions for research in territorial and regional fields are: do also those rules regulate a territorial system? Can we state that a territorial system could be modelled as a network and that it is similar to a B-A model? In the case of weighted networks can we use the approaches proposed by Jost and Joy (2002) and by Barrat et al. (2004b)? Our findings show similar patterns among territorial systems and complex networks. Thus it is reasonable to think that the complex network paradigm, applied to regional and urban systems, can describe an evolution in time of regional phenomena. In the case of B-A model the network evolution has been described by the variation in the number of nodes and their preferential attachment. For weighted networks, both models proposed by Jost and Joy (2002) and Barrat et al. (2004b) aimed at simulating the systems' evolution based on two coupled mechanisms: the topological growth and the weights' dynamics. Thus, they provided two models that took into account the interplay between topology and traffic in a network. The time evolution of a territorial system needs some rudiments to be fully modelled:

- The minimum amount of time for a window observation
- To fix the network features that can describe its evolution

Serrano et al. (2007) studied the world trade network. In order to take into account the network evolution in a 40 year time period, they used three observations: 1960, 1980, and 2000. Actually to better understand the change of such real world phenomena one may need to consider observations over a shorter period of time.

In the case of the SMCN, the analyst is able to study the dynamics of a complex system starting from census data – the origin destination table (ISTAT 2001a) describing commuting movements among Sardinian towns – available for 1981, 1991, and 2001 (every decade). The SMCN, as a municipal transportation network, presents a fairly high degree of robustness and a very weak tendency to increase the number of nodes. Thus it is pointless to apply a complex network model, like the B-A, to the case of commuting (as modelled in this chapter). On the other side, it might be worth investigating the patterns of the fluctuations, in time, of network measures, such as the degree, the clustering coefficient, the weight, and the strength.

In the case of commuter movement in a European lagging behind region such as Sardinia, it is reasonable to guess that, even in a mid-century, some topological properties – that is, the number of nodes – of the network will change only slightly. By contrast, we expect that the same traffic properties will vary – that is, number of commuters moving in the whole network and weight measuring the commuter flow conveyed on each single edge.

### *17.3.3 Comparative Analysis through Complex Network Theory*

The complex network theory starts from the assumption that we are not able to model the single interactions between each pair of elements of a large system. Hence we study these systems taking into account their overall properties. The power laws give a description of scale invariance features and have been used as universal rules. The concept of "universality" stems from the assumption that there are properties for a large class of systems that are independent of the dynamical details. According to the concepts introduced in the field of statistical mechanics, in physics and mathematics these laws control complex systems and phase transitions. Also we can find the power law and universality behaviour in economical systems, social networks, and fluctuation of goods (Barabási 2000).

In this theoretical framework we aim at modelling a territorial system as a network in order to analyse real phenomena and support urban and regional planning at different scales of observation (local, regional, and national).

Our first studies follow this path: we took into account a bounded regional system (Sardinia is an island) and gave a description of dynamics of socio-economical features by using the commuter movements between Sardinian towns. This modelling approach has highlighted both some hidden properties and trivial ones.

Territorial systems are influenced by many relevant variables so that we often define them as complicated systems: are they also complex systems? Are they composed by interacting elements characterised by universal laws?

We plan to test the assumption that spatial systems – even in different geographical settings – display similar characteristics obeying to constant rules that, when properly proved, may lead to the definition of universal laws. We have begun to compare two similar – that is, insular – regional commuter systems of Sardinia and Sicily. While the results confirm that the two commuter systems are similar, we are aware of the need to check our thesis out studying other dissimilar regional systems.

We will study some non-insular systems from another part of Italy. Our inspection will cover northern Italy which has a better economical situation and also a better transportation system than the south. These new case studies should confirm or confute our thesis: there are hidden universal laws that underlie the network of commuter movements we have observed in insular Italy.

Forthcoming inspections will look upon the influence of our modelling approach at different space scales.

It would be very interesting also to verify whether similar statistical properties – that is, power law behaviour, small-world effect, and presence of hubs – may be observed at a larger scale of observation. If this is the case, the same phenomena recovered in a regional commuter network should be found also for a national system. The opposite result might mean that, in Italy, regional commuting networks are typical, obey to local characteristics and rules and do not parallel the general features of the national system.

### *17.3.4 Community Detection on Commuter Networks*

Very promising research on CNA concerns the detection of communities. A community in a network may be defined as a set of nodes that present a high number of internal links and few links toward the nodes of other communities. Many authors have proposed methods and algorithms to develop community detection on networks, while others have discussed theoretical and operative limits (Danon et al. 2005; Newman 2004, 2006; Palla et al. 2005; Radicchi et al. 2004; Guimerà et al. 2003a, b; Rosvall and Bergstrom 2007; Fortunato and Barthélemy 2007; Arenas et al. 2007).

Research on communities has also involved the realms of regional science, geography and planning, for many purposes. Cluster analysis and other related methods have been considered and tested throughout by many scholars, since the pioneering studies of Berry (1964) and Fischer (1979). Recently, the interest has revamped among geographers for regionalization methods also provoked by the advances in geographic information science (Noronha and Goodchild 1992).

We believe network community detection may contribute to seeking finer methods for classifying homogeneous sub regions. This may be achieved integrating traditional clustering methods by explicitly taking into account the relational properties of the entities (the nodes). The other way around, territorial settings may provide us with a useful real world case study for testing the versatility of community detection over complex networks.

Motivated by these background reflections, we will be studying the communities of the commuting networks above and comparing them to the current relevant administrative bodies. Hopefully the pattern of productive basins, constructed on the self-organised daily movements of students and workers, will inspire decision makers to calibrate the boundaries of administrative subdivisions accordingly.

## 17.4 Conclusion and Outlook Remarks

In this chapter, we have presented some retrospective thoughts on recent results obtained by applying complex network analysis (CNA) to the study of commuter networks sited in insular Italy.

These encouraging results have led us to reflect on a possible research agenda that comprehends four main research areas: (a) the integration between GIS and CNA, (b) the study of the evolution of networks in time, (c) the analysis of comparable and non-comparable networks, and (d) the detection of communities on networks.

We will direct our efforts in such a way that the research agenda above will serve as a guideline for our future works, under the assumption that integration is definitely needed between CNA and the more operational theories of regional science, policy making and planning.

**Acknowledgments** A.C., A.D. M., M.C. and G.D. acknowledge financial support by MIUR under project PON-CyberSar. This work makes use of results produced by the Cybersar Project managed by the Consorzio COSMOLAB, a project co-funded by the Italian Ministry of University and Research (MUR) within the Programma Operativo Nazionale 2000–2006 "Ricerca Scientifica, Sviluppo Tecnologico, Alta Formazione" per le Regioni Italiane dell'Obiettivo 1 (Campania, Calabria, Puglia, Basilicata, Sicilia, Sardegna) – Asse II, Misura II.2 "Società dell'Informazione", Azione a "Sistemi di calcolo e simulazione ad alte prestazioni". More information is available at http://www.cybersar.it.

## References

Albert R, Barabási AL (2002) Statistical mechanics of complex networks. Rev Mod Phys 74:47–97

Arenas A, Fernadez A, Gomez S (2007) Multiple resolution of the modular structure of complex networks. ArXiv0703218v1: http://lanl.arxiv.org/abs/physics/0703218v1

Barabási AL (2000) Linked: how everything is connected to everything else and what it means, Plume

Barabási AL, Albert R (1999) Emergence of scaling in random networks. Science 286:509–512

Barrat A, Barthélemy M, Pastor-Satorras R et al. (2004a) The architecture of complex weighted networks. Proc Natl Acad Sci 11:3747–3752

Barrat A, Barthélemy M, Vespignani A (2004b) Modeling the evolution of weighted networks. arXiv:cond-mat/04006238 v1

Batty M (2001) Cities as small worlds. Editorial. Environ Plan B: Plan Des 28 637–638

Berry BJL (1964) Approaches to regional analysis: a synthesis. Ann Assoc Am Geogr 54:2–11

Campagna M, Caschili S, Chessa A et al. (2007) Modeling commuters dynamics as a complex network: the influence of space. In: Proceedings of the 10th international conference on computers in urban planning and urban management (CUPUM), Iguassu Falls, July, 11–13 2007

Centro Interregionale (2007) Progetto DBPrior10k: strati prioritari di interesse nazionale. Centro Interregionale, Regione Autonoma della Sardegna, Assessorato Enti Locali Finanze e Urbanistica, Servizio per la Pianificazione Territoriale e Cartografia

Chiricota Y, Melançon G, Phan Quang TT et al. (2008) Visual exploration of (French) commuter network, In: Proceedings of geovisualization of dynamics, movement, and change agile'08 satellite workshop Spain, accessible at http://hal-lirmm.ccsd.cnrs.fr/lirmm-00272786/en/

Chowell G, Hyman JM, Eubank S et al. (2003) Scaling laws for the movement of people between locations in a large city. Phys Rev E 68:066102

Colizza V, Barrat A, Barthelemy M et al. (2007) Modeling the worldwide spread of pandemic influenza: baseline case and containment interventions. PLoS Med 4(1):e13 doi:10.1371/journal.pmed.0040013

Crucitti P, Latora V, Porta S (2006) Centrality measures in spatial networks of urban streets. Phys Rev E 73:036125

Danon L, Díaz-Guilera A, Duch J, Arenas A (2005) Comparing community structure identification. J. Stat. Mech.: P09008

De Montis A, Barthélemy M, Chessa A et al. (2007) The structure of interurban traffic: a weighted network analysis. Environ Plan B: Plan Des 34(5):905–924

De Montis A, Caschili S, Campagna M et al. (2008) Time evolution of complex networks: commuting systems in insular Italy. In: 48th Congress of the European Regional Science Association, Liverpool, UK, pp 27–31, August 2008

De Montis A, Chessa A, Caschili S et al. (2009) Modelling commuting systems through a complex network analysis: a study of the Italian islands of Sardinia and Sicily. J Transp Land Use (forthcoming)

Fischer MM (1979) Regional Taxonomy: a comparison of some hierarchic and non-hierarchic strategies, Reg Sci Urban Econ 10:503–537

Fortunato S, Barthélemy M (2007) Resolution limit in community detection. Proc Natl Acad Sci 104(1):36–41

Gastner MT, Newman MEJ (2004) The spatial structure of networks. Cond-mat 0407680

Gorman SP, Kulkarni R (2004) Spatial small worlds: new geographic patterns for an information economy. Environ Plan B: Plan Des 31:273–296

Guimera R, Mossa S, Turtschi A et al. (2003a) Structure and efficiency of the World-wide Airport network. Cond-mat 0312535

Guimerà R, Danon L, Díaz-Guilera A et al. (2003b) Self-similar community structure in a network of human interactions. Phys Rev E 68(6) 065103

Italian National Institute of Statistics (Istat), (1991) 13° Censimento generale della popolazione e delle abitazioni, Matrice origine destinazione degli spostamenti pendolari della Sardegna (13th General census of population and houses, Origin destination matrix of the commuting movements of Sardinia)

Italian National Institute of Statistics (ISTAT) (2001a) 14° Censimento generale della popolazione e delle abitazioni, Matrice origine destinazione degli spostamenti pendolari della Sicilia (14th General census of population and houses, Origin destination matrix of the commuting movements of Sicily)

Italian National Institute of Statistics (ISTAT) (2001b) 14° Censimento generale della popolazione e delle abitazioni, (14th General census of population and houses)

Jiang B, Claramunt C (2004) Topological analysis of urban street networks. Environ Plan B: Plan Des 31:151–162

Jost J, Joy MP (2002) Evolving networks with distance preferences. Phys Rev E 66(3):036126

Latora V, Marchiori M (2003) Economic small-world behavior in weighted networks. Eur Phys J B 32:249–263

Latora V, Marchiori M (2002) Is the Boston subway a small-world network?. Physica A 314: 109–113
Newman MEJ (2004) Fast algorithm for detecting community structure in networks. Phys Rev E 69:066133
Newman MEJ (2006) From the cover: modularity and community structure in networks. PNAS 103(23):8577–8582
Newman MEJ (2003) Structure and function of complex networks. SIAM review 45:167–256
Noronha VT, Goodchild MF (1992) Modeling interregional interaction: implications for defining functional regions. Ann Assoc Am Geogr 82:86–102
O'Kelly ME (1998) A geographer's analysis of hubs-and-spoke networks. J Transp Geogr 6: 171–186
Palla G, Derényi I, Farkas I et al. (2005) Uncovering the overlapping community structure of complex networks in nature and society. Nature 435(7043):814–818
Pastor-Satorras R, Vespignani A (2004) Evolution and structure of the Internet. Cambridge University Press, Cambridge, USA
Porta S, Crucitti P, Latora V (2008) Multiple centrality assessment in Parma campus: a network analysis of paths and open spaces. Urban Design Int 13:41–50
Radicchi F, Castellano C, Cecconi F et al. (2004) Defining and identifying communities in networks. Proc Natl Acad Sci USA 101:2658–2663
Reggiani A, Signoretti S, Nijkamp P et al. (2009) Network measures in civil air transport: a case study of Lufthansa. In: Naimzada AK, Stefani S, Torriero A (eds) Networks, Topology and Dynamics. Theory and Applications to Economic and Social Systems, Lecture Notes in Economics and Mathematical Systems, vol 613. Springer, Berlin, pp 257–282
Rosvall M, Bergstrom CT (2007) An information-theoretic framework for resolving community structure in complex networks. Proc Natl Acad Sci 104(18):7327–7331
Schintler LA, Gorman SP, Reggiani A et al. (2005) Complex network phenomena in telecommunication systems. Networks Spat Econ 4:351–370
Sen P, Dasgupta S, Chatterjee A et al. (2003) Small world properties of the Indian railway network. Phys Rev E 67:036106
Serrano MA, Boguña M, Vespignani A (2007) Patterns of dominant flows in the world trade web. arXiv:0704.1225v1, 10 april 2007
Strano E, Cardillo A, Iacoviello V, Latora V et al. (2007) Street centrality vs. commerce and service locations in cities: a kernel density correlation case study in Bologna. Italy, physics/ 0701111
Watts DJ, Strogatz SH (1998) Collective dynamics of 'small-world' networks. Nature. 393: 440–442

# Chapter 18
# Spatial and Commuting Networks

## A Unifying Perspective

**Roberto Patuelli, Aura Reggiani, Peter Nijkamp, and Franz-Josef Bade**

## 18.1 Introduction

A wide literature is devoted to the study of the relevance of space, encompassing several fields and disciplines, such as geography, economics, epidemiology, environmental and regional sciences. For example, space-time modelling has been a relevant focus of research in spatial economics starting from Hägerstrand (1967) and Wilson (1967, 1970). While the former paid attention to the modelling of spatial diffusion phenomena, the latter unified movements of spatial flows under the umbrella of statistical and information theory, by means of spatial interaction models. In these models, the relevance of spatial structure emerged in the associated cost/impedance functions. In parallel, starting from Zipf (1932) and Simon (1955), the importance of spatial structures (homogeneous or heterogeneous) has been discussed extensively in the literature, by focusing on the relationships between urban growth, agglomeration economies, and commuting costs (see, among others, Krugman 1991; Rossi-Hansberg and Wright 2006). A point of concern is that, in these spatial (growth and interaction) models, the effects of spatial topology and connectivity are only implicitly included, but never explicitly considered and discussed.

Tied to the spatial topology and connectivity issue is the network concept, which received a great deal of attention in social sciences and spatial economics in the past decades. Examples are the popular ideas of social complex networks (Barabási 2003; Vega-Redondo 2007), the network economy (Shapiro and Varian 1999), and the knowledge economy (Cooke 2001). Networks are based on the existence of interactions – at multiple levels/layers – between agents, giving rise to synergy effects. The effects of these interactions are often investigated and modelled in the

R. Patuelli (✉)
Institute for Economic Research (IRE), University of Lugano, Lugano, Switzerland; The Rimini Centre for Economic Analysis, Rimini, Italy

A. Reggiani and P. Nijkamp (eds.), *Complexity and Spatial Networks*, Advances in Spatial Science,
DOI: 10.1007/978-3-642-01554-0_18, 

form of, for example, network externalities or spillover effects (Yilmaz et al. 2002). The labour market literature is no exception: spatial matching processes have been widely studied in a social network framework (Montgomery 1991), as well as commuting, which has been modelled in both urban and regional contexts (for example, see van Nuffel and Saey 2005; Russo et al. 2007; Reggiani and Bucci 2008). In addition, network-based results can be tied to widely used econometric techniques (see, for instance, the relation between topological accessibility and spatial weights matrices, discussed in Mackiewicz and Ratajczak 1996).

The commuting literature has long been interested in problems of urban shape and regional networks of cities, in particular with regard to monocentricity and polycentricity (Button 2000). Cases of the latter are found at increasingly larger spatial scales, leading to the idea of "network cities" (Batten 1995), in which horizontal city-relations emerge (Wiberg 1993; van der Laan 1998), also because of improved transportation infrastructure and accessibility. In this framework, network modelling approaches to the analysis of commuting flows are worth noting. Russo et al. (2007) analyse commuting flows in Germany to identify "entrepreneurial cities" in Germany. Van der Laan (1998) and van Nuffel and Saey (2005) investigate the emergence of multinodality in the Netherlands and in the Flanders, respectively. In particular, van der Laan finds that increasing horizontal relations emerge for regions with modern economic structures, while the hierarchical status quo is preserved for peripheral, less advanced regions.

In line with the above developments, the present chapter investigates, for the case of Germany, the relevance of the volume and distribution of commuting flows, as well as of the commuting network's connectivity and topology. We aim to assess how network topology and its changes over time affect the geographic commuting system and its hierarchies. The reason for studying the commuting network in a connectivity perspective is inspired by the idea that the distribution of commuting can help explain other relevant economic phenomena, such as the convergence or divergence of labour market indicators (see for example Patacchini and Zenou 2007) or production levels. In this regard, the value added of network analysis is that it allows inspecting – in an intuitive fashion – commuting-induced topology and accessibility. Therefore, we aim to further inspect the connectivity perspective, to improve our understanding of the spatial-economic perspective.

The chapter is structured as follows: Sect. 18.2 briefly reviews recent developments in the field of network analysis. Sect. 18.3.1 illustrates a preliminary spatial analysis of commuting flows in Germany, with reference to the "open" cities (that is, to the cities with high propensity to mobility), while Sect. 18.3.2 presents the results of the network modelling experiments, by focusing on the network connectivity properties. Sect. 18.4 presents then a comparative multicriteria analysis that synthesises the dynamics of the different hierarchies – concerning the German "open" districts – emerging from the spatial and network approach. Finally, Sect. 18.5 concludes the chapter with some final remarks and directions for future research.

## 18.2 Spatial and Network Analysis: Recent Perspectives

Recent developments in spatial analysis call for a better understanding of the influence of space in the dynamics of economic growth patterns (for example, agglomeration economies). Relationships between agglomeration economies, fractal patterns, and rank size rules can be found, among others, in Batty (2005), and Chen (2004), while spatial equilibrium models consisting of a system of monocentric cities (city network) have also been adopted (see, for example, Abdel-Rahman 2003). However, these models have rarely embedded network concepts.

Here below we briefly discuss recent developments in network analysis and, in particular, their implications for regional networks. The focus is on recent works published by Barabási and Albert (BA) (1999), which radically changed the pre-existing frameworks for the analysis of large networks, by developing the concept of "scale-free (SF) networks", and by providing a model that helps explaining their (topological) properties.

SF networks are usually discussed vis-à-vis "random networks" (see, for example, the conventional Poisson random graph, Erdös and Renyi 1960). SF networks – first formalised by Price (1965, 1976) – are characterised by the presence of a few nodes ("hubs") with a high number of connections ("links") to other nodes (a high "degree"), and by a vast majority of nodes exhibiting a low number of links. The term "scale-free" refers to the statistical properties deriving from the above characteristics (see Newman 2003) and implies a great heterogeneity of the degree distribution.

The probability distribution of the nodes' degree $x$ (its "degree distribution") for SF networks tends to decay following a power function:

$$\Pr(X = x) \sim x^{-a}. \tag{18.1}$$

For large $x$, the value of the exponent $a$ in SF networks converges to 3 (Bollobás et al. 2001). A direct relation follows between the power law and Zipf's law (Zipf 1932), a distribution relating the degree of the nodes to their rank (Adamic 2000). According to Zipf, the relation between these two variables is as follows:

$$x \sim r^{-b}, \tag{18.2}$$

where $r$ is the rank of the node concerned. The value of the exponent $b$ is expected to be 1. Following from the mathematical relation of the Pareto distribution (which can be interpreted as a rank size rule) and power-law distributions (Adamic 2000), the relation between (18.1) and (18.2) is given by:

$$a = 1 + 1/b. \tag{18.3}$$

On the basis of the above considerations, we apply – in our empirical experiments – (18.2) (in logarithmic terms), by then extrapolating the value of $a$ according to (18.3).

In contrast to SF models, random networks (RNs) belong to a long-established class of networks (Rapoport 1957; Erdös and Renyi 1960). In an RN, the links between nodes in the network are expected to arise randomly. As a result, the probability of a node having degree $x$, $\Pr(X=x)$, follows, for a large-enough number of nodes, a Poisson distribution, implying a homogeneous distribution of connections. Consequently, most of the nodes have a similar number of links and importance.

In our empirical application, we test whether the German commuting network shows SF or RN characteristics, that is, if it is heterogeneous or homogeneous. Consistently with (18.2), we adopt, in the RN case, the exponential equation (18.4), where the degree of the nodes $x$ is sorted in decreasing order:

$$x = ke^{-\beta r}. \tag{18.4}$$

By synthesizing, the empirical evidence of rank size rules in urban economics, biology, and other fields is strictly related to the underlying connectivity network properties expressed by the associated power law. In other words, the rank size rule advocated in spatial economic science and the power law advocated in social sciences can be considered as two sides of the same coin, and hence interpreted in a unifying perspective.[1]

The above analytical frameworks are tested, for the case of the German commuting network, in Sect. 18.3.2, subsequently to a preliminary spatial analysis.

## 18.3 Case Study: Dynamics of German Commuting

### *18.3.1 Spatial Analysis: The "Open Cities"*

Before analysing the network properties of spatial commuting patterns, we will synthesise the characteristics of the German database from a regional/spatial perspective.

The data employed in our analysis refer to the registered residence and workplace of all dependent employees in Germany, at two points in time: 1995 and 2005. The data are aggregated in 439 German administrative districts, called *Kreise* (NUTS-3), and were collected by the Federal Employment Services (*Bundesanstalt für Arbeit*, BA) within social security services.[2] We can then form an origin-destination (OD) matrix, of dimension 439 × 439, which has, for each cell ($i$, $j$), the number of employees living in district $i$ and working in district $j$. In addition, we classify the German districts with regard to their levels of urbanisation and

[1] See also Chap 19 by Reggiani, in this volume.

[2] Since the data are directly gathered at the single firm level, it is reasonable to expect low and non-systematic measurement errors.

surrounding agglomeration[3] (BBR– *Bundesanstalt für Bauwesen und Raumordnung*) (Böltgen and Irmen 1997).

As indicators of the propensity to mobility of the districts, we employ indicators of incoming and outgoing mobility, which we refer to as inward and outward openness (authors' adaption from van der Laan 1998). The inward openness of a district indicates to which extent it attracts outside workers, and is computed, for a generic district $j$, as the ratio between the number of employees of the district $j$ residing in other districts, and the total number of employees of district $j$:

$$\sum_{i \neq j} e_{ij} \Big/ \sum_{i} e_{ij}.$$

Similarly, the outward openness can be defined as the percentage of residents who commute outside of their district, and is computed as:

$$\sum_{j \neq i} e_{ij} \Big/ \sum_{j} e_{ij}.$$

As a synthetic indicator of mobility (openness), we compute the average of inward and outward openness. This synthetic openness measure represents the capacity of a district to be "mobile" and, consequently, "active". Van der Laan (1998, p. 238) identifies high values of openness as possible signs of a "multi-nodal urban region".

In Fig. 18.1, central cities (CBDs) and highly urbanised districts mainly emerge as the most "active" in both 1995 and 2005. The Munich *Landkreis* results as the most "open". The higher concentration of population and economic activities (located within or in the surroundings of the main cities) – or a mobile population exploring new work opportunities – might explain this result (van Oort 2002). Notable exceptions – with low openness values – are Berlin and the CBD of Munich, due to their larger areas, which tend to contain commuting with the district boundaries. Over the ten year period we observe a generalised increase in the propensity to mobility, while a bigger positive variation can be found for the Berlin area.

In this context, it could be interesting to explore whether the most "open" cities seen above are also connected together in a city-network pattern. In summary, given the mobility characteristics of the districts, it might be relevant to explore how these are affected by the underlying connectivity networks, also in the light of the findings supporting multinodality, recently presented in the literature (Batten 1995; van Nuffel and Saey 2005). The next section investigates this aspect.

[3]The districts are classified as follows: (1) central cities in regions with urban agglomerations; (2) highly urbanised districts in regions with urban agglomerations; (3) urbanised districts in regions with urban agglomerations; (4) rural districts in regions with urban agglomerations; (5) central cities in regions with tendencies towards agglomeration; (6) highly urbanised districts in regions with tendencies towards agglomeration; (7) rural districts in regions with tendencies towards agglomeration; (8) urbanized districts in regions with rural features; and (9) rural districts in regions with rural features.

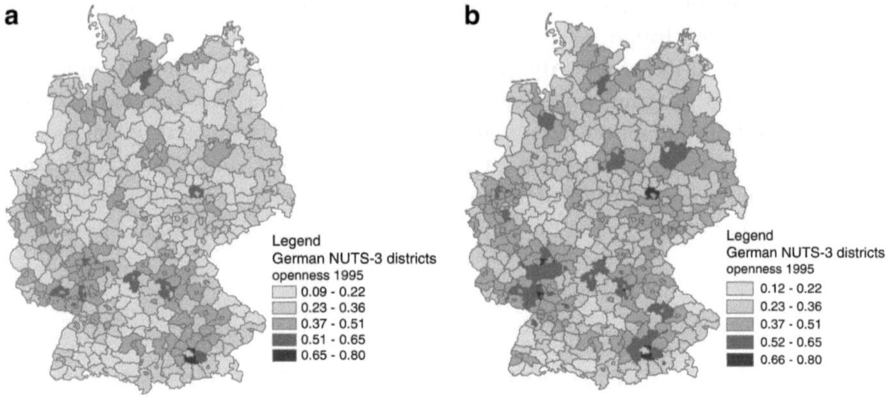


**Fig. 18.1** Maps of openness of districts, 1995 and 2005 (Patuelli et al. 2009)

**Table 18.1** $R^2$ values and exponents for power and exponential interpolations of incoming connections (indegree) and inward openness, 1995 and 2005

| Year | Indegree | | Inward openness | |
|---|---|---|---|---|
| | Power law | Exponential | Power law | Exponential |
| 1995 | 0.7002 | 0.9739 | 0.8027 | 0.9871 |
| (exponent) | (0.2442) | (0.0022) | (0.4623) | (0.0039) |
| 2005 | 0.6046 | 0.9316 | 0.7820 | 0.9859 |
| (exponent) | (0.2589) | (0.0025) | (0.4000) | (0.0034) |

*Source*: Patuelli et al. (2009)

## 18.3.2 Network Analysis: The "Connected" Cities

### 18.3.2.1 Connectivity Distribution

An initial analysis of the network underlying the commuting activities can be carried out by considering the statistical distribution of the data. In order to identify the (network) attractiveness and the propensity to mobility of the districts, we propose two exploratory approaches, based on the so-called indegree and on the inward openness. First, the number of inward connections per district (*indegree*) is examined, that is, from how many districts commuters come. From this viewpoint, which regards the logical topology of the commuting network, it is relevant *if* there is (any) commuting between two districts $i$ and $j$, whatever its extent. Secondly, we examine the inward openness of the districts (as defined above). In this case we consider the commuting inflows, that is, the weights tied to the links. In this case, the total inflows of each district are standardised by the number of jobs available in-place.

We next interpolate our data with a power function and an exponential function (see (18.2) and (18.4)). Table 18.1 shows the resulting $R^2$ coefficients and the values of the function exponents. For the case of the indegree distribution, an exponential distribution fits well the degree decay, although with a sharp cut-off at the end, and its exponent also remains extremely low in time. The $R^2$ for the power function is lower and also decreasing over time. On the other hand, its coefficient is more meaningful from an economic point of view. Transforming the indegree power-law coefficient according to (18.3), we obtain coefficients much greater than 3, suggesting random network characteristics (that is, a homogeneous pattern). Overall, these findings suggest the existence of a highly interconnected (logical) commuting network. However, the ambiguity between exponential and power law suggests that no clear agglomeration-pattern can be inferred in the case of the indegree distribution.

As for the indegree distribution, the distribution of the inward openness remains fairly stable in the 2 years considered, and the exponential function better interpolates the data. However, the power function also has a high $R^2$. In addition, the exponent values for the power interpolation are now higher (0.40–0.46), which implies transformed power-law coefficients greater than[4] 3. Overall, this preliminary data exploration shows that the exponential function is a better fit to both the indegree and the inward openness distributions, thus suggesting – according to these variables – an equilibrated network. This result is indeed in agreement with the associated rank size rule (18.2), since power-law coefficients smaller than 1 indicate an even spatial distribution of the two variables at hands (indegree and openness) (Brakman et al. 2001).

#### 18.3.2.2 Network Indices

After exploring the data and their distribution, we provide a set of synthetic indices, which describe three principal aspects exploring the network under different perspectives: (a) centralisation; (b) clustering; and (c) variety/dispersion.

Network centralisation is an aggregate assessment of the degree of inequality of a network. It may be computed on the basis of individual node centrality measures. The "centrality" of a node may be seen as a measure of its structural importance. The centrality index presented here may be called *indegree centralisation*, and is based on the concept of relative degree centrality of nodes, which measures the "visibility" of a node. This concept can be linked to the one of "hub" (Latora and Marchiori 2004), since the most visible nodes can be considered as hubs. The index only considers direct connections (indirect connections can only be considered if the transportation infrastructure is included in the analysis), and, in our case, only inward connections are considered (hence, the denomination "indegree centralisation"), in order to show the nodes' attractiveness for outside workers. Relative

[4]Our result would vary if we imposed a minimum threshold on the flows associated with each network link. A threshold set at three would support a finding of scale-free characteristics of the commuting network.

indegree centrality ($ric_i$) is computed, for each node $i$, as the ratio between the observed and the maximum possible number of connections of a node ($n - 1$):

$$ric_i = \text{indegree}_i/(n-1),$$

where $n$ is the total number of nodes. Consequently, the aggregate network indegree centralisation (*NIC*) index is computed, similarly to Freeman (1979), as:

$$NIC = \sum_{i \in N} (ric^* - ric_i)\backslash(n-2),$$

where $ric^*$ is $\max_i (ric_i)$.

The second index we compute refers to network clustering. Network clustering coefficients have been used extensively in network analysis (see, for example, Watts and Strogatz 1998) in order to determine the level of interconnectedness of networks. In order to compute a clustering coefficient for a node, we need to define its neighbourhood, which is given – if first order relations only are considered – by the nodes directly connected to the node concerned. A clustering coefficient for node $i$ is then computed as the ratio of the number of links existing between its neighbours and the maximum number of links that may exist between the same (neighbours): $c_i = l_i/l_i^*$, where $l_i$ and $l_i^*$ are the actual and possible number of links in node $i$'s neighbourhood, respectively. A synthetic network clustering coefficient is then computed as the average of the single nodes' coefficients. Clearly, if $n$-order neighbours are considered, a node's neighbourhood is represented by all the nodes that can be reached in $n$ hops.

As a third index, in order to assess the variety/dispersal of the nodes, we use an entropy indicator. Entropy is a concept derived from information theory (Shannon 1948) and widely used in spatial-economic science (Wilson 1967, 1970). Entropy is employed here as an indicator of the probability that the flows observed are generated by a "stochastic spatial allocation process" (Nijkamp and Reggiani 1992, p. 18). Higher entropy levels indicate that the flows are more homogeneous and dispersed over the network. The indicator $E$ is computed as:

$$E = -\sum\nolimits_{ij} p_{ij} \ln p_{ij},$$

where

$$p_{ij} = t_{ij}/O_i.$$

In $p_{ij}$, $t_{ij}$ is the number of commuters between districts $i$ and $j$, while $O_i$ is the outflows of district $i$.

Table 18.2 presents the results obtained for the German commuting network for the three indices described above, for the years 1995 and 2005. Though without dramatic changes, the network shows two distinct trends over 10 years. On the one hand, the network becomes less centralised, while the entropy increases. These

**Table 18.2** Descriptive indices for the German commuting network, 1995 and 2005

| Indices | 1995 | 2005 |
|---|---|---|
| Indegree centralisation | 0.33 | 0.31 |
| Clustering | 0.59 | 0.63 |
| Entropy | 8.23 | 8.38 |

*Source*: Patuelli et al. (2009)

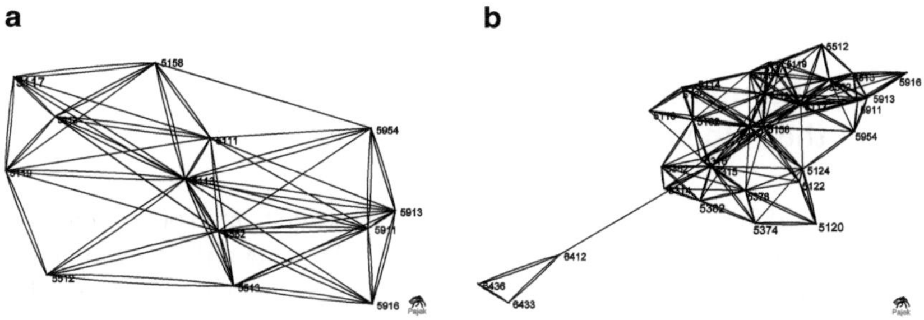


**Fig. 18.2** "4-cores" in the commuting network: (**a**) 1995; (**b**) 2005 (Patuelli et al. 2009)

results imply a more distributed structure of the network. On the other hand, the clustering coefficient of the network grows, suggesting a tendency towards greater interconnectivity. These results seem to confirm the findings emerging in our spatial analysis (Sect. 18.3.1), highlighting the network's tendency towards a multinodal structure (van der Laan 1998).

A graphical representation of the tendency towards greater interconnectivity in the commuting network can be obtained, for 1995 and 2005, on the basis of the "$k$-core" concept (Fig. 18.2), again from an *inward connections* viewpoint. A $k$-core is a subgraph (or more) in which each node has a minimal degree (in our case, *indegree*) of $k$, that is, each node in the $k$-core has connections with at least $k$ other nodes in the subgraph (Holme 2005). For a more meaningful computation and a readable graph, we select a subsample of the data, consisting of the flows above the arbitrary threshold of 1,000 individuals per OD pair. We find – for both 1995 and 2005 – $k$-cores of level 4, comprising first 13 and then 33 districts.

For the year 1995, the small core of 13 districts identifies a local and heavily interconnected network, headed by Düsseldorf and Dortmund, showing intense horizontal (local) relations. The fact that other districts do not appear in the 4-core does not mean that they have no reciprocal flows of commuters with the core districts. Simply, these other nodes do not feature the minimum levels of interconnectedness and flows of the core nodes, although they can show several flows much greater than 1,000 individuals. Frankfurt is a clear example. If we consider the year 2005, a larger graph of 33 districts is found. Here, the Düsseldorf/Dortmund cluster increases, and it represents most of the core. But it is noteworthy to cite the function of Frankfurt, which now acts as a hub, connecting the Frankfurt (code 6412) local cluster to the main Düsseldorf/Dortmund cluster.

Overall, the results of the network analysis seem to confirm the multinodal structure of the German commuting network (especially at the local level), while also suggesting an increased connectivity among the major centres – as centrality decreases over time – and, consequently, a tendency towards two layers of multi-nodality: (a) at the local level; and (b) at the regional level (the city-network level). As also seen by van Nuffel and Saey (2005, p. 326) and by van der Laan (1998, p. 244), these relations between the main centres do not overshadow local links – which still carry most of the mobility – but complement them.

## 18.4 Multidimensional Synthesis: The Network of the "Open" and Connected Cities

As a final step of this research endeavour, it is worthwhile to map out the hierarchies of the districts and their persistence over time, in order to identify the main relevant centres from both a spatial and a network viewpoint. We aim to offer a "synthetic" measure of the multiple spatial and connectivity dimensions observed above, by using a multidimensional method well known in the spatial-economic literature, known as multicriteria analysis (MCA). The synthetic assessment of the district characteristics – from the spatial and the connectivity perspectives – allows us to define a dominance rank of the districts concerned, and to investigate the changes which occurred in this rank over the period 1995–2005.

In order to look at the most representative districts only, we select a subsample of districts ("alternatives") to be employed in our MCA, using a synthetic connections-flows (CF) index, computed, for each district $i$, as

$$(CF)_i = [C_i/\max_i(C_i)]^*[F_i/\max_i(F_i)],$$

where $C_i$ and $F_i$ are the number of incoming connections (the indegree) and the inward openness of district $i$, respectively. The index is the product of the two normalised indicators $C_i$ and $F_i$, and is constrained from 0 to 1. It aims to provide a balanced assessment of the openness and connectedness of the districts, that is, from the conventional spatial interaction perspective and from the network perspective, respectively. On the basis of the CF index, we select 26 districts (listed in Table 18.3), which appear among the top 30 districts for both 1995 and 2005. Such a large group of "open" districts (26 of 30) over a 10-year period suggests an overall stability of the upper tier of the districts, according to the CF index. The districts selected, with a few exceptions, are urban districts – that is, central cities of type 1 and 5.

We carry out the MCA[5] on the basis of two aggregate assessment criteria (macro-criteria): spatial mobility (inward and outward openness) and connectivity (relative indegree centrality and clustering coefficients). We proceed in two steps: first,

[5]We employ the regime multicriteria method Hinloopen and Nijkamp (1990). In detail, three scenarios have been considered: (a) equal weights to all criteria; (b) ascending weights; and (c) descending weights. A final MCA of the rankings obtained provides the final results. We assume the hypothesis of no correlation between the criteria employed in the MCA.

by carrying out an MCA for each macro-criterion[6] and, second, by carrying out a final MCA which synthesises the two previous analyses.

With respect to the MCA based on spatial-economic indicators (spatial mobility macro-criterion), the results (presented in Table 18.3) show that Munich (*Landkreis*) steadily occupies the first position. Moreover, the ranking of the top districts is rather stable over the period considered. The results of the second MCA, based on the connectivity macro-criterion, provide – in 1995 – a different ranking, as the main cities are dominant. As seen earlier for the *k*-core analysis, Düsseldorf emerges from a network perspective. Further large cities, such as Frankfurt, Stuttgart and Munich, follow. We can also note that, with the exception of Munich, the districts that headed the spatial MCA rankings only perform intermediately in the connectivity MCA.

The final MCA results, synthesising the two preceding analyses, can be summarised along a few main observations. The district of Munich (*Landkreis*) – which also happens to be the richest German district according to per capita GDP – emerges as the most dominant for both 1995 and 2005, while a reshuffling in the rank of the districts can be observed over the 10-year period. Other districts seem to emerge. In particular, these are: Wiesbaden (from 7th to 2nd), Mannheim (14th to 6th), Frankfurt (12th to 8th), Stuttgart (15th to 11th), Düsseldorf (18th to 13th) and Karlsruhe (21st to 14th). The progress observed for these districts is mainly due to the connectivity macro-criterion. In other words, their high clustering coefficients show that the above districts are oriented towards stronger agglomeration patterns, in addition to their openness.

The findings summarised here lead us to propose a reinterpretation (or integration) in an economic sense of the concept of hub (for conventional hub definitions, see Barabási 2003), on the basis of a node's capacity of not only attracting connections from many other nodes, but also of generating an increased propensity to mobility. This double role by a few main nodes may drive the network towards multinodal characteristics.

However, although the districts emerging in the above analysis are the most "open" and "active", they still cannot be considered as the main "attractors". If we want to explore this characteristic, we then have to use, in the CF index computation, different variables (such as inflows or workplaces), in order to detect the relevance of the destinations, as the attraction models in the transport literature suggest.[7]

---

[6]The two macro-criteria employed here clearly identify two different types of phenomena: Spearman's correlation between the rankings resulting from the spatial and connectivity MCAs is equal to –0.369 for 1995 and to –0.311 for 2005. This is confirmed by the cross-correlations between the spatial and the connectivity criteria, which range – in absolute values – from 0.066 to 0.501.

[7]In this context, had inflows and outflows been employed as criteria within the spatial mobility macro-criterion, a ranking similar to the one obtained for the connectivity macro-criterion would have emerged.

**Table 18.3** Multicriteria analysis for the "open" and connected districts: results for 1995 and 2005

| Districts | Spatial results[a] | | Districts | Connectivity results[b] | | Districts | Final results[c] | |
|---|---|---|---|---|---|---|---|---|
| | 1995 | 2005 | | 1995 | 2005 | | 1995 | 2005 |
| 09184 Munich | 1 | 1 | 05111 Düsseldorf | 1 | 1 | 09184 Munich | 1 | 1 |
| 06436 Main-Taunus-Kreis | 2 | 2 | 06412 Frankfurt am Main | 2 | 2 | 06436 Main-Taunus-Kreis | 2 | 4 |
| 09661 Aschaffenburg | 3 | 4 | 08111 Stuttgart | 3 | 4 | 06411 Darmstadt | 3 | 3 |
| 06413 Offenbach am Main | 4 | 3 | 09184 Munich | 4 | 7 | 07315 Mainz | 4 | 5 |
| 06411 Darmstadt | 5 | 5 | 09564 Nuremberg | 5 | 8 | 08221 Heidelberg | 5 | 9 |
| 07314 Ludwigshafen am Rhein | 6 | 6 | 05314 Bonn | 6 | 9 | 05314 Bonn | 6 | 7 |
| 08221 Heidelberg | 7 | 8 | 08222 Mannheim | 7 | 6 | 06414 Wiesbaden (Landeshauptstadt) | 7 | 2 |
| 07315 Mainz | 8 | 7 | 06414 Wiesbaden (Landeshauptstadt) | 8 | 3 | 09562 Erlangen | 8 | 15 |
| 09662 Schweinfurt | 9 | 15 | 06436 Main-Taunus-Kreis | 9 | 11 | 08121 Heilbronn | 9 | 16 |
| 08121 Heilbronn | 10 | 9 | 08212 Karlsruhe | 10 | 5 | 07314 Ludwigshafen am Rhein | 10 | 18 |
| 09461 Bamberg | 11 | 12 | 06411 Darmstadt | 11 | 10 | 08421 Ulm | 11 | 12 |
| 08421 Ulm | 12 | 11 | 07315 Mainz | 12 | 13 | 06412 Frankfurt am Main | 12 | 8 |
| 09562 Erlangen | 13 | 10 | 09562 Erlangen | 13 | 12 | 06413 Offenbach am Main | 13 | 10 |
| 06611 Kassel | 14 | 16 | 08221 Heidelberg | 14 | 15 | 08222 Mannheim | 14 | 6 |
| 07111 Koblenz | 15 | 13 | 08421 Ulm | 15 | 14 | 08111 Stuttgart | 15 | 11 |
| 06414 Wiesbaden (Landeshauptstadt) | 16 | 14 | 08121 Heilbronn | 16 | 20 | 06611 Kassel | 16 | 17 |
| 05314 Bonn | 17 | 17 | 09663 Wuerzburg | 17 | 22 | 09661 Aschaffenburg | 17 | 20 |
| 09362 Regensburg | 18 | 20 | 07314 Ludwigshafen am Rhein | 18 | 21 | 05111 Düsseldorf | 18 | 13 |
| 09161 Ingolstadt | 19 | 24 | 06413 Offenbach am Main | 19 | 16 | 09663 Wuerzburg | 19 | 24 |
| 09663 Wuerzburg | 20 | 19 | 06611 Kassel | 20 | 17 | 07111 Koblenz | 20 | 22 |
| 08222 Mannheim | 21 | 18 | 09161 Ingolstadt | 21 | 18 | 08212 Karlsruhe | 21 | 14 |
| 06412 Frankfurt am Main | 22 | 22 | 09362 Regensburg | 22 | 19 | 09564 Nuremberg | 22 | 19 |
| 08111 Stuttgart | 23 | 21 | 07111 Koblenz | 23 | 24 | 09461 Bamberg | 23 | 25 |
| 05111 Düsseldorf | 24 | 25 | 09661 Aschaffenburg | 24 | 23 | 09161 Ingolstadt | 24 | 23 |
| 08212 Karlsruhe | 25 | 26 | 09461 Bamberg | 25 | 25 | 09362 Regensburg | 25 | 21 |
| 09564 Nuremberg | 26 | 23 | 09662 Schweinfurt | 26 | 26 | 09662 Schweinfurt | 26 | 26 |

*Source*: Patuelli et al. (2009)

[a]Spatial criteria: inward and outward openness

[b]Connectivity criteria: relative indegree centrality and clustering coefficient

[c]Final MCA: uses as criteria the spatial and connectivity results

## 18.5 Conclusions

This chapter has attempted to provide a novel analysis of commuting data and their trends, investigating both the spatial distribution of work mobility and the underlying logical commuting network. We have analysed data on journey-to-work trips for 439 German districts, for the years 1995 and 2005.

From a spatial perspective, we searched for the most mobile and "open" centres, with a particular focus on the openness of different typologies of districts. From the network perspective, we first considered the distributional properties of mobility indicators such as inward openness and indegree. We then computed aggregate indicators showing the evolution of the commuting network structure. Overall, we found evidence of the presence, in addition to a local and strongly interconnected network, of a *regional* network, which, however, does not overshadow established local patterns (see, for example, the results of the $k$-core analysis).[8]

In order to provide a unifying perspective, we synthesised the two (spatial and network) analyses by carrying out a multicriteria analysis (MCA). The MCA allowed us to observe, through a systematic assessment of the various indicators computed that the German districts are stable at the spatial mobility level, that is, with regard to their hierarchies. In addition, the results of the connectivity-based MCA show that the clustering coefficient indicator appears to influence most network connectivity, as suggested by Watts and Strogatz (1998).

A number of further research directions can be traced, in order to push further (or to fully exploit) the multidisciplinarity of the analytical approach proposed here. From the theoretical viewpoint, a deeper investigation of the influence of distance, travel time and accessibility, as well as of labour market characteristics, on commuting would be commendable. In this regard, and in order to better understand the relationship between the spatial economy and its underlying interaction networks, further research should frame our approach within more extensive regional labour market theoretical models (for example, the one developed by Blanchard and Katz (1992)). A further investigation of local commuting networks and agglomeration economies could be sought by integrating power-law-based and Zipf's-based evidence. Behavioural analysis at a micro-level (or taking into account different socio-economic groups) would also be fruitful, in order to test the aggregate behaviour.

From the methodological viewpoint, additional topological characteristics, such as betweenness-based centrality measures, should be investigated by means of a joint network/physical infrastructure analysis. Moreover, incorporating physical infrastructure would allow us to fully exploit network analysis tools, and to inspect widely relevant policy issues, such as infrastructure criticalities and bottlenecks.

[8] If high-degree nodes were found to be also connected to each other, then highly interconnected clusters could emerge, possibly leading, according to Holme (2005), to a core-periphery network structure (Chung and Lu 2002). Most importantly, Holme shows that transportation networks (more generically, geographically-embedded networks) tend to share this characteristic.

An integration of spatial and network-based measures into spatial econometric interaction models (see, for example, Griffith in this volume) should also be sought, in particular in order to investigate the relationship between clustering and network autocorrelation.

From the empirical viewpoint, the study of pre- and post-unification commuting networks in Germany, as well as of alternative geographical settings (for example, islands; see Chap 17 by De Montis et al. in this volume) and aggregation levels, could provide much needed information on the different long-run evolution of commuting networks.

All in all, the integrated "space-network" approach seems to offer novel pathways for the analysis of commuting and the associated interacting economic activities.

**Acknowledgments** The present chapter is adapted from a previous paper by the same authors titled: "The Evolution of the Commuting Network in Germany: Spatial and Connectivity Patterns", published in *Journal of Transport and Land Use* (2009) (available at http://www.jtlu.org). The authors wish to thank Pietro Bucci for his help in the empirical application regarding the connectivity analysis and multicriteria evaluation.

## References

Abdel-Rahman HM (2003) The city network paradigm: new frontiers (Working paper No. 2003–10). University of New Orleans, Department of Economics and Finance, New Orleans

Adamic LA (2000) Zipf, power-laws, and pareto – a ranking tutorial. Retrieved 16 April, 2007, from www.hpl.hp.com

Barabási A-L (2003) Linked: the new science of networks. Perseus Publishing, Cambridge

Barabási A-L, Albert R (1999) Emergence of scaling in random networks. Science 286:509–12

Batten DF (1995) Network cities: creative urban agglomerations for the 21st century. Urban Stud 32(2):313–28

Batty M (2005) Cities and complexity: understanding cities with cellular automata, agent-based models, and fractals. MIT, Cambridge

Blanchard OJ, Katz LF (1992) Regional evolutions. Brookings Pap Econ Activity 1:1–75

Bollobás B, Riordan O, Spencer J et al. (2001) The degree sequence of a scale-free random graph process. Random Struct Algorithm 18:279–90

Böltgen F, Irmen E (1997) Neue siedlungsstrukturelle regions- und kreistypen. Mitteilungen und Informationen der BfLR H. 1, S. 4–5 (in German)

Brakman S, Garretsen H, van Marrewijk C (2001) An introduction to geographical economics. Cambridge University Press, Cambridge

Button K (2000) Where did the 'New Urban Economics' go after 25 Years? In A. Reggiani (ed) Spatial economic science. Springer, Berlin, pp 30–50

Chen H-P (2004) Path-dependent processes and the emergence of the rank size rule. Ann Reg Sci 38(3):433–49

Chung F, Lu L (2002) The average distances in random graphs with given expected degrees. PNAS 99(25):15879–82

Cooke P (2001) Regional innovation systems, clusters, and the knowledge economy. Ind Corp Change 10(4):945–74

Erdös P, Renyi A (1960) On the evolution of random graphs. Publication of the Mathematical Institute of the Hungarian Academy of Science, vol. 5

Freeman LC (1979) Centrality in social networks: conceptual clarification. Soc Network 1:215–39
Hägerstrand T (1967) Innovation diffusion as a spatial process. University of Chicago Press, Chicago
Hinloopen E, Nijkamp P (1990) Qualitative multiple criteria choice analysis. Qual Quant 24:37–56
Holme P (2005) Core-periphery organization of complex networks. Phys Rev E 72:046111
Krugman P (1991) Geography and trade. MIT, Cambridge
Latora V, Marchiori M (2004) A measure of centrality based on the network efficiency, from http://arxiv.org/abs/cond-mat/0402050
Mackiewicz A, Ratajczak W (1996) Towards a new definition of topological accessibility. Transp Res B: Methodological 30(1):47–79
Montgomery JD (1991) Social networks and labour-market outcomes: toward an economic analysis. Am Econ Rev 81(5):1408–18
Newman MEJ (2003) The structure and function of complex networks. SIAM Rev 45(2):167–256
Nijkamp P, Reggiani A (1992) Interaction, evolution and chaos in space. Springer, Berlin
Patacchini E, Zenou Y (2007) Spatial dependence in local unemployment rates. J Econ Geogr 7(2):169–91
Patuelli R, Reggiani A, Nijkamp P et al. (2009) The dynamics of the commuting network in Germany: spatial and connectivity patterns. J Transp Land Use (forthcoming)
Price DJdS (1965) Networks of scientific papers. Science 149:510–5
Price DJdS (1976) A general theory of bibliometric and other cumulative advantage processes. J Am Soc Inf Sci 27:292–306
Rapoport A (1957) Contribution to the theory of random and biased nets. Bull Math Biol 19(4):257–77
Reggiani A, Bucci P (2008) Accessibility and impedance forms: empirical applications to the german commuting network. Paper presented at the 55th Annual North American Meetings of the Regional Science Association International, New York, November
Rossi-Hansberg E, Wright MLJ (2006) Establishment size dynamics in the aggregate economy. Am Econ Rev 97(5):1639–66
Russo G, Reggiani A, Nijkamp P (2007) Spatial activity and labour market patterns: a connectivity analysis of commuting flows in Germany. Ann Reg Sci 41(4):789–811
Shannon CE (1948) A mathematical theory of communication. American Telephone and Telegraph Co, New York
Shapiro C, Varian HR (1999) Information rules. Harvard Business School Press, Boston
Simon HA (1955) Aggregation of variables in dynamical systems: DTIC Research Report AD0089516
van der Laan L (1998) Changing urban systems: an empirical analysis at two spatial levels. Reg Stud 32(3):235–47
van Nuffel N, Saey P (2005) Commuting, hierarchy and networking: the case of flanders. Tijdschrift voor Economische en Sociale Geografie 96(3):313–27
van Oort F (2002) Agglomeration, economic growth and innovation. Spatial analysis of growth- and r&d externalities in the Netherlands. Erasmus Universiteit, Tinbergen Institute Thesis No. 260, Rotterdam
Vega-Redondo F (2007) Complex social networks. Cambridge University Press, Cambridge
Watts DJ, Strogatz SH (1998) Collective dynamics of small-world networks. Nature 363:202–204
Wiberg U (1993) Medium-sized cities and renewal strategies. Pap Reg Sci 72(2):135–143
Wilson AG (1967) A statistical theory of spatial distribution models. Transp Res 1:253–269
Wilson AG (1970) Entropy in urban and regional modelling. Pion, London
Yilmaz S, Haynes KE, Dinc M (2002) Geographic and network neighbors: spillover effects of telecommunications infrastructure. J Reg Sci 42(2):339–60
Zipf GK (1932) Selected studies of the principle of relative frequency in language. Harvard University Press, Cambridge

# Part D
# Epilogue

# Chapter 19
# From Complexity to Simplicity

## Interdisciplinary Synthesis and Future Perspectives

**Aura Reggiani**

## 19.1 Conceptual Background

"*Near is beautiful*" was argued by Miller (2004, p. 248) in his essay on "Tobler's First Law and Spatial Analysis". The awareness has also grown that relations among things that are near can generate complex spatio-temporal phenomena. The simplicity of Tobler's law[1] invokes reflections on the complexity[2] of interacting phenomena and the 'simple' laws which have been articulated in the scientific literature when attempting to 'decode' these phenomena. Certainly, from a spatial economic viewpoint, Tobler's law is consistent with the minimum cost-distance principle. In addition, Miller sheds light on the meaning of 'near' and 'distant': near is central to the space-economy, it is a more flexible and powerful concept than is often appreciated, and it could be expanded to include both space and time. Thus, not only (near or distant) space, but also the time component is fundamental in the analysis of the interacting economic phenomena. In parallel with Tobler, Hägerstrand (1967) pointed to the relevance of joint space-time diffusion processes, and Wilson (1967) linked spatial interaction with statistical information principles and entropy laws. An associated microeconomic foundation of spatial interaction modelling was subsequently developed by Anas (1983) on the basis of random utility theory (McFadden 1974). Later on, Nijkamp and Reggiani (1992) linked dynamic entropy with (dynamic) spatial interaction models.

A. Reggiani
Department of Economics, Faculty of Statistics, University of Bologna, Bologna, Italy

[1] "*Everything is related to everything else, but near things are more related than distant things*" (Tobler, 1970, in Miller, 2004, p. 284; see also the introductory Chap 1 by Reggiani and Nijkamp in this volume).

[2] We can identify and describe in a meaningful way the fundamental feature of complexity as follows (Bossomaier and Green 2000, p. 5): "*The essence of complexity is the outcome should not be obvious from the simple building blocks*". For a discussion on the concept of complexity, see also Chap 5 by Kulkarni et al., Chap 6 by Couclelis, Chap 12 by Donaghy, and Chap 14 by Seel and Waters, in this volume.

A. Reggiani and P. Nijkamp (eds.), *Complexity and Spatial Networks*, Advances in Spatial Science,

DOI: 10.1007/978-3-642-01554-0_19, 

The clear methodological interrelationships between the above-mentioned theories and models call for further reflections on the complexity of space-time phenomena and the *simplicity of the laws* describing these phenomena. The primary idea of complexity concerns the mapping of a system's non-intuitive behaviour, particularly the evolutionary patterns of connections among interacting components of a system whose long-run behaviour is hard to predict (Casti 1979). But, particularly at a dynamic level, it is noteworthy that May's law (May 1976) – describing the evolution of a population in discrete terms by means of a simple logistic equation – shows irregular and chaotic (and thus unpredictable) characteristics for certain values of the parameters and initial conditions. Also the 'complex' interacting evolution of two species can be described by the 'simple' Lotka–Volterra equations, whose analytical form is based on two interrelated logistic equations, and, surprisingly, the dynamic logistic equation turns out to be to be the dynamic form (under a certain condition of the utility function) of the associated logit model, and hence of the related spatial interaction model of the Wilson type (Reggiani 2004).

The recent enormous interdisciplinary interest in network[3] concepts, analysis, and modelling – arising from the study of complex interconnected dynamic systems – again underlines the 'simplicity law'. Networks often show common behaviour, based on their topological characteristics, and this behaviour is mainly derived from exponential/power forms, which are strongly related to the equations that govern spatial interaction (see Table 19.4 in Sect. 19.2). In other words, the topological properties of a network can give useful insights into: how the network is structured; which are the most 'important' nodes/agents; and how network topology can influence the conventional spatial economic laws (such as equilibrium theory, spatial interaction theory, etc.). However, this topology structure is again expressed by very simple laws, and in most cases these laws can be interpreted in a spatial economic framework. For example, if we find – in certain complex network typologies – the well-known Albert and Barabási (1999) power-law model, we may infer a rank-size rule of the Zipf (1949) type; or, if we calculate the redundancy of a node, in order to test the 'structural' holes (Burt 1992) in a network, we end up with a function strictly related to the entropy concept – concerning spatial economic systems – developed by Wilson (1970).

By considering whether different network topologies/typologies affect the evolutionary trajectories of complex spatial systems, the following methodological questions can be considered as fundamental points of concern:

- How do network structures affect interaction?
- How do changes in networks lead to changes in equilibrium structure?
- Are utilities a function of network structures?
- Do the functional forms of utility functions depend on the network structures?

[3]Networks can be interpreted as complex interconnected space-time systems, given the nonlinear characteristics of the network structure. For a brief review on complexity and networks, see among others, Chap 1 in Reggiani and Nijkamp (2006).

In this framework, it is still an open research issue which specific and novel contributions network analysis can offer to spatial economic analysis, and – vice versa – whether the solidity of spatial economic laws needs to be reconsidered in the light of recent advances in complexity and network theory. Hence, a dual analysis is necessary, in order to explore potential connections between these two approaches. In this respect, a synthesis of preliminary reflections is presented in the subsequent sections.

## 19.2 Spatial Economic Analysis and Network Analysis: A Dual Perspective

Spatial structures matter, according to different distributions of the centres/industries (on the basis of the Zipf's/Gibrat's law[4]), as witnessed by an extensive literature,[5] starting from Simon (1955) to Krugman (1996), Gabaix (1999), Duranton (2002), and Batty (2005), among others. These authors investigate the economic meaning of Zipf's/Gibrat's law, in particular exploring how agglomeration economies can be consistent with the city-size distribution and its growth (see also Rossi-Hansberg and Wright 2006).

Topological structures matter, according to different types of connectivity (Barabási and Oltvai 2004). It should be noted that, recently, there have been a great number of contributions which deal with topology, connectivity and networks in economic and social sciences (Goyal 2007; Friesz 2007; Naimzada et al. 2009; Vega-Redondo 2007; Vervest et al. 2009). By referring to Barabási's work, in essence, the statistical distribution of the links between centres/nodes can be expressed as follows:

- (a) Poisson distribution[6]: *random network*
- (b) Power distribution[7]: *scale-free network (with the possibility of hubs[8])*

---

[4]The proposition established by Gibrat (1931) is that the proportionate growth process (that is, the growth rate of a city's population does not depend on the site of the city) gives rise to the log-normal distribution, and not to the power distribution (that is, Zipf's law; see Eeckhout 2004).

[5]See also Chap 3 by Benguigui et al. as well as Chap 10 by Frenken, in this volume.

[6]Poisson distribution:

$$P(k) \propto e^{-\langle k \rangle} \frac{\langle k \rangle^k}{k!},$$

where $P(k)$ is the probability that a chosen node – in a certain network – has exactly $k$ links. Random networks are also called exponential, because the probability that a node is connected to $k$ other sites decreases exponentially for large $k$.

[7]Power law distribution: $P(k) \propto k^{-\gamma}$, where $\gamma$ is the exponent.

[8]Hubs are the preferential nodes/attractors in a network (hub: a single vertex with a large number of connections). A hub configuration/hierarchy exists for $2 \leq \gamma < 3$, according to Barabási and Oltvai (2004).

It would be interesting therefore to look at the similarities and differences between these two fields of analysis (spatial economics and networks), in the light of their fundamental *simple* laws (Table 19.1).

Surprisingly, the rank size coefficient equal to 1 means that Zipf's law holds. On the right side of Table 19.1, the power law coefficient equal to 2 means a hub-and-spoke network, according to Barabási and Oltvai (2004). By combining these two findings we might test whether the largest cities in Zipf's law are the hubs in the network connectivity of cities, for example, by considering the existence of commuting (whatever the volume) to be a link.

Further, we can certainly infer – in spatial economics – an analytical relationship between Zipf's law, the rank-size rule, and Wilson's (1967) spatial interaction models (SIMs)[9] (see also Brackman et al. 2001).

In addition, SIMs emphasise the homogeneity or heterogeneity of centres by means of different forms of the associated (negative) impedance cost functions $f(t_{ij})$ (see also Olsson 1980; Parr 1985; Richardson 1969). To summarise, the following association between decay form and spatial structure seems to emerge, as shown in Table 19.2.

It should be noted that the topological/connectivity structure has not been explicitly taken into account in SIMs, even though it is certainly "hidden" in the flow and time/cost matrix.

For example, recent empirical experiments concerning commuting in Germany show how accessibility functions based on the above deterrence forms (a) and (b)

**Table 19.1** Dual analysis

| Spatial economic analysis | Network analysis |
|---|---|
| Spatial structure | Topological structure |
| Statistical distribution of city population (Rank-size rule) | Statistical distribution of nodes (with $k$ links) |
| Rank-size coefficient (minus or greater than 1) | Power law coefficient (minus or greater than 2) |
| Homogeneity vs. Heterogeneity | Homogeneity vs. Heterogeneity |

**Table 19.2** The association between decay form and spatial structure (the coefficients $\beta$ and $\gamma$ represent the time/cost-sensitivity parameters)

| Deterrence form | Spatial structure |
|---|---|
| a) $f(t_{ij}) = e^{-\beta t_{ij}}$ (Exponential-decay) | Homogeneity of centres (regular/isotropous space) |
| b) $f(t_{ij}) = t_{ij}^{-\gamma}$ (Power-decay) | Heterogeneity of centres (agglomeration economies) |

[9]In general an (unconstrained) SIM reads as follows:

$$T_{ij} = KO_iD_jf(t_{ij}),$$

where the flows $T_{ij}$ represent the flows (commuting, trade, ideas, etc.) from origin $i$ to the destination $j$. They are a function of the outflows $O_i$ and of the inflows, $D_j$, as well as of the deterrence function $f(t_{ij})$; $t_{ij}$ is the travel time (or travel cost) between $i$ and $j$; the parameter $K$ is a scaling factor.

can give rise to completely different hierarchical orders and patterns. In particular, accessibility based on the exponential-decay function shows smooth patterns (likely due to regular connectivity structures), while accessibility based on the power-decay function displays high levels of discrepancy (likely due to the presence of "privileged" nodes (hubs), to which all the other nodes prefer to be attached). In other words, the exponential-decay function in accessibility seems to capture the hypothesis of a homogeneous network, while the power-decay function in the accessibility seems to grasp the hypothesis of a hub network (Reggiani 2008).

The various analytical forms of the impedance function applied to interaction/movement and the statistical distribution of the nodes seem therefore to constitute *two sides of the same coin*. In more general terms, we can consider a "methodological" synthesis scenario, as in Table 19.3, which highlights that economic variables might be better interpreted by jointly exploring the associated connectivity structure (and vice versa).

The related theoretical foundations, approaches, and functional forms are summarised in Table 19.4. It should also be noted here that the theoretical foundations and emerging approaches in spatial economics (such as gravity models/spatial interaction models/logit models) are analytically compatible and thus intrinsically connected, as mentioned in the previous section.

Furthermore, Table 19.4 indicates the use of the exponential and the power function in both disciplines (spatial economics and network analysis).

Much interesting "debate" on the exponential versus the power form in modelling interacting activities has already taken place in the past literature. It should be noted that the SIM embraces not only the exponential form but also the power form,

**Table 19.3** Synthesis scenario: two sides of the same coin

| Spatial Economic analysis | Network analysis |
|---|---|
| (Complex) interactions between nodes | (Complex) interactions between nodes |
| Focus on the related economic variables | Focus on the related links |
| Focus on the economic meaning of the functional forms | Focus on the connectivity patterns of the functional forms |

**Table 19.4** An interdisciplinary synthesis: relevance of exponential/power forms in spatial economics and network analysis

| Theoretical foundations | Emerging approaches | Functional forms |
|---|---|---|
| | Spatial economics | |
| Newton's law (1960s) (macro-level) | Gravity model | Power form (decay function) |
| Entropy maximization (macro-level) (1970s) | Spatial interaction model | Exponential/power form (decay function) |
| Utility maximization (micro-level) (1980s) | Logit model | Exponential form (utility function) |
| | Network analysis | |
| Complex networks (micro-macro level) (1990s) | Random/scale-free network | Exponential/power form (statistical distribution) |

if, by maximizing an entropy function, the cost constraint is expressed in a logarithmic form. In this context, Haynes (1974) stressed the advantage of the exponential-decay form for modelling both human and animal activities as opposed to the power model.

On the other hand, Bak (1996) and Pumain et al. (2006), among others, indicated the relevance and ubiquity of the power law in nature, as well as in urban economics, and hence its role as a "transversal tool" to many sciences.

In the spatial economic field, the essential analytical difference is that the exponential-decay exhibits a straight line in its semi-logarithmic transformation, while the power-decay is transformed into a straight line by considering its logarithmic transformation. All in all, the difference between these two functions consists of semi-logarithmic versus logarithmic axes. Consequently, it seems that either the semi-logarithmic or the logarithmic expression[10] of two variables under analysis is able to identify and fit – at an aggregate level – the interacting phenomena between these two variables.

In the network field, we might also observe a strict relationship between the power law and the exponential distribution, by means of what are called "combinations of exponentials" (Newman 2005).

It is interesting to note here that, in the evolution of the economic network a "complexity" view is based on the interaction represented by heterogeneous agents who behave boundedly rational. In this context, strategy choice might follow an evolutionary selection principle of the dynamic logit type.[11] Here the $\beta$ parameter, interpreted as the "intensity of choice", indicates the "random" or "preferential" behaviour of agents (from zero to very large values, respectively). The same happens in the context of the evolution of the social network, in the presence of noise.[12] Here, gradual adjustment and learning in games is modelled by the logit form, where the $\beta$ parameter indicates the sensitivity of agents' adjustment to their local environment. Again the $\beta$ parameter, for very low values close to zero, modulates the actions chosen with the same probability (that is, random network).

All in all, the role of the $\beta$ parameter in economic network analysis is relevant (see Table 19.5).

**Table 19.5** The association between the $\beta$ parameter in the exponential logit form and the economic network of agents

| $\beta$ Parameter in the exponential logit form | Economic networks (agents) |
|---|---|
| ($\beta = 0$) | Random behaviour of agents |
| ($\beta = \infty$) | Preferential behaviour of agents |

[10] Of course, the combination of exponential and power in the decay function is also possible, as, for example, in the Tanner function (for example, see March, 1971).
[11] See Chap 7 by Hommes.
[12] See Chap 8 by Ehrahardt et al.

The relationship in Table 19.5 shows a clear connection with the exponential-decay form in Table 19.2. There the "space" is expressed by the time/cost interaction matrix between the zones, while here it is "hidden" in the utility of the agents. Again (the dynamic) logit form turns out to be a simple model which "decodes" complexity and seems to be the "core model" in different disciplines.

Empirical applications concerning these matters show the possibility of mapping out: "simple" logit/spatial interaction models by means of a multi-agent simulation approach for the transport mode choice[13]; and dynamic spatial interactions between urban form and transport,[14] between population and spatial employment,[15] and between commuting and spatial employment.[16]

The methodological considerations outlined above, mainly concerning the "ubiquity" of the exponential/power form in several disciplines, call for reflections on the underlying economic meanings and theories (see Tables 19.4 and 19.5), and hence for synergy and cross-fertilization between these three scientific fields (spatial economics, network economics and network science). For example, it is necessary to have more understanding of the linkages between the theories in spatial economics (for example, between maximization of entropy at the aggregate level and maximization of random utility at the micro-level) and of the theories underlying network analysis. Consequently, a series of research issues need to be investigated in this context.

## 19.3 Towards a Research Agenda

### *19.3.1 Methodological Issues*

Some methodological questions can now be highlighted with reference to the synthesis presented in Tables 19.1–19.4, and hence to possible research topics connecting spatial economics and network analysis:

- *Agglomeration of economic activities and connectivity:* Are the centres – more important from the economic viewpoint or more "open" to innovation, growth and mobility – also the more connected? And vice versa: If the infrastructure network is randomly or scale-free connected, is this also the case for the associated economic variables/activities?
- *Utility models and topological structures:* Do topological structures (for example, random and scale-free networks) influence individual/aggregate utility/choice functions? And hence, which choice models/utility functions are associated with

[13] See Chap 13 by Grether et al.
[14] See Chap 4 by Medda et al.
[15] See Chap 15 by Schintler and Galiazzo.
[16] See Chap 16 by Griffith, Chap 17 by De Montis et al., and Chap 18 by Patuelli et al.

random/scale-free networks? If we conjecture that a logit model underlies a random network (depending on the $\beta$ parameter), likewise a "nested logit" model might underlie a scale-free network.
- *Entropy and constraints:* Entropy seems to be fundamental in the three fields of analysis (that is, spatial economics, network economics, and network science), since it can be considered as an "order parameter"[17] for identifying the concentration of the network. However, further reflection on the maximization of entropy, cost constraints, and emerging equilibrium models is necessary.
- *Structural holes:* Structural holes (Burt 1992) are the missing nodes connecting the networks. They can be filled by hubs, but how can they be analytically identified? Which is the related utility function?
- *Interdisciplinary integration*: Finally, would be it possible to consider an integration of the two field of analysis: for example, by considering different types of decay functions in the preferential attachment in a scale-free network?

The above research issues are obviously not exhaustive. Essentially, they indicate that a strong interdisciplinary effort is needed, in order to capture and map out the dynamics and complexity of spatio-temporal patterns and phenomena, and understand their relevance for scenario analyses and policy strategies.

### *19.3.2 Policy Issues*

The interest – in both disciplines (that is, spatial economics and network science) – concerning the analysis and modelling of network connectivity and evolution, also reflects the relevance of this theme from a policy viewpoint. For example, decision tools that strongly influence the topology and dynamics of the network might include the cost/utility functions of enlarging the network and/or the cost of adding a new node/link.

Thus new policy issues that can be raised in this context are, among others:

- New strategies in relation to innovative scenarios in connectivity networks (for example, as a result of the death/emergence of hubs, new clustering and organizations, new locations of firms, etc.).
- The construction of (dynamic) optimization problems based on collective utility functions (for example, max interaction/entropy) subject to some constraints (for example, random/scale free network indicators, in addition to individual or aggregate cost functions). It is interesting to note here the relevance of the constraints.[18]
- The 'empirical' research of the $\beta$ parameters (previously discussed) in order to extrapolate network behaviour, in the light of suitable forecasting.

---

[17] See also Chap 2 by Wilson in this volume.
[18] See also Chap 9 by Ricottilli, and Chap 11 by Friesz et al., in this volume.

- The identification of the (economic and social) accessibility with each type of network (random, preferential, etc.).
- The evaluation of new (dynamic) emerging scenarios (new configurations of connections and interactions, new forms of accessibility, etc.).

The research agenda summarised above might be a platform from where to depart for both original theoretical and empirical research, with the aim of jointly exploring the two fields of analysis (spatial economics and networks). In this context, the role of micro-behavioural attitudes deserves particular attention, and there is also a need for "dynamic" data provision, together with appropriate statistical tests for the verification of the underlying spatial economic and network processes/patterns. Meta-analysis on the transferability of results might then lead to the inference of the final "simple" law(s) that can unify complexity in the findings.

## References


Albert R, Barabási AL (1999) Emergence of scaling in random networks. Science 286:509–512

Anas A (1983) Discrete choice theory, information theory and the multinomial logit and gravity models. Transp Res B 17:13–23

Bak P (1996) How nature works. Springer, Berlin

Barabási A-L, Oltvai ZN (2004) Networks biology: understanding the cell's functional organization. Nat Rev Genet 5(2):101–113

Batty M (2005) Cities and complexity: understanding cities with cellular automata, agent-based models, and fractals. MIT Press, Cambridge

Bossomaier TRJ, Green DG (2000) Introduction. In: Bossomaier TRJ, Green DG (eds) Complex SYSTEMS, Cambridge University Press, Cambridge, pp 2–9

Brakman S, Garretsen H, van Marrewijk C (2001) An introduction to geographical economics: trade, location and growth. Cambridge University Press, Cambridge

Burt R (1992) The structural holes: the social structure of competition. Harvard University Press, Cambridge

Casti J (1979) Connectivity, complexity, and catastrophe in large-scale systems. Wiley, New York

Duranton G (2002) City size distribution as a consequence of the growth process. London School of Economics and Political Science, Centre of Economic Performance, London

Eeckhout J (2004) Gibrat's law for (all) cities. Am Econ Rev 94(5):1429–1451

Friesz T (2007) Network science, nonlinear science and infrastructure systems. Springer, NY

Gabaix X (1999) Zipf's law and the growth of cities. Am Econ Rev 89(2):129–132

Gibrat R (1931) Les inégalités économiques. Librairie du Recueil Sirey, Paris

Goyal S (2007) Connections: an introduction to the economics of networks. Princeton University Press, Princeton

Hägerstrand T (1967) Innovation diffusion as a spatial process. (Trans: Pred A). The University of Chicago Press, Chicago

Haynes RM (1974) Application of exponential distance decay to human and animal activities. Geogr Ann Ser B 2:90–104

Krugman P (1996) Confronting the mystery of urban hierarchy. J Jpn Int Econ 22(2/3):339–352

March L (1971) Urban systems: a generalised distribution function. In: Wilson AG (ed) Urban and regional planning, Pion, London

May R (1976) Simple mathematical models with very complicated dynamics. Nature 271:459–467

McFadden D (1974) Conditional logit analysis of qualitative choice behaviour. In: Zarembka P (ed) Frontiers in econometrics, Academic Press, New York, pp 105–142

Miller HJ (2004) Forum: on Tobler's first law of geography. Ann Assoc Am Geogr 94(2): 284–289
Naimzada AK, Stefani S, Torriero A (2009) Networks, topology and dynamics. Lecture notes in economics and mathematical systems, vol 613, Springer, Berlin
Newman MEJ (2005) Power laws, pareto distributions and Zipf's law. Contemp Phys 46(5): 323–351
Nijkamp P, Reggiani A (1992) Interaction, evolution and chaos in space. Springer, Berlin
Olsson G (1980) Birds in egg/eggs in bird. Pion Limited, London
Parr J (1985) The form of regional density function. Urban Stud 22:289–303
Pumain D, Paulus F, Vacchiani-Marcuzzo C et al. (2006) An evolutionary theory for interpreting urban scaling laws. Cybergeo, Systèmes, Modélisation, Géostatistiques vol 343:http://www.cybergeo.eu/index2519.html
Reggiani A (2004) Evolutionary approaches to transport and spatial systems. In: Hensher DA, Button KJ, Haynes KE, Stopher PR (eds) Handbook of transport geography and spatial systems, Elsevier, Amsterdam, pp 237–252
Reggiani (2008) Networks and accessibility structures: German commuting patterns. In: Becker U, Böhmer J, Gerike R (eds) How to define and measure access and need satisfaction in transport, University of Dresden, Dresden Institute for Transportation and Environment (DIVU), Dresden, Issue 7, pp 95–108
Reggiani A, Nijkamp P (2006) Spatial dynamics, networks and modelling. Edward Elgar, Cheltenham
Richardson HW (1969) Elements of regional economics. Penguin Books, Harmondsworth
Rossi-Hansberg E, Wright MLJ (2006) Urban structure and growth. Federal Reserve Bank of Minneapolis. Research Department Staff Report 381
Simon HA (1955) Aggregation of variables in dynamical systems. DTIC Research Report AD0089516
Tobler W (1970) A computer movie simulating urban growth in the Detroit region. Econ Geogr 46 (2):234–240
Vega-Redondo F (2007) Complex social networks. Cambridge University Press, Cambridge
Vervest PHM, van Liere DW, Zheng L (eds) (2009) The network experience. Springer, Berlin
Wilson AG (1967) A statistical theory of spatial distribution models. Transp Res 1:253–269
Wilson AG (1970) Entropy in urban and regional modelling. Pion, London
Zipf GK (1949) Human behaviour and the principle of least effort. Addison-Westley Press, Cambridge

MIX
Papier aus verantwortungsvollen Quellen
Paper from responsible sources
FSC® C105338

FSC
www.fsc.org

If you have any concerns about our products,
you can contact us on
**ProductSafety@springernature.com**

In case Publisher is established outside the EU,
the EU authorized representative is:
**Springer Nature Customer Service Center GmbH**
**Europaplatz 3, 69115 Heidelberg, Germany**

Printed by Libri Plureos GmbH
in Hamburg, Germany